Quantum AI Systems (QAIS)

Quantum AI Systems (QAIS)

Theory, Architectures, and Applications

Volume I

FOUNDATIONS OF QUANTUM AI SYSTEMS

A Systems-Level Architecture for Quantum AI

QuSciTech Press

Paperback ISBN-13: 979-8-9948369-0-3
Hardcover ISBN-13: 979-8-9948369-1-0

First Edition, Volume I
2026

Published by QuSciTech Press
P.O. Box 270014
Las Vegas, Nevada 89127, USA

Printed in the United States of America.

Permanent Scholarly Record and Citation

Series Archival DOI:

https://doi.org/10.5281/zenodo.18955800

The official archival record for the Quantum AI Systems series, including publication metadata, companion resources, and public repository references, is maintained through the Zenodo record above and associated QuSciTech repositories.

Companion laboratory environments, instructional systems, and supplemental computational resources are provided separately through the QuSciTech instructional platform and associated GitHub repositories.

Computational and generative visualization tools may be used during figure refinement and production workflows; however, all scientific interpretation, architectural design, systems engineering, and technical validation remain under direct author control.

For permissions, academic inquiries, and publication correspondence:
contact@quscitech.com

Dedication

To the pioneers of quantum information science and artificial intelligence, whose work revealed that information is physical and intelligence emerges through structure, interaction, and constraint.

And to those who will shape the future of Quantum AI Systems.

"We are no longer satisfied with insights into particles, fields, forces, and geometry. Today we seek to understand information, computation, and complexity as fundamental features of the physical world."

— Adapted from themes developed by John Archibald Wheeler and later information-theoretic physics researchers

Foreword

Quantum artificial intelligence cannot be understood by combining quantum computing and artificial intelligence only at the algorithmic level. Its development requires a systems perspective connecting physical quantum behavior, information representation, computation, learning, communication, sensing, control, governance, and deployment.

The *Quantum AI Systems (QAIS, pronounced QWAS): Theory, Architectures, and Applications* series provides that perspective. It treats quantum-enabled intelligence as the product of coordinated quantum and classical components operating under physical, computational, informational, and institutional constraints. This approach shifts attention from isolated algorithms to the complete architectures required to transform quantum capabilities into useful and trustworthy systems.

Volume I, *Foundations of Quantum AI Systems*, establishes the basis for the entire series. Quantum states, gates, circuits, measurement, coherence, hardware, control systems, software frameworks, and classical coordination are not merely introductory topics. They define what every subsequent QAIS capability can represent, compute, learn, communicate, sense, and control.

This foundational emphasis matters because higher-level claims about quantum learning or intelligence remain incomplete unless they can be traced to the physical operations and supporting infrastructure that make them possible. Volume I therefore establishes the vocabulary, operational principles, and systems context needed to study quantum AI with technical discipline.

By beginning with first principles and connecting them to an architectural framework, Volume I prepares the transition from quantum behavior to representation and reasoning in Volume II and supports the broader progression of the QAIS series.

Preface

I developed Volume I to address a recurring problem in the study of quantum artificial intelligence: quantum physics, computation, hardware, software, and artificial intelligence are often presented as separate subjects even though practical quantum-enabled systems depend on their coordinated operation.

This separation can make it difficult to understand how a physical quantum state becomes an information-bearing resource, how gates and circuits transform that resource, how measurement produces classical evidence, and how hardware and software coordinate the complete execution process. The central intellectual problem is therefore not simply how quantum mechanics works, but how its principles become usable components of an intelligent system architecture.

Volume I approaches this problem from first principles and then progressively expands toward implementation. Chapter 1 establishes the physical and conceptual foundations of quantum information. Chapter 2 develops the operations through which quantum states are prepared, transformed, and measured. Chapter 3 connects those operations to hardware platforms, control systems, classical processors, software frameworks, and execution environments.

Throughout the volume, mathematical ideas are linked to physical meaning and architectural function. Figures clarify relationships among states, operations, layers, and system components. Spotlights and Sidebars connect foundational concepts to research, engineering, and emerging applications.

The purpose is not to reduce quantum mechanics to a simplified computing metaphor. It is to provide a technically grounded pathway from quantum behavior to quantum-system realization. This foundation enables readers to evaluate the representations, reasoning processes, learning architectures, communication systems, sensing interfaces, and governance structures developed in the remaining volumes.

Acknowledgments

The development of *Quantum AI Systems (QAIS): Theory, Architectures, and Applications, Volume I—Foundations of Quantum AI Systems* rests on contributions from generations of physicists, mathematicians, computer scientists, electrical engineers, systems engineers, quantum-hardware researchers, software developers, and artificial intelligence researchers.

I gratefully acknowledge the universities, research laboratories, government agencies, professional associations, and standards organizations whose research, education, collaboration, and technical guidance continue to advance quantum mechanics, quantum information science, quantum computing, artificial intelligence, and systems engineering.

I also recognize the quantum-hardware developers, control engineers, software engineers, framework maintainers, and open-source communities whose processors, simulators, programming environments, documentation, and educational resources have expanded opportunities for experimentation and analysis.

The QuSciTech Labs initiative contributed to the conceptual and instructional development of the QAIS framework and its treatment of quantum foundations, operational principles, infrastructure, and quantum–classical system integration.

Finally, I thank the students, educators, researchers, engineers, practitioners, colleagues, and readers whose questions, critiques, experiments, and technical observations have contributed to this work.

How to Use This Book

Volume I supports sequential study, formal instruction, research preparation, professional development, and focused technical reference. No prior study of the remaining QAIS volumes is required.

First-time readers should study Chapters 1 through 3 in order. Chapter 1 establishes the physical and conceptual foundations of quantum information and introduces the architectural framing through which those foundations are interpreted within Quantum AI Systems (QAIS). Chapter 2 develops the operational grammar through which quantum states are prepared, transformed, correlated, measured, and verified. Chapter 3 connects these principles and operations to quantum technology, infrastructure, execution environments, ecosystem development, institutional coordination, and workforce requirements.

Readers already familiar with quantum mechanics may use Chapters 1 and 2 as a structured review before proceeding to the infrastructure, ecosystem, and systems-integration material in Chapter 3. Engineering and information-technology professionals may begin with the architectural and infrastructure discussions in Chapter 3 and return to earlier sections when additional physical, mathematical, or operational context is needed.

Each chapter combines introductory framing, technical development, figures, Technical Spotlights, case-study analysis using the **Quantum AI Learning and Inference System (QALIS, pronounced *QWAL-iss*)** and **Cryptographically Relevant Quantum Computing–Large Language Models (CRQC–LLM)** lenses, practical applications, a chapter conclusion, a Quick Self-Check, and references. Quick Self-Check answers are provided in Appendix B, while foundational terminology is consolidated and tracked in Appendix C. Additional appendices provide mathematical foundations, curriculum guidance, laboratory resources, and expanded treatment of the Quantum Stack.

QAISKit extends the volume through laboratory activities, simulations, architectural exercises, assessments, and implementation resources for quantum and hybrid quantum–classical environments. These resources support the transition from conceptual understanding to experimentation, architectural analysis, and system-level application. Instructors may use the complete three-chapter sequence or organize selected topics into specialized instructional, workforce-development, or research modules.

Note to the Reader

Quantum AI Systems and the associated Quantum Stack are proposed conceptual frameworks for organizing the study, design, and evaluation of quantum-enabled intelligent systems. They are analytical and architectural reference models, not formally adopted standards or mandatory implementation structures.

The framework draws upon established principles from quantum mechanics, quantum information science, artificial intelligence, computer science, electrical engineering, and systems engineering. Some higher-level relationships—including integrated quantum–classical intelligence, cross-layer coordination, and scalable QAIS realization—remain developing areas requiring further formalization and empirical validation.

Actual implementations may combine layers, distribute functions differently, or use terminology specific to a hardware platform, software environment, or organization. The models and examples in this volume support explanation and investigation; they do not guarantee quantum advantage, hardware scalability, fault-tolerant operation, or present-day realization of complete QAIS architectures.

Readers should evaluate implementation and performance claims against current experimental evidence, technical documentation, applicable standards, and the physical limitations of the technologies involved. Technically grounded corrections may be submitted to:

press@quscitech.com

Architectural Progression of the Book

The QAIS series follows a cumulative structure in which each chapter provides capabilities required by those that follow. It progresses from quantum foundations throughepresentation, learning, communication, and sensing to governance, deployment, and evolution.

Volume I—Foundations of Quantum AI Systems

Volume I establishes the physical, computational, and infrastructural foundations used throughout the series.

Chapter 1: Foundations of Quantum AI Systems

Chapter 1 introduces quantum states, qubits, superposition, interference, entanglement, and measurement as the physical basis of QAIS.

Chapter 2: Operations and Scientific Framework of QAIS

Chapter 2 converts these principles into state preparation, gates, circuits, evolution, coherence, and measurement operations.

Chapter 3: Evolution, Ecosystem, and Workforce of QAIS

Chapter 3 connects quantum operations to hardware, control systems, classical processors, software frameworks, and execution environments.

Volume II—Quantum Representation and Reasoning

Volume II uses the foundations of Volume I to develop structured information representations and reasoning processes.

Chapter 4: Encoding Classical Data in Quantum Systems

Chapter 4 explains how data, features, concepts, relationships, and problem structures are encoded for quantum and hybrid processing.

Chapter 5: Quantum Reasoning and Decision Architectures

Chapter 5 applies quantum transformations, probability, correlation, measurement, and classical interpretation to encoded information.

Volume III—Quantum Learning and System Architectures

Volume III extends representation and reasoning into adaptive quantum–classical systems.

Chapter 6: Quantum Advantage in QAIS

Chapter 6 evaluates quantum capabilities against classical baselines using performance, resource, scalability, and verification criteria.

Chapter 7: Quantum Machine-Learning Architectures

Chapter 7 organizes parameterized circuits, quantum neural networks, kernel methods, generative models, and hybrid workflows into trainable architectures.

Chapter 8: Optimization and Control in Quantum AI Systems (QAIS)

Chapter 8 develops the objectives, parameter updates, feedback, stability, and control mechanisms through which learning architectures adapt.

Volume IV—Quantum Information and Communication Systems

Volume IV extends adaptive QAIS capabilities across distributed systems and physical environments.

Chapter 9: Quantum Information Metrics and Model Evaluation

Chapter 9 establishes measures of uncertainty, correlation, distinguishability, fidelity, channel behavior, and model performance.

Chapter 10: Quantum Communication Architectures

Chapter 10 applies these measures to quantum channels, entanglement distribution, teleportation, repeaters, routing, synchronization, and security.

Chapter 11: Quantum Sensing Architectures

Chapter 11 connects QAIS to physical environments through measurement, sensing, metrology, signal integration, feedback, and interpretation.

Volume V—Quantum Governance, Resilience, and Deployment

Volume V integrates the preceding capabilities into governed, resilient, and sustainable operational systems.

Chapter 12: Capstone QAIS Integration

Chapter 12 combines the preceding layers through cross-layer coordination, bounded propagation, verification, interoperability, resilience, and system evaluation.

Chapter 13: The Future of Quantum AI Systems

Chapter 13 extends the integrated architecture into deployment, scaling, institutional adoption, governance, long-term stability, and responsible evolution.

Together, the thirteen chapters trace a continuous progression from quantum behavior and controlled operations to representation, reasoning, learning, communication, sensing, integration, governance, and deployment.

Table of Contents

List of Figures

Guide to Special Elements

This book includes several specialized elements designed to support learning, application, and architectural understanding:

Equations

Formal mathematical expressions that define quantum states, transformations, optimization processes, and information-theoretic measures. These equations provide the analytical foundation for QAIS, enabling precise modeling of representation, propagation, and system behavior under constraint.

Figures

Visual representations of system architectures, quantum processes, and computational models that illustrate the structure and behavior of QAIS across layers and applications.

Tables

Structured summaries of key concepts, comparisons, system architectures, and design trade-offs.

Technical Spotlights

Focused explorations of core concepts, design principles, and implementation insights that provide deeper technical interpretation of key ideas.

Case Studies

Applied scenarios and architectural anchors illustrating how QAIS concepts operate across domains, from foundational principles to large-scale system behavior.

Hands-On Quantum Labs

Practical exercises designed to support experimentation with quantum AI concepts and hybrid quantum–classical workflows. Additional labs are available at https://quscitech.com

List of Abbreviations and Acronyms

AI — Artificial Intelligence
API — Application Programming Interface
ASIN — Amazon Standard Identification Number
BQC — Blind Quantum Computing
CNOT — Controlled-NOT Gate
CPTP — Completely Positive Trace-Preserving (Map)
CRQC — Cryptographically Relevant Quantum Computing
CRQC-LLM — Cryptographically Relevant Quantum Computing–Large Language Model
DEV — Development Environment
DOI — Digital Object Identifier
GPU — Graphics Processing Unit
HHL — Harrow–Hassidim–Lloyd (Algorithm)
IBM — International Business Machines
IoT — Internet of Things
LLM — Large Language Model
ML — Machine Learning
NISQ — Noisy Intermediate-Scale Quantum
NLP — Natural Language Processing
PDF — Portable Document Format
PQC — Post-Quantum Cryptography
QAIS — Quantum AI Systems *(*pronounced *QWAS)*
QALIS — Quantum AI Learning and Inference System (pronounced QWAL-iss)
QC — Quantum Computing
QEC — Quantum Error Correction
QFT — Quantum Fourier Transform
QIS — Quantum Information Science
QKD — Quantum Key Distribution
QPU — Quantum Processing Unit
QRAM — Quantum Random Access Memory
RAM — Random Access Memory
SIA — System Integrity Assurance

Chapter 1

—

Foundations of Quantum AI Systems

Architectural Position

Defines the physical and mathematical foundations underlying QAIS. Within the Quantum Stack, this chapter establishes the Layer 1 substrate, specifying the principles of superposition, entanglement, and measurement that constrain all higher-level system behavior. These foundational laws determine the admissible structure of quantum states and the limits of all subsequent transformation, representation, and computation processes.

Core Principle

QAIS systems are defined by how information is represented, propagated, and stabilized under constraint.

Abstract

Chapter 1 establishes the physical foundation of QAIS, hybrid architectures in which classical data are encoded into quantum states and processed through quantum–classical learning pipelines. Model behavior is jointly governed by representational geometry in Hilbert space and the constraints of quantum hardware. The chapter reframes core Quantum Information Science (QIS) phenomena—superposition, entanglement, coherence, and measurement disturbance—as architectural primitives that define QAIS structure, behavior, and bounded propagation. Qubits are introduced as coherence-limited resources, entanglement as a connector for distributed processing, and measurement as the mechanism for inference and verification. These principles are unified into system-level design rules for encoding, stability, and decision-making. Chapter 1 serves as the conceptual gateway, linking quantum physics to the hybrid architectures and learning systems developed in subsequent chapters.

Keywords

Quantum AI Systems; Quantum Information Science; qubits; superposition; entanglement; coherence; measurement disturbance; quantum physics; quantum computing; information as a physical resource; control surface; system behavior; bounded propagation; QAIS architectures; system-level design

Terminology Clarification: *QIS refers to quantum information science, the foundational scientific field upon which this volume builds. QAIS refers to Quantum AI Systems, the architectural and systems-engineering framework developed throughout this series. QIS supplies the underlying physical and informational foundations; QAIS translates those foundations into integrated system architectures.*

1.1 Introduction — Physical Law to Architecture

Architectural Role: *Establishes the physical substrate of intelligence, defining the quantum mechanical constraints (superposition, entanglement, measurement) that bound all representation, computation, and inference.*

Architectural Spine: *Quantum States → Superposition → Entanglement → Coherence → Measurement → System Architecture*

Chapter Framing: *Chapter 1 establishes the interpretive and architectural lens for the book by positioning quantum phenomena as system-level constraints that govern representation, propagation, and stability in QAIS. Rather than treating quantum mechanics as a computational abstraction, the chapter frames it as the physical basis of an executable system grammar in which information, transformation, and measurement define the limits of intelligent behavior. Within this perspective, QAIS is introduced as a system-level discipline integrating quantum computation and artificial intelligence, where representation functions as a control surface and system behavior emerges under physical constraint. The Quantum Stack is presented as a vertically integrated control hierarchy that organizes this transition from physical substrate to actionable intelligence, while the QALIS and CRQC-LLM frameworks provide evaluative structure for analyzing hybrid quantum–classical systems throughout the book.*

Quantum Stack Layer: L1 *— Quantum Physics → Establishing the physical laws governing quantum states, superposition, entanglement, and measurement.*

Unlike classical computational systems, where information is often treated as an abstract symbolic entity independent of its substrate, QAIS treats information as physically embodied within evolving quantum systems. Under this perspective, representation, transformation, propagation, and observation are inseparable from the physical processes governing coherence, interference, entanglement, and measurement.

This shift carries important architectural consequences. System behavior no longer emerges solely from algorithmic instruction sequences, but from the interaction of constraint, feedback, propagation, and contextual measurement across multiple operational layers. As QAIS architectures scale, challenges involving robustness, verification, coherence preservation, and control increasingly appear not as isolated technical problems, but as interconnected manifestations of deeper structural relationships within the system itself.

Accordingly, this text approaches QAIS as a systems discipline in which intelligent behavior emerges through the coordinated evolution of physical information structures operating under bounded operational conditions.

QAIS begins at the physical substrate where quantum behavior originates. In the Quantum Stack model, this corresponds to Layer 1 — Quantum Physics, where superposition, entanglement, and measurement disturbance [1] define the laws governing how information can exist, evolve, and be interrogated.

Chapter 1 establishes this layer not as background science but as the foundation upon which every QAIS architecture, protocol, and workflow is built. Phenomena traditionally explored in QIS are reframed here as design primitives: operational resources that determine how correlations are encoded, how coherence is preserved, and how information behaves when embedded within intelligent systems.

In this chapter, "operational" refers to physical relevance rather than executable procedure; formal operational definitions are introduced in Chapter 2.

In QAIS, information is inseparable from the physics of its carrier. The qubit is not merely a representational unit but a coherence-sensitive resource whose lifetime constrains how long quantum advantages can be sustained. Entanglement functions as an architectural connector for distributed systems rather than a conceptual curiosity. Measurement is elevated from a textbook postulate to a system rule that governs how quantum evolution becomes classical decision. This reframing transforms quantum laws into engineering constraints, shaping model design, verification strategies, and resilience boundaries.

Building on this foundation, QAIS represents a shift from treating quantum information as an isolated theoretical construct to treating it as the substrate of intelligent architectures. Whereas QIS examined superposition, entanglement, coherence, and measurement disturbance as physical effects, QAIS treats them as system-level design rules for representation, inference, coordination, and resilient decision-making.

Modern challenges in scalable inference, secure communication, distributed coordination, and intelligent control require architectures grounded in the physics that governs information itself.

To formalize that progression, this chapter introduces the Quantum Stack, a layered framework tracing the path from fundamental physics to engineered intelligence. Industry and standards-based models, including those adopted by IBM, NIST, and IEEE, describe the progression from qubits and control electronics to compilers, circuits, and algorithmic workflows.

---

[1] **Measurement Disturbance:** The principle that quantum measurement irreversibly alters the measured state. Unlike classical observation, quantum measurement participates in state determination.

These models accurately capture computation but do not extend to the layers required for intelligence; they stop before addressing how quantum systems encode structure, support learning, or maintain trust under operational stress.

The QAIS Quantum Stack extends this progression beyond programmable quantum computation by explicitly introducing Hybrid Quantum-Classical Computing & AI as the integration layer through which quantum computational resources are coordinated with classical processing, data systems, optimization, simulation, learning, inference, and orchestration. QAIS then forms the culminating layer, composing these capabilities into governed systems for learning, reasoning, decision-making, verification, resilience, and deployment.

This chapter's thematic arc follows that progression: quantum phenomena become system-defining resources that determine how information behaves, persists, and enables intelligence under quantum law. The Quantum Stack provides the structural logic for that transformation; QAIS provides its system expression.

Scope and Method of This Volume

This book is structured as a systems monograph. Each chapter advances a cumulative architectural argument, where physical constraint, operational grammar, representational encoding, learning architecture, and governance discipline are developed in sequenced progression. Concepts introduced earlier are not revisited descriptively but activated structurally in later chapters.

Guiding Principle 1 — Architecture Precedes Operation

Each chapter in this volume begins with a Guiding Principle that frames the intellectual obligation of the material that follows. The Guiding Principle is not a summary. It is a structural commitment. It defines what must be established before the chapter advances to technical development. For Chapter 1, the governing commitment is that architecture precedes operation.

QAIS systems must be architected before they are operationalized. This chapter therefore establishes architectural doctrine rather than execution grammar. Physical law precedes workflow, and structural decomposition precedes implementation.

QAIS begins at the physical substrate where quantum behavior originates. Superposition, entanglement, coherence, and measurement disturbance are not treated as isolated phenomena; they are formalized as design primitives. These primitives define the boundaries within which representation, inference, coordination, and verification may occur. Information is inseparable from the physics of its carrier.

Consequently, architecture must be grounded in coherence limits, correlation structure, and governed measurement boundaries before operational claims are made.

Chapter 1 defines the QAIS architecture in three complementary dimensions:

- **Conceptual Architecture.** The Quantum Stack (Figure 1.1.1) establishes the vertical ascent from quantum physics to intelligent system deployment. It clarifies where QAIS resides within the progression from substrate law to engineered intelligence.
- **Layered System Architecture.** The QAIS System-Layer Model
- Figure 1.2.2 decomposes QAIS internally across substrate, operational control, encoding and metrics, learning architectures, and hybrid deployment. This layered structure replaces conceptual aggregation with disciplined cross-layer integration.
- **Constraint-Aware Architectural Doctrine.** The dual-lens evaluative contract Figure 1.2.1 and the Cross-layer constraint–capability flow Figure 1.2.3 formalize the principle that capability and constraint are co-instantiated. Superposition becomes representational breadth; entanglement becomes structural connectivity; coherence becomes runtime budget; measurement becomes governed boundary. These are architectural constants, not implementation details.

Equally important is what Chapter 1 does not attempt. It does not define execution pipelines, gate-level operational grammar, control-loop orchestration, or deployment schematics. Those elements belong to Chapter 2, Operations & Scientific Framework of QAIS, where architecture transitions into executable structure.

The discipline established here is sequential and intentional:

- Physics defines possibility space.
- Architecture defines structural organization.
- Operation activates governed workflow.

This progression governs not only Chapter 1, but the structural rhythm of the book as a whole. Each subsequent chapter begins by establishing its governing architectural commitment before advancing into operational or technical detail, ensuring that structural clarity precedes procedural implementation throughout the volume.

With this foundation established, the discussion now turns to the Quantum Stack itself: the layered architectural framework tracing the progression from quantum physical law to intelligent system architecture.

1.1.1 The Quantum Stack: Physics to Architectures

The foundational structure of QAIS is best understood through the **Quantum Stack**, a layered conceptual and architectural model that traces the disciplined progression from physical law to intelligent-system deployment. More than a taxonomy of technological components, the stack establishes an ordered hierarchy of dependencies in which each layer inherits constraints from the layer below and enables increasingly sophisticated capabilities above. Physics defines the space of what is possible; engineering stabilizes and controls that space; information science organizes it; computation activates it; and intelligence integrates it into purposeful system behavior.

Unlike traditional quantum computing stacks, which typically culminate in algorithm execution, the QAIS Quantum Stack extends the ascent further—into representational architecture, learning systems, hybrid governance, and resilience engineering. It does not stop at programmable computation. It asks how programmable quantum structure becomes disciplined intelligent behavior under physical and adversarial constraints.

QIS organizes engineered capabilities—such as quantum-state preparation, coherent control, entanglement, interference, measurement, error management, information transmission, and precision detection—into structured protocols for computation, communication, and sensing. Quantum computing implements the computational subset of these protocols within programmable systems that encode candidate states or solutions as quantum states and transform their probability amplitudes through controlled operations, entanglement, and interference.

Hybrid Quantum-Classical Computing & AI then integrates programmable quantum computation with classical processing, optimization, simulation, machine learning, inference, data services, and orchestration. This layer provides the integration bridge through which specialized quantum computational capabilities participate in coordinated hybrid intelligent workflows.

At the highest layer lies QAIS, where quantum resources are no longer treated as isolated computational accelerators but are assembled into architectures for learning, optimization, inference, coordination, and interpretable system behavior. This layered progression mirrors the intellectual architecture of the book itself: each chapter ascends one or more layers of the stack, demonstrating how interference-driven physical processes become governed, system-level capability.

The conceptual foundation of QAIS requires a structured framework that connects fundamental physical principles with higher-order system behavior. No single layer can be understood fully in isolation because each depends on the capabilities provided by the layers beneath it and operates within the constraints those layers impose. The framework must therefore be viewed as an integrated hierarchy through which physical phenomena are progressively engineered into computational, learning, and intelligent-system capabilities. Figure 1.1.1 presents this hierarchy from a unified architectural perspective, illustrating how capabilities emerge upward through the stack while lower-layer constraints shape what higher layers can realistically achieve.

The Structural Logic of the Quantum Stack

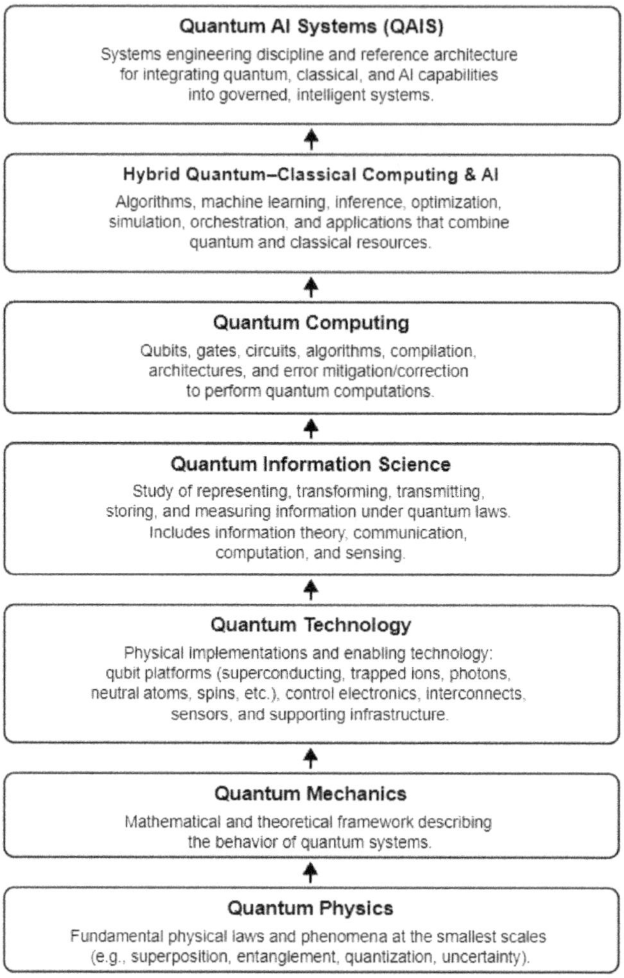

Figure 1.1.1 — The Quantum Stack: From Physics to QAIS.

This conceptual architecture figure presents the seven-layer Quantum Stack, beginning with quantum physics and ascending through quantum mechanics, quantum technology, quantum information science, quantum computing, hybrid quantum–classical computing and AI, and ultimately QAIS at the top. The stack illustrates how fundamental quantum phenomena are progressively formalized, physically realized, organized as information-processing resources, transformed through programmable quantum computation, integrated with classical computing and AI, and ultimately composed into architectures supporting learning, optimization, inference, reasoning, decision-making, and governed intelligent operation. The layers are conceptual and architectural rather than hardware-specific.

The Quantum Stack should be read as a constraint-aware architectural ascent, not a hardware bill of materials. Its layers represent ordered transitions in abstraction, control, and system accountability. Each layer introduces a new form of organization while remaining bounded by the structural conditions imposed from below. Progression is therefore cumulative, not substitutive: higher layers do not replace lower ones; they operationalize the lower layers. If Figure 1.1.1 defines *where* QAIS resides in the architectural ascent, the next question is *how does that ascent prepare QAIS structurally*?

The Quantum Stack is not passive background. It is preparatory architecture. Each layer establishes preconditions that must remain vertically coherent for intelligent behavior to be sustainable. Coherence budgets, calibration discipline, synchronization tolerance, measurement staging, and verification placement are inherited upward.

The following Technical Spotlight makes this pre-operational preparation explicit. It presents the Quantum Stack not as a static classification diagram, but as a vertical conditioning framework through which physical possibilities and constraints are progressively translated into engineered control, information protocols, programmable workflows, and validated intelligent architectures before a QAIS system is operationalized or operational performance claims are made.

The stack encodes a disciplined transformation:

- **From natural law to formal rule**— translating observed physical behavior into mathematical structures governed by unitary evolution and quantum-state-space geometry.
- **From formal rule to engineered stability**— making abstract quantum dynamics physically realizable, stable, and controllable through precision fabrication, calibration, timing, control, and environmental isolation.
- **From engineered stability to information structure and protocol**— organizing stable, controllable quantum states and operations into representations and repeatable protocols for storing, transforming, transmitting, and measuring quantum information across computation, communication, and sensing contexts.
- **From information structure and protocol to programmable quantum transformation**— implementing quantum-information processes through executable gates, circuits, algorithms, compilation, state preparation, controlled interference, information exchange, and staged measurement within defined runtime constraints.
- **From programmable quantum transformation to hybrid computational intelligence**— coordinating quantum computation with classical computation, artificial intelligence, optimization, simulation, inference, data processing, and orchestration to form integrated quantum–classical computational workflows.
- **From hybrid computational intelligence to QAIS architecture**—integrating these hybrid capabilities into governed systems that support learning, optimization, inference, reasoning, decision-making, verification, resilience, and operational deployment.

These transitions are not merely technical escalations; they are shifts in responsibility. Quantum physics governs physical possibility. Quantum mechanics formalizes quantum behavior. Quantum technology governs physical realization and controllability. Quantum information science governs the representation, transformation, transmission, storage, and measurement of quantum information. Quantum computing governs programmable quantum execution. Hybrid quantum–classical computing and AI govern the integration and orchestration of quantum computation with classical computation, data processing, optimization, simulation, learning, and inference. QAIS governs the systems-level composition, coordination, verification, resilience, governance, and deployment of these integrated capabilities.

At every layer, enablement and limitation co-emerge. Superposition expands representational capacity while demanding phase coherence and precision control. Entanglement strengthens relational structure while tightening synchronization tolerance and error sensitivity across subsystems.

Measurement produces actionable classical evidence while irreversibly collapsing amplitude structure and constraining future evolution.

Coherence sustains structured interference yet imposes finite runtime windows that bound architectural depth. Constraint is not downstream correction; it is co-present with capability at each ascent step.

Because these tensions accumulate upward, architectural layers cannot be optimized independently. A gain in expressive power at the representational level must remain supportable within coherence budgets and calibration tolerances. An expansion of distributed coordination must remain sustainable under synchronization drift and verification overhead.

The stack therefore encodes vertical dependency: instability or misalignment at lower layers propagates upward as architectural fragility.

QAIS represents the culminating layer of this cumulative conditioning. It is the point at which physical, informational, computational, and hybrid capabilities are composed into governed intelligent-system architectures operating under inherited physical and operational constraints.

In this sense, the Quantum Stack is not merely descriptive scaffolding. It is a constraint-aware architectural progression for understanding how quantum-enabled intelligent systems emerge, are evaluated, and remain bounded by the capabilities and limitations inherited from lower layers.

Technical Spotlight — Preparing QAIS Through the Quantum Stack

The Quantum Stack provides the architectural logic through which QAIS is understood as an integrated systems doctrine rather than a collection of techniques. Each layer defines both the constraints and the affordances inherited by the layers above it, ensuring vertical coherence across the progression from physics to intelligent deployment.

QAIS does not transcend its physical substrate; instead, its representational capacity, operational feasibility, and resilience posture are conditioned and delimited by the governing principles of quantum mechanics.

Conceptual Grounding

Quantum physics establishes the physical phenomena and boundaries within which quantum information can exist, constraining coherence duration, correlation structure, and disturbance sensitivity. Quantum mechanics provides the mathematical and theoretical framework governing how quantum states are represented and evolve through unitary dynamics, measurement, and controlled operations.

Quantum technology physically realizes these principles through qubit platforms, control electronics, interconnects, operating environments, timing precision, and calibration discipline. Quantum information science organizes these physical capabilities around the representation, transformation, transmission, storage, and measurement of information under quantum laws. Quantum computing then makes selected information-processing transformations programmable through qubits, gates, circuits, algorithms, compilation, measurement, and error-management mechanisms.

Hybrid quantum–classical computing and AI provide the next level of integration by coordinating quantum computational resources with classical processors, data systems, optimization methods, simulation environments, machine-learning and inference mechanisms, and orchestration workflows. This layer establishes the computational bridge through which specialized quantum capabilities participate in larger hybrid intelligent processes.

At the highest layer, QAIS composes these integrated capabilities into governed systems supporting learning, optimization, inference, reasoning, decision-making, verification, resilience, and operational deployment. In this framing, intelligence emerges not as an abstraction detached from physics, but as systems-level architecture built upon vertically integrated physical, informational, computational, and hybrid capabilities.

System Relevance (QAIS-Layer Scoped)

The Quantum Stack ensures that QAIS is never interpreted independently of physical constraint. Every QAIS architecture depends on:

- coherence windows,
- calibration precision,
- noise behavior,
- operational measurement checkpoints,
- verification boundaries,
- hybrid quantum–classical interface stability.

These factors define the design space within which quantum-intelligent behavior can emerge. They determine what forms of learning and inference are physically sustainable and what resilience measures must be embedded structurally rather than retrofitted.

Cross-Disciplinary Orientation

Physicists interpret the stack as a progression from fundamental quantum phenomena toward controlled information processing. Engineers treat it as a readiness hierarchy in which device reliability, calibration, synchronization, control precision, and interface stability condition the viability of higher-level architectures.

AI specialists focus increasingly on the Hybrid Quantum–Classical Computing & AI and QAIS layers, where quantum computational capabilities are coordinated with classical processing, data systems, optimization, learning, inference, and orchestration before being composed into governed intelligent systems. Cybersecurity experts examine relevant concerns throughout the stack for exposure surfaces, integrity failures, verification bottlenecks, interface vulnerabilities, and dependencies through which lower-level limitations can propagate into higher-level system risk.

- **QALIS:** QALIS principally realizes learning-and-inference capability through the upper portion of this progression, drawing on hybrid quantum–classical computing and AI capabilities and incorporating them into disciplined learning, inference, adaptation, and operational decision-making processes. Within the QAIS System-Layer Model of Figure 1.2.2, its principal realization spans the Ecosystem and Infrastructure, Architectures and Learning Models, and Applications and Hybrid Deployment layers.

-
- **CRQC–LLM:** CRQC–LLM provides a complementary security, resilience, integrity, and verification perspective across relevant concerns in the progression. Within the QAIS System-Layer Model of Figure 1.2.2, it is represented as a cross-layer overlay that coordinates relevant cryptographic-security, LLM-integrity, monitoring, orchestration, verification, mitigation, resilience, provenance, and governance concerns across all six system layers, as appropriate to each layer. Its scope represents oversight and traceability rather than direct component control.

Together, these constructs ensure that capability and constraint are evaluated jointly: QALIS emphasizes the realization of learning and inference capability, while CRQC–LLM emphasizes relevant security, integrity, resilience, verification, and governance concerns that must remain traceable across the architecture.

Section 1.1 Summary — Physical Law to Architecture

With the Quantum Stack formally established, the architectural terrain of QAIS is now defined in structural terms. The seven-layer ascent from quantum physics through quantum mechanics, quantum technology, quantum information science, quantum computing, hybrid quantum–classical computing and AI, and ultimately QAIS has been clarified, and the conditioning effect of each layer on those above it has been made explicit. What remains is not further decomposition, but disciplined interpretation and evaluation.

Section 1.2 — Case Study Orientation and Alignment introduces the complementary analytical constructs QALIS and CRQC–LLM. QALIS provides the capability-oriented learning-and-inference perspective through which hybrid quantum–classical and AI resources are organized toward intelligent behavior. CRQC–LLM provides a complementary perspective for examining relevant cryptographic-security, LLM-integrity, verification, resilience, provenance, mitigation, and governance concerns across the architecture. These constructs ensure that the progression defined by the Quantum Stack does not remain merely descriptive; they establish complementary lenses through which subsequent architectural mechanisms can be evaluated for both capability and constraint.

The progression therefore shifts from structural mapping to methodological evaluation. Having established the conceptual progression that culminates in QAIS, the chapter now formalizes how QAIS mechanisms are to be interpreted, engineered, evaluated, and stress-tested under physical constraint, operational limitation, and relevant adversarial conditions.

1.2 Case Study Orientation and Alignment

Architectural Framing: Systems Architecture

Section Framing: This section establishes the evaluative architecture that governs how representation, propagation, and stability are interpreted across QAIS. Representation is treated as a control surface whose structure must be analyzed under both constructive and adversarial conditions. Propagation is framed as system behavior subject to constraint, verification, and potential destabilization across layers. Stability is therefore enforced as bounded propagation through dual-lens evaluation, ensuring that capability and constraint remain co-instantiated throughout the system lifecycle. This section defines the methodological discipline that regulates all subsequent architectural development.

Chapter 1 introduced the Quantum Stack as the architectural ascent from physical law to intelligent system deployment. That structural progression now requires an interpretive discipline. Without such discipline, the layered ascent risks becoming descriptive rather than governed. Section (1.2) therefore establishes the evaluative contract that will regulate every architectural claim made in this volume.

Two persistent analytical constructs are introduced: **QALIS** and **CRQC–LLM**. These are not illustrative examples inserted for narrative convenience. They are methodological lenses that operate across the upper layers of the Quantum Stack and remain active throughout the entire book. Together, they formalize the defining tension of QAIS: intelligent capability must emerge under physical constraint, and every expansion of expressive power must be evaluated simultaneously for resilience under adversarial and operational stress.

From this point forward, each mechanism introduced in QAIS will be interpreted in two lights:

- As a pathway to scalable quantum-intelligent capability, and
- As a surface requiring synchronization discipline, verification placement, and governance alignment.

This dual orientation is structural, not rhetorical. It defines how QAIS is studied, engineered, and evaluated.

1.2.1 QALIS — Constructive Evaluation Framework

QALIS (Quantum AI Learning and Inference System) is the constructive evaluation framework used throughout the book to show how QAIS capabilities are assembled in practice. It models how quantum primitives are assembled into disciplined, deployable intelligent systems.

QALIS begins with the physical substrate but does not remain at the level of phenomenon. Superposition, entanglement, coherence, and measurement are treated as architectural variables. These primitives are composed into learning pipelines, inference structures, sensing workflows, and coordinated decision architectures whose stability depends on disciplined cross-layer alignment.

Under QALIS, superposition is not merely the coexistence of computational alternatives; it becomes representational breadth within a structured inference pipeline. Entanglement is not mystical correlation; it becomes a structural connector that binds distributed subsystems into unified representational spaces. Coherence is not simply a fragile physical property; it becomes a runtime budget governing how long structured evolution may proceed before degradation invalidates inference.

Measurement is not an abstract projection rule; it becomes a governed boundary where amplitude geometry resolves into classical evidence within verification-aware workflows.

The constructive logic of QALIS proceeds in architectural sequence:

- State preparation must be coherence-aware and calibration-aligned.
- Controlled transformations must preserve phase integrity and amplitude structure.
- Interference must be engineered deliberately to bias computational pathways.
- Measurement placement must be staged to preserve interpretability and trust boundaries.

Each stage is treated not as isolated operation but as a component within a cross-layer system discipline.

QALIS therefore demonstrates how quantum information becomes scalable intelligent infrastructure. It provides the constructive grammar for assembling quantum primitives into learning and inference architectures that are physically grounded, verification-aware, and deployment-conscious.

Importantly, QALIS does not assume advantage. Advantage must be architecturally earned through coherence budgeting, synchronization discipline, and measurement governance. Expressive capacity expands only within bounded physical envelopes. Capability emerges not from isolated algorithmic speedup, but from disciplined composition under constraint.

In this way, QALIS models the constructive side of QAIS: the systematic ascent from amplitude structure to coordinated intelligence.

1.2.2 CRQC-LLM — Stress-Test Evaluation Framework

If QALIS models constructive assembly, CRQC-LLM (Cryptographically Relevant Quantum Computing-Large Language Models) is the structured stress-test framework used to evaluate resilience, verification boundaries, and adversarial pressure in QAIS architectures.

CRQC-LLM operates as the complementary stress-test framework within QAIS. It examines how the same primitives that enable quantum advantage simultaneously expand fragility surfaces under scale, automation, cryptographic pressure, and adversarial reasoning.

CRQC-LLM combines two conceptual forces:

1. Cryptographically relevant quantum capability, particularly where quantum computation threatens classical security assumptions; and
2. Automated reasoning systems capable of orchestrating adversarial exploration across architectural surfaces.

This combination forces QAIS architectures to confront resilience requirements at the same level of seriousness as capability claims.

Under CRQC-LLM, superposition is evaluated not only for representational breadth but also for phase sensitivity under noise and perturbation. Entanglement is evaluated not only as coordination infrastructure but as synchronization surface vulnerable to drift or degradation.

Coherence is examined not merely as runtime resource but as a narrowing constraint envelope susceptible to adversarial timing exploitation. Measurement is treated as a verification checkpoint whose misplacement may leak trust or distort inference.

CRQC-LLM therefore probes QAIS architectures along multiple axes:

- coherence decay under accumulated noise,
- synchronization instability across distributed subsystems,
- verification boundary misplacement at hybrid interfaces,
- cryptographic disruption under scalable quantum computation,
- and governance failure under automation and scale.

The stress-test framework does not negate QALIS. It ensures that constructive design is never interpreted independently of resilience accounting. It exposes brittleness before scale amplifies it. It treats fragility not as downstream implementation error, but as a co-emergent property of expressive capacity.

Within QAIS, adversarial interrogation is not external to architecture. It is structurally embedded as a methodological requirement.

1.2.3 The Dual-Lens Evaluation Contract

The relationship between QALIS and CRQC–LLM is formalized in Figure 1.2.1, *Dual Systems-Evaluation Lenses in QAIS*. Together, these constructs define the evaluative contract governing the remainder of the volume.

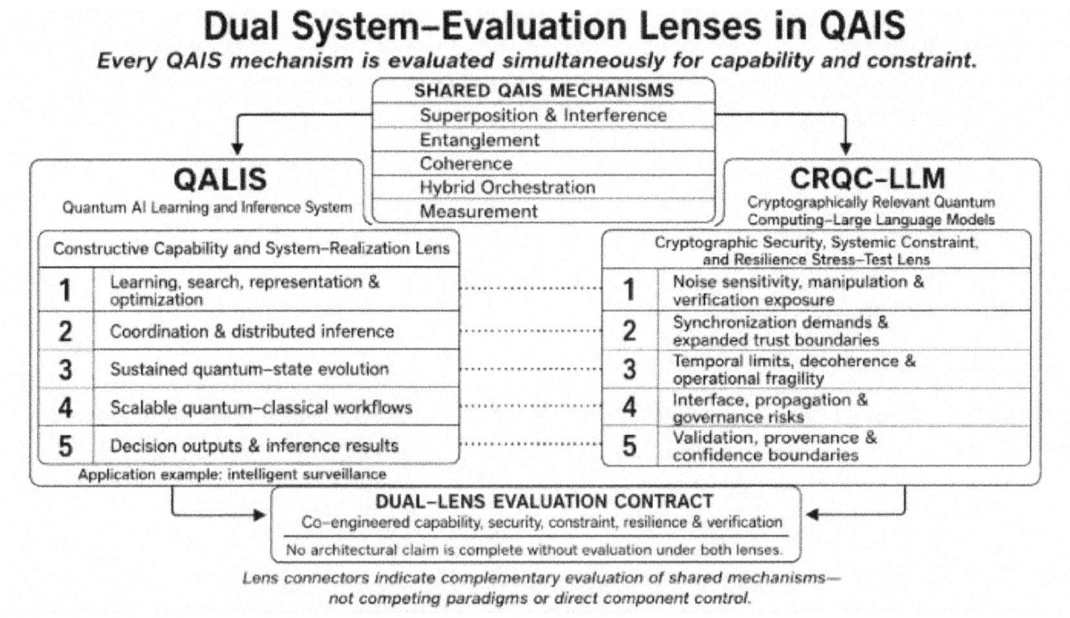

Figure 1.2.1 — Dual System-Evaluation Lenses in QAIS.

This conceptual framing figure presents a dual-lens framework through which the same QAIS mechanisms are evaluated simultaneously for constructive capability and systemic constraint. QALIS provides the constructive capability and system-realization lens through which quantum and hybrid mechanisms are assembled into scalable learning, inference, coordination, optimization, and decision-support capabilities. CRQC-LLM provides the complementary cryptographic-security, systemic-constraint, resilience, and verification lens through which those same mechanisms are examined for fragility, adversarial exposure, synchronization demands, cryptographic risk, governance obligations, and verification-boundary limitations.

The framework is built around five shared QAIS mechanisms:

- Superposition and interference
- Entanglement
- Coherence
- Hybrid orchestration
- Measurement

Each mechanism is interpreted through both lenses. Under QALIS, superposition and interference support learning, search, representation, and optimization; under CRQC–LLM, they introduce noise sensitivity, manipulation risks, and verification exposure. Entanglement enables coordination and distributed inference while expanding synchronization requirements and trust boundaries. Coherence sustains structured quantum-state evolution while imposing temporal limits and susceptibility to decoherence. Hybrid orchestration enables scalable quantum–classical workflows while creating interface, propagation, security, and governance risks. Measurement produces decision and inference outputs while establishing validation, provenance, and confidence boundaries.

Evaluating a QAIS architecture therefore requires simultaneous consideration of what a mechanism enables and what conditions constrain its reliable, secure, and governable use. Neither perspective alone provides a complete characterization of system behavior. The two lenses are not competing paradigms; they are complementary forms of analysis applied to the same architectural mechanisms.

Under this evaluative contract:

- **QALIS** interprets shared QAIS mechanisms as composable architectural resources for constructing scalable learning, inference, coordination, optimization, and decision-support systems.
- **CRQC–LLM** interprets those same mechanisms as cryptographically relevant security, resilience, constraint, and verification surfaces that must withstand coherence decay, adversarial pressure, synchronization drift, cryptographic exposure, infrastructure fragility, and governance failure.

No architectural claim is complete until it has been evaluated under both lenses.

This dual-lens discipline operationalizes a central principle of Chapter 1: the mechanisms that enable capability also establish the boundaries within which that capability can remain stable, secure, verifiable, and governable. The amplitude structure that enables interference-driven computation also increases sensitivity to disturbance and manipulation. Entanglement that enables distributed coordination also expands synchronization demands and trust boundaries. Coherence that sustains structured evolution also imposes strict temporal limits. Measurement that produces usable results also creates validation, provenance, and confidence obligations.

Capability and constraint are therefore co-instantiated variables. Every constructive architectural decision produces corresponding requirements for verification, resilience, security, governance, and operational control. Design and stress evaluation must advance together.

By embedding both interpretations at the foundational stage, Chapter 1 prevents subsequent analysis from drifting toward capability inflation or treating resilience, security, and verification as afterthoughts. The resulting contract requires every QAIS mechanism to be evaluated for both its contribution to intelligent system behavior and the conditions necessary to sustain that behavior throughout the system lifecycle.

Integration Across the Book

QALIS and CRQC–LLM are not confined to Chapter 1. Their complementary perspectives structure analysis throughout the subsequent chapters:

- **In Chapter 2,** QALIS guides the constructive organization of operational grammar and measurement workflows, while CRQC–LLM examines staging, control, verification, and measurement-boundary risks.
- **In the encoding chapters,** QALIS evaluates representational capability and computational utility, while CRQC–LLM examines fidelity thresholds, manipulation exposure, and representational instability.
- **In the learning-architecture chapters,** QALIS addresses model construction, learning, inference, and optimization, while CRQC–LLM evaluates coherence budgets, inference-depth limitations, adversarial exposure, and verification requirements.
- **In the communication and sensing chapters,** QALIS evaluates distributed coordination and information use, while CRQC–LLM examines synchronization discipline, cryptographic exposure, provenance, and trust boundaries.
- **In the governance and resilience chapters,** the two lenses converge through verification placement, security controls, accountability, lifecycle oversight, and cross-layer resilience.

QALIS and CRQC–LLM are therefore not case studies in the episodic sense. They are persistent methodological instruments that structure the book's analytical progression. QALIS asks how a mechanism can be assembled into useful learning and inference capability. CRQC–LLM asks whether that capability can remain secure, resilient, verifiable, and governable under cryptographically relevant and operationally realistic conditions.

With the evaluative contract established, Section 1.2.4 differentiates QAIS, QALIS, and CRQC–LLM architecturally. QAIS is presented as the overarching system discipline and architectural framework; QALIS is identified as a learning-and-inference system realization and constructive evaluation lens; and CRQC–LLM is identified as a specialized cryptographic-security, constraint, resilience, and verification construct.

The QAIS System-Layer Model is then introduced to formalize how capability emerges—and how constraint propagates—through cross-layer integration.

The progression therefore moves from interpretive orientation to structural consolidation. Having defined how QAIS mechanisms must be evaluated, the chapter next clarifies how QAIS is internally organized as a governed, cross-layer architecture operating under physical, informational, operational, security, and institutional constraints.

1.2.4 Differentiating QALIS, QAIS, and CRQC–LLM

QAIS is the systems-level discipline studied throughout this volume. QAIS defines how quantum resources, operational primitives, engineered architectures, and hybrid workflows integrate to produce scalable quantum-enabled intelligent systems spanning sensing, communication, optimization, learning, reasoning, and structured decision-making. QAIS is not a subsystem or a single implementation; it is the architectural and operational domain within which intelligent quantum–classical systems are designed, analyzed, evaluated, and deployed.

Within this discipline:

- **QALIS (Quantum AI Learning and Inference System)** is the constructive evaluation framework used throughout the book to examine how QAIS capabilities emerge in practice. Through the QALIS lens, superposition, entanglement, coherence, interference, and measurement are evaluated as enabling resources that support learning workflows, inference pipelines, adaptive control loops, and distributed coordination mechanisms. QALIS emphasizes capability realization, architectural coherence, and governed system performance within QAIS environments.
- **CRQC–LLM (Cryptographically Relevant Quantum Computing–Large Language Model)** is the structured stress-test evaluation framework used to examine the resilience, verification boundaries, security posture, and systemic robustness of QAIS architectures. Through the CRQC–LLM lens, identical mechanisms are evaluated under adversarial, failure-inducing, and propagation-sensitive conditions to expose synchronization limits, coherence fragility, interface vulnerabilities, interpretability challenges, and governance constraints. CRQC–LLM emphasizes architectural stress, bounded propagation, and resilience analysis within QAIS environments.

Together, these frameworks reinforce the chapter's governing principle: quantum phenomena become system-level resources whose value must be evaluated simultaneously in terms of capability and constraint. QALIS provides the constructive lens for examining how QAIS architectures succeed, while CRQC–LLM provides the stress-test lens for examining how those same architectures fail, degrade, or require additional governance and resilience controls. Neither framework replaces QAIS itself; both operate as complementary evaluative perspectives within the broader QAIS discipline.

1.2.5 QAIS System-Layer Architecture

From Substrate to Intelligent Deployment

Layer-Reference Note: To distinguish the six-layer QAIS System-Layer Architecture from the seven-layer conceptual Quantum Stack, references to **"layer" or "layers" within this section refer to the QAIS System-Layer Architecture and its corresponding System-Layer Model unless explicitly identified as a "Quantum Stack Layer."** References to the conceptual progression will use the explicit designation **"Quantum Stack Layer."**

A coherent understanding of QAIS requires decomposing system functionality into distinct yet interacting layers, each responsible for specific roles within the overall architecture. This layered decomposition enables systematic analysis of how information is represented, transformed, and controlled across the system.

Figure 1.2.2 lays out this structure within a formal framework, providing a foundation for reasoning about system organization, interaction, and cross-layer dependencies.

The solid annotation denotes the principal QALIS realization span across the Ecosystem and Infrastructure, Architectures and Learning Models, and Applications and Hybrid Deployment layers. The dashed connector pathways denote CRQC–LLM cross-layer oversight and traceability across relevant concerns in all six layers. These scope connectors do not indicate direct component control. Together, these annotations distinguish between the principal QALIS realization span and CRQC–LLM cross-layer oversight and traceability within distributed QAIS environments.

The QAIS System-Layer Model complements the Quantum Stack introduced in Section 1.1.1. Whereas the Quantum Stack traces the progression from physical principles to QAIS as a discipline, the System-Layer Model provides an internal architectural view, detailing how functionality is distributed and coordinated across layers within a QAIS system.

Positioned at the close of this introductory sequence, the model reinforces a central thesis: QAIS capability emerges progressively from quantum physical principles through structured operational and architectural integration, culminating in deployable intelligent systems.

The Organizational Logic of QAIS

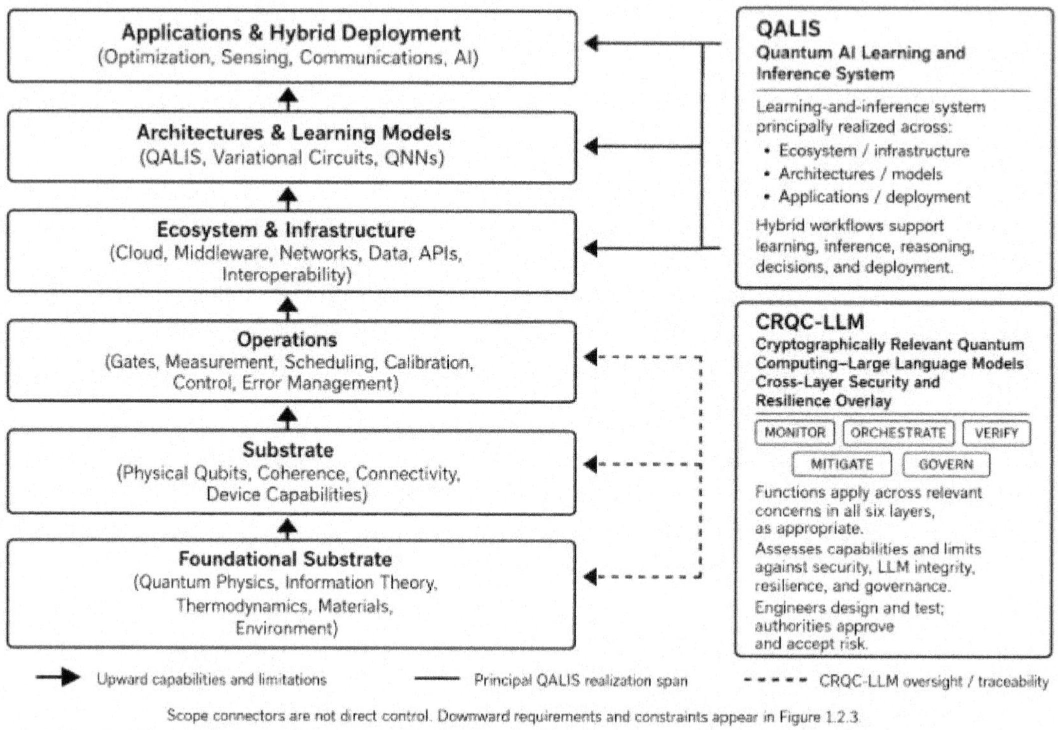

Figure 1.2.2 — QAIS System-Layer Model with QALIS and CRQC–LLM Cross-Layer Roles.

This figure presents the canonical six-layer QAIS System-Layer Model, progressing from the Foundational Substrate through the Substrate, Operations, Ecosystem and Infrastructure, Architectures and Learning Models, and Applications and Hybrid Deployment layers. he solid upward arrows represent the upward progression of capabilities and limitations from lower to higher layers toward deployable hybrid intelligence. QALIS is shown as a distinct learning-and-inference system principally realized across the Ecosystem and Infrastructure, Architectures and Learning Models, and Applications and Hybrid Deployment layers. CRQC–LLM is represented separately as a cross-layer overlay that specifies and coordinates cryptographic-security, LLM-integrity, monitoring, orchestration, verification, mitigation, resilience, and governance mechanisms across relevant concerns in all six layers, as appropriate to each layer. The connectors indicate architectural scope and traceability rather than direct control. Downward requirements and constraints, together with their reconciliation against upward capabilities and limitations, are detailed in Figure 1.2.3.

The Cross-layer Roles of QALIS and CRQC–LLM

Neither QALIS nor CRQC–LLM constitutes an additional numbered layer of the QAIS System-Layer Model. They are distinct constructs with different architectural roles.

QALIS—the Quantum AI Learning and Inference System—is a learning-and-inference system principally realized across the Ecosystem and Infrastructure, Architectures and Learning Models, and Applications and Hybrid Deployment layers.

It organizes the quantum, classical, data, model, and orchestration capabilities required to support learning, inference, adaptation, and operational decision-making, while interacting with verification and governance mechanisms provided by the broader QAIS architecture and, where relevant, coordinated through CRQC–LLM.

CRQC–LLM—the Cryptographically Relevant Quantum Computing–Large Language Model construct—is represented separately as a specialized cross-layer overlay. It specifies and coordinates mechanisms for monitoring, orchestrating, verifying, mitigating, and governing cryptographic-security, LLM-integrity, resilience, provenance, and related risks across relevant concerns in all six QAIS layers. These functions apply differently according to the responsibilities and constraints of each layer.

CRQC–LLM does not replace the systems-engineering process, independently control the layers, or perform every form of system validation. Engineers, specialized tools, system owners, security authorities, and governance bodies retain responsibility for design, implementation, testing, approval, and risk acceptance. CRQC–LLM supports these actors by coordinating relevant evidence, controls, traceability, verification, and mitigation across layer boundaries.

By distributing capability through integration across layers rather than isolating intelligence within a single component, QAIS reinforces the progression from physical constraint to governed intelligent deployment. Intelligent performance is not attributed to one breakthrough mechanism. It results from disciplined coordination among physical resources, operational control, infrastructure services, representational methods, learning architectures, and deployment governance.

Each layer receives capabilities and limitations from the layers below while translating higher-level requirements into more specific constraints for the layers beneath it. Coherence budgets, fidelity thresholds, latency limits, synchronization tolerances, security requirements, and verification criteria collectively determine which forms of intelligent behavior are physically and operationally sustainable. Capability therefore emerges through the systematic alignment of physics, engineering discipline, architectural intent, and governance across the complete QAIS lifecycle.

Where QAIS Resides in the System-Layer Model

QAIS is not confined to a single numbered layer or only to the upper portion of the architecture. The QAIS reference architecture encompasses all six layers:

layer 1 — Foundational Substrate. Defines the physical, thermodynamic, causal, material, environmental, and information-theoretic boundaries within which QAIS can exist.

layer 2 — Substrate. Realizes usable quantum resources through physical qubits, devices, coherence resources, connectivity mechanisms, control interfaces, and operating environments.

layer 3 — Operations. Converts substrate resources into controlled quantum behavior through state preparation, gates, calibration, scheduling, measurement, compilation, control, and error-management processes.

layer 4 — Ecosystem and Infrastructure. Integrates quantum and classical resources through computing platforms, middleware, networks, data services, APIs, security mechanisms, interoperability services, and orchestration environments.

layer 5 — Architectures and Learning Models. Composes infrastructure and quantum capabilities into learning, inference, optimization, and reasoning architectures, including QALIS, variational circuits, quantum neural networks, and hybrid quantum–classical models.

layer 6 — Applications and Hybrid Deployment. Operationalizes QAIS architectures within real-world environments supporting optimization, sensing, communication, learning, inference, decision support, and other hybrid quantum–classical applications.

All six layers are required to describe a complete Quantum AI System.

The lower layers establish physical feasibility, quantum resources, operational control, and executable behavior. The upper layers provide infrastructure integration, architectural composition, learning and inference capabilities, and operational deployment.

The Foundational Substrate defines the governing physical, thermodynamic, causal, and information-theoretic boundaries. The Substrate realizes quantum resources through selected qubit technologies, devices, materials, and operating environments. Operations converts those resources into controlled transformations through calibration, gates, scheduling, measurement, compilation, and error-management processes.

At the Ecosystem and Infrastructure layer, quantum and classical resources are connected through cloud platforms, middleware, networks, data services, security mechanisms, APIs, and orchestration environments. This layer provides the operational foundation through which higher-level learning architectures can access and coordinate distributed resources.

QAIS becomes most visibly associated with intelligent system composition at the Architectures and Learning Models layer. Here, quantum and classical resources are assembled into variational circuits, quantum neural networks, inference pipelines, hybrid learning models, and systems such as QALIS.

At the Applications and Hybrid Deployment layer, QAIS becomes operationally manifest within enterprise and domain-specific environments. Quantum computing, classical AI, data, sensing, optimization, communication, security, and governance capabilities are integrated into deployable workflows. Architectural discipline must withstand environmental noise, resource constraints, scaling pressure, adversarial conditions, and regulatory obligations at this layer.

QAIS therefore spans all six layers. Its physical feasibility is established in the lower layers, its integration is enabled through the middle layers, and its intelligent and operational identity becomes most visible in the upper layers.

The Organizational Logic of QAIS

The QAIS System-Layer Model demonstrates that intelligent quantum systems do not emerge at one abstraction boundary. They result from disciplined alignment across substrate physics, operational control, infrastructure integration, architectural composition, learning and inference, and deployment governance. Each layer enables and constrains the layers above while responding to requirements imposed from above.

At the foundational and substrate levels, quantum mechanics and the selected physical implementation establish the resource envelope, including coherence lifetimes, entanglement capabilities, connectivity, fidelity, measurement limits, and environmental requirements. Operational control translates these resources into executable transformations through calibrated gates, noise-aware scheduling, measurement, and error management.

Ecosystem and infrastructure services make those transformations accessible through computing platforms, middleware, networks, APIs, data services, security controls, and orchestration mechanisms.

Encoding and representational decisions determine how information is mapped into classical and quantum state spaces, influencing expressiveness, trainability, robustness, resource consumption, and verification. Architectural composition then combines these elements into coordinated learning and inference pipelines.

Deployment places those pipelines within organizational, regulatory, mission, and ecosystem conditions that determine scalability, resilience, trust, and acceptability.

By making these relationships explicit, the model reveals:

- how physical and information-theoretic limits propagate upward into architectural feasibility;
- how calibration, coherence, fidelity, latency, and synchronization constrain learning and inference;
- how encoding and model choices affect scalability, reliability, and verification;
- how mission and governance requirements propagate downward into engineering specifications; and
- how demonstrated capabilities and limitations must be reconciled with required performance and assurance criteria.

The QAIS System-Layer Model therefore functions as an architectural dependency and traceability map rather than merely a descriptive taxonomy. It exposes vertical dependencies and discourages isolated optimization. Coherence limits become architectural boundary conditions. Encoding strategies affect resource demand, state-space structure, noise exposure, measurement design, and verification placement. Security and governance requirements influence interfaces, trust boundaries, data handling, measurement staging, provenance, and deployment approval.

QAIS systems emerge only when capability, resilience, security, verification, interoperability, and governance are co-engineered across layers. Cross-layer co-engineering is performed by interdisciplinary engineering and governance teams through the iterative reconciliation of two flows:

Upward capabilities and limitations ⟷ ***Downward requirements and constraints***

If a lower layer cannot satisfy a downward requirement, the limitation is reported upward. Engineers must then redesign the architecture, improve or replace the technology, modify the allocation of quantum and classical resources, introduce mitigation, formally adjust the requirement, accept authorized residual risk, or reject the deployment.

CRQC–LLM supports this process for relevant cryptographic, LLM-integrity, security, resilience, provenance, and governance concerns.

Through its Monitor, Orchestrate, Verify, Mitigate, and Govern functions, it helps coordinate evidence and controls across the architecture without replacing the engineers, specialized validation tools, system owners, or authorized decision-makers.

Figure 1.2.3 illustrates this bidirectional relationship from a cross-layer perspective. The upward flow communicates the capabilities and limitations that the implemented layers can provide. The downward flow communicates the requirements and constraints that lower layers must satisfy. Their iterative reconciliation determines whether the proposed QAIS architecture is feasible, verifiable, secure, resilient, and suitable for deployment.

Cross-Layer Constraint and Capability Flow in QAIS

While Figure 1.2.2 decomposes QAIS into architectural layers, Figure 1.2.3 shows how capability, constraint, orchestration, security, and resilience propagate across and between those layers. Intelligent performance emerges only when upward capabilities and limitations and downward requirements and constraints are continuously co-engineered.

The figure also distinguishes the architectural scopes of QALIS and CRQC–LLM. QALIS represents the principal learning-and-inference realization spanning the upper layers of the architecture, while CRQC–LLM provides a specialized cross-layer security and resilience overlay. These scope relationships represent architectural dependencies and responsibilities rather than direct control of every connected component.

QAIS Architecture Model—Architectural Validation and Design Rationale for the QAIS Cross-Layer Capability, Constraint, and Orchestration Model

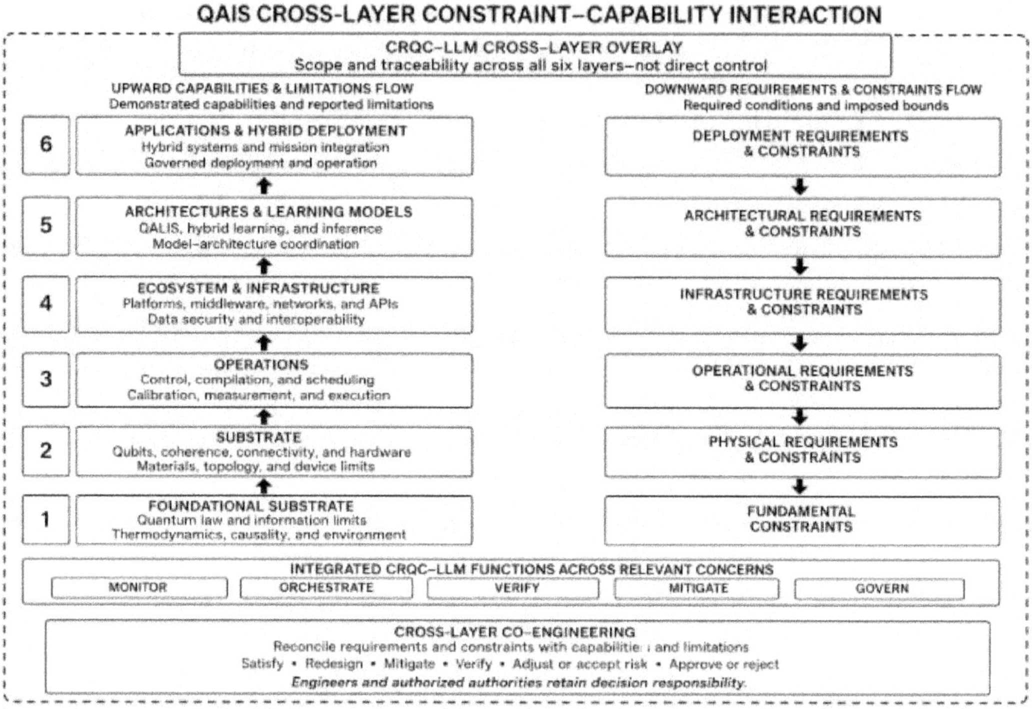

Figure 1.2.3 — Cross-Layer Capability, Constraint, and Orchestration Flow in QAIS.

This figure presents the canonical six-layer QAIS architecture, progressing from foundational physical and informational principles through quantum resources, operational control, ecosystem infrastructure, learning architectures, and hybrid intelligent deployment environments. The model emphasizes that intelligent quantum capability emerges through coordinated cross-layer integration rather than within any single architectural layer.

Upward capability flow illustrates how physical resources and operational mechanisms progressively enable infrastructure, learning, inference, and application capabilities. Downward constraint flow demonstrates how deployment objectives, architectural requirements, and operational demands propagate toward implementation while ultimately remaining bounded by physical and information-theoretic limits.

The QALIS scope connector identifies the principal architectural span through which quantum-AI learning and inference capabilities are assembled across ecosystem infrastructure, architectures and learning models, and applications and hybrid deployment. The CRQC–LLM overlay identifies the cross-layer security, cryptographic intelligence, verification, mitigation, resilience, and governance functions required to manage cryptographically relevant risks across the QAIS lifecycle. These connectors denote architectural scope, dependency, and responsibility rather than direct command or control.

Figure Interpretation

This figure should be interpreted as a cross-layer control and co-engineering architecture for QAIS, illustrating how capabilities, constraints, orchestration mechanisms, learning and inference functions, security controls, and governance processes interact throughout the system lifecycle. Rather than depicting a simple hierarchical stack, the model emphasizes that intelligent quantum-enabled behavior emerges through coordinated interaction among physical resources, operational controls, infrastructure services, learning architectures, security overlays, and deployment environments.

The upward and downward arrows represent architectural flows. The QALIS and CRQC–LLM connectors identify the layers across which those respective systems depend on, integrate with, or provide specialized functions. They do not imply direct control over every component within those layers.

Upward Capability Flow—Enablement

The upward capability flow represents the progressive accumulation of enabling functionality across the QAIS architecture:

1. **Foundational Substrate layer → Substrate layer:** Quantum mechanics, information-theoretic limits, thermodynamics, materials, and environmental conditions establish the physical and informational foundations upon which all quantum behavior depends.
2. **Substrate layer → Operations layer:** Physical qubits, coherence resources, connectivity mechanisms, and hardware capabilities provide the resources required to execute quantum operations.
3. **Operations layer → Ecosystem and Infrastructure layer:** Gates, measurement, scheduling, calibration, control systems, and error-mitigation mechanisms transform hardware resources into usable computational services.
4. **Ecosystem and Infrastructure layer → Architectures and Learning Models layer:** Quantum hardware, cloud platforms, middleware, networks, data services, and interoperability mechanisms create the environment necessary for scalable learning architectures and intelligent-system development.
5. **Architectures and Learning Models layer → Applications and Hybrid Deployment layer:** QALIS, variational circuits, quantum neural networks, and hybrid quantum–classical learning architectures enable intelligent applications, decision-support capabilities, and operational deployment environments.

The upward flow therefore illustrates how increasingly sophisticated capabilities emerge from lower-layer resources and enabling mechanisms. QALIS depends on this accumulated capability; it does not exist independently of the infrastructure, operational services, and physical resources that support it.

QALIS Architectural Scope

QALIS—Quantum AI Learning and Inference System—occupies the principal learning-and-inference realization span of QAIS. It integrates capabilities across the following upper layers:

- **Ecosystem and Infrastructure:** Provides computing services, middleware, data resources, networking, interoperability, and access to quantum and classical resources.
- **Architectures and Learning Models:** Provides quantum, classical, and hybrid models for learning, inference, representation, optimization, and reasoning.
- **Applications and Hybrid Deployment:** Operationalizes learning and inference capabilities within domain-specific and enterprise environments.

QALIS is therefore not an isolated layer or a stand-alone quantum computer. It is a system realization that depends on coordinated services and capabilities distributed across multiple QAIS layers.

CRQC–LLM Overlay—Cross-Layer Security and Resilience

CRQC–LLM—Cryptographically Relevant Quantum Computing–Large Language Models—provides a specialized cross-layer security and resilience overlay within QAIS. It supports the identification, interpretation, coordination, and mitigation of cryptographically relevant risks arising across quantum, classical, AI, data, infrastructure, and governance environments.

Its cross-layer functions include:

- **Monitor:** Observe security posture, cryptographic dependencies, system health, operational state, performance metrics, environmental conditions, model behavior, and contextual information across relevant layers.
- **Orchestrate:** Distribute risk information, dependencies, constraints, policies, alerts, cryptographic-transition requirements, and coordination directives throughout the architecture.
- **Verify:** Evaluate correctness, consistency, interoperability, security control effectiveness, cryptographic posture, compliance, and policy conformance.
- **Mitigate:** Adapt system behavior to address security degradation, cryptographic exposure, synchronization drift, operational disruption, adversarial activity, resilience threats, and performance instability.
- **Govern:** Support security oversight, policy enforcement, compliance assurance, auditability, lifecycle management, cryptographic transition governance, and organizational accountability.

These functions establish a security- and resilience-oriented coordination framework. They complement the broader orchestration mechanisms of QAIS rather than replacing them.

Downward Constraint Flow—Limitations and Requirements

The downward constraint flow represents the propagation of requirements and limitations from higher-level objectives toward lower layers:

1. **Deployment Constraints → Architectural Constraints:** Application requirements for performance, reliability, scalability, security, compliance, and cost restrict the available architectural design choices.
2. **Architectural Constraints → Infrastructure Constraints:** Architectural decisions establish infrastructure requirements involving networking, interoperability, service availability, bandwidth, data access, and resource allocation.
3. **Infrastructure Constraints → Operational Constraints:** Infrastructure capabilities affect scheduling, calibration accuracy, control precision, measurement fidelity, latency, availability, and error-management performance.
4. **Operational Constraints → Physical Constraints:** Operational demands encounter hardware limitations such as coherence time, crosstalk, thermal effects, connectivity restrictions, control fidelity, and environmental noise.
5. **Physical Constraints → Fundamental Constraints:** All physical implementations remain bounded by quantum uncertainty, no-cloning restrictions, thermodynamic limits, information-theoretic constraints, material limitations, and environmental conditions.

The downward flow emphasizes that higher-level objectives must remain compatible with lower-level physical realities and available resources. These flows are iterative: discovered physical or operational limitations may require higher layers to revise their architectures, models, policies, or deployment objectives.

Cross-Layer Co-Engineering

The Cross-Layer Co-Engineering foundation represents the system-wide activities required to preserve long-term architectural integrity:

- **Alignment:** Ensure that objectives, requirements, capabilities, and constraints remain consistent across all layers.
- **Coordination:** Synchronize activities among distributed quantum, classical, AI, networking, data, and governance components.
- **Verification:** Validate correctness, consistency, interoperability, security, and operational conformance.
- **Resilience:** Preserve stability, adapt to changing conditions, and recover from disruption or failure.

- **Governance:** Enforce policies, accountability, compliance requirements, risk controls, and lifecycle oversight.

Viewed collectively, the model demonstrates that capability creation, constraint management, learning and inference, orchestration, security, resilience, verification, and governance cannot be optimized independently within isolated layers. Intelligent quantum-enabled behavior emerges through continuous co-engineering across the entire QAIS architecture.

Technical Spotlight—Cross-Layer Control in QAIS

The organizational logic of QAIS is not hierarchical in a purely linear sense; it is cybernetic. Each layer enables capabilities above it while remaining subject to requirements imposed from above and limitations inherited from the layers below. Substrate coherence budgets determine the feasible depth of quantum circuits. Operational noise constrains representational fidelity. Encoding decisions influence architectural scalability and composability. Infrastructure and governance readiness determine whether deployment is technically sustainable, secure, compliant, and operationally permissible.

In this framing, optimization within a single layer is insufficient. QAIS demands cross-layer equilibrium. Capability without constraint awareness produces fragility. Constraint without architectural compensation produces stagnation.

The system must therefore be engineered as a whole, with physical law, operational control, representational design, architectural composition, security, resilience, and deployment governance treated as coupled variables rather than independent modules.

Figure 1.2.3 should consequently be read not as a static taxonomy but as a dynamic control and co-engineering architecture in which capability accumulates upward, requirements propagate downward, and feedback moves throughout the system. Intelligent performance emerges only when these directional pressures and feedback relationships are coherently aligned.

Case Study — Architectural Realization Alignment

Intent: To formalize QAIS as a cross-layer control and co-engineering architecture rather than merely a stacked abstraction model.

- **QALIS** interprets the ascent through the layers as constructive system assembly. Physical and operational capabilities enable infrastructure services, structured learning, coordinated inference, and scalable intelligent behavior through disciplined systems integration.

- **CRQC-LLM** interprets the architecture through a specialized cryptographic-security and resilience perspective. Synchronization drift, coherence decay, adversarial perturbation, encoding instability, cryptographic exposure, governance failure, and infrastructure fragility can propagate across interacting layers. Intelligent-system security and stability therefore depend on continuous cross-layer monitoring, verification, coordination, and mitigation rather than isolated optimization within individual abstraction domains.

Together, these realizations reinforce a central QAIS design principle: intelligent capability, systemic constraint, security, and resilience are inseparable variables that must be co-engineered throughout the full system lifecycle.

Reframing Information as an Architectural Resource

marks the book's first decisive transition in scope: from QIS as the study of quantum information, its Quantum Physics, and its information-processing consequences to QAIS as a system discipline that transforms those foundations into architectural rules. QIS established that superposition, interference, entanglement, coherence, and measurement effects are physically real, mathematically describable, and potentially exploitable for information processing.

QAIS extends this foundation by treating these mechanisms not merely as phenomena to observe, but as resources to engineer, constraints to budget, and verification surfaces to govern. QAIS therefore asks not only what quantum information is, but also how it must be represented, prepared, protected, transformed, measured, and sustained to support learning, inference, coordination, optimization, and trustworthy decision-making under realistic operational limits.

This reframing clarifies why QAIS cannot be reduced to "quantum computing plus AI." Within QAIS, information is physical and therefore:

- lifetime-limited by coherence;
- correlation-structured through entanglement;
- phase- and amplitude-dependent in its representation;
- interface-constrained by measurement at the quantum–classical boundary; and
- integrity-sensitive because disturbance is a consequence of physical law, not merely an implementation defect.

These properties function as architectural constraints. They shape how computational and learning pipelines are composed, where trust and verification boundaries must be placed, how control and measurement must be staged, and which design trade-offs are physically and operationally admissible. They also determine how scaling, synchronization, error management, and hybrid integration must be disciplined so that expressive capacity does not exceed physical feasibility.

By elevating quantum information to architectural status, QAIS requires representation, control flow, security, resilience, verification, and governance to be co-designed from system inception. Information is no longer treated merely as a payload transported through neutral infrastructure. It becomes part of the governing substrate that determines which structures can be stabilized, which correlations can be trusted, which transformations can be executed reliably, and which decisions can be operationally justified under constraint.

Figure 1.2.4 consolidates this transition by organizing the architectural inputs into three categories:

- foundational quantum phenomena and resources, including superposition, interference, entanglement, and coherence;
- measurement as the quantum–classical resolution boundary; and
- hybrid orchestration as the system-integration mechanism coordinating quantum and classical resources.

These inputs are evaluated through the complementary perspectives supplied by QALIS and CRQC–LLM before being incorporated into integrated QAIS architectures. QALIS examines how the inputs can be assembled into constructive learning, inference, coordination, optimization, and decision-support capabilities.

CRQC–LLM examines the associated cryptographic-security exposures, systemic constraints, resilience requirements, and verification obligations.

The model does not imply that QIS ends where QAIS begins. Rather, QAIS operationalizes QIS principles within engineered, governed, secure, and deployable system frameworks.

Placed after the Quantum Stack in Figure 1.1.1 and the QAIS System-Layer Model, Figures 1.2.2 and 1.2.4 perform a specific organizing function by connecting progression across the layers to system-level design constraints. The Quantum Stack establishes the position of QAIS within the progression from physics to intelligent application. The System-Layer Model characterizes the internal decomposition and cross-layer relationships of QAIS. Figure 1.2.4 then explains why quantum information must be treated not as an abstract payload, but as a governed architectural resource subject to physical, informational, operational, security, and institutional constraints.

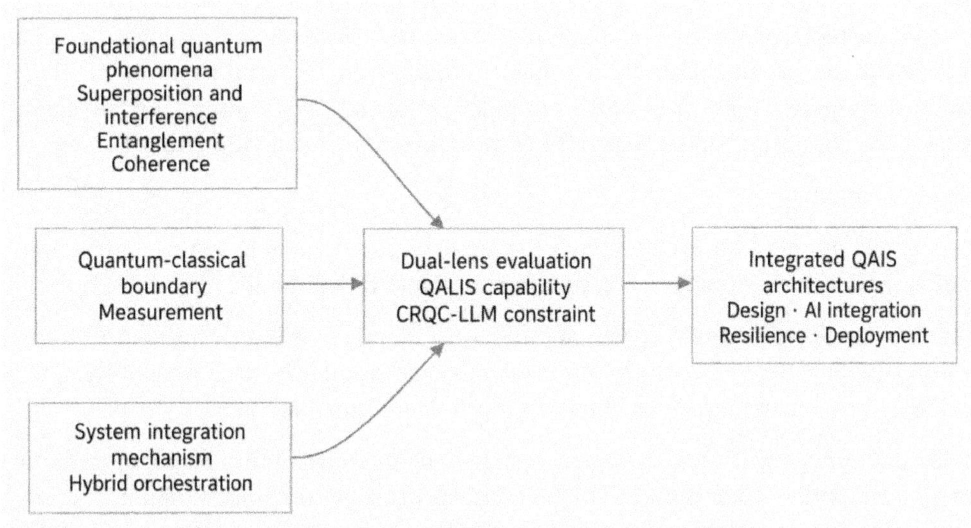

Figure 1.2.4 — From QIS to QAIS as a System Discipline.

Foundational quantum phenomena and resources, measurement as the quantum–classical boundary, and hybrid orchestration as the system-integration mechanism are evaluated through complementary QALIS and CRQC–LLM perspectives. QALIS examines constructive capability and system realization, while CRQC–LLM examines cryptographic security, systemic constraint, resilience, and verification. Their combined evaluation informs the design, AI integration, resilience, and deployment of integrated QAIS architectures.

Technical Spotlight—Information as a System Resource

This spotlight elaborates on the transition from the foundational study of quantum information to its disciplined use within intelligent system architectures. Quantum behavior becomes not only a source of computational potential but also a source of architectural requirements, operating limits, security exposures, and verification responsibilities.

Conceptual Grounding

QIS demonstrates that information can be encoded quantum mechanically through amplitudes, phases, correlations, and coherence. QAIS accepts that foundation but changes the primary engineering question from existence to architecture. Quantum properties become design resources that must be composed into pipelines with explicit assumptions about coherence budgets, correlation structures, noise, control fidelity, timing, and readout constraints.

Within QAIS, information cannot be separated from the processes through which it is prepared, encoded, protected, transformed, transmitted, measured, verified, and resolved into classical evidence. A quantum state is therefore not treated as an abstract data container. It is a physically instantiated and temporally bounded information structure whose usefulness depends on the conditions under which it can be maintained and manipulated.

System Relevance

Treating information as an architectural resource requires QAIS to make essential system properties explicit rather than leaving them as implementation details.

Coherence becomes a temporal resource budget. It bounds feasible circuit depth, constrains synchronization strategies, limits the duration of learning and inference workflows, and helps determine whether distributed coordination is physically sustainable.

Entanglement functions as a correlation resource connecting qubits or subsystems. It can support coordinated computation, distributed protocols, sensing, communication, and verification. At the same time, it introduces fragility, synchronization demands, validation overhead, and expanded trust boundaries that must be managed architecturally.

Superposition and interference provide structured possibility spaces through which algorithms can transform probability amplitudes. Their usefulness depends on precise state preparation, controlled evolution, phase relationships, and measurement design. The same properties that enable computational advantage also create sensitivity to noise, drift, control error, and adversarial manipulation.

Fidelity, entropy, and trace distance cease to be merely retrospective diagnostic quantities. Within a systems architecture, they become forward-looking design thresholds used to determine whether state preparation, transformation, transmission, or inference remains sufficiently accurate, stable, and trustworthy under noise.

Measurement operates as a quantum–classical resolution boundary. It converts accumulated quantum-state structure into classical evidence and therefore constrains how inference results can be expressed, interpreted, validated, and acted upon. Because measurement generally changes the state being measured, its placement, timing, basis, and repetition strategy must be designed as part of the architecture.

Hybrid orchestration coordinates quantum processors, classical computing resources, AI models, middleware, data services, control systems, and human or automated decision processes. It transforms individual quantum operations into usable system workflows, but it also introduces interface dependencies, latency, propagation risks, security boundaries, and governance obligations.

These realities explain why QAIS architectures cannot treat resilience, security, or verification as afterthoughts. Verification checkpoints, calibration discipline, provenance controls, error-management mechanisms, and governance responsibilities must be embedded structurally within the system rather than appended as external features after implementation.

Cross-Disciplinary Orientation

Reframing information as a governed architectural resource aligns disciplines that have historically approached quantum systems from different perspectives.

Physicists recognize a transition from demonstrating quantum effects to producing reproducible, controllable, and measurable operational behavior that can be relied upon within larger systems.

Engineers interpret the same transition as formalized constraint management. Coherence lifetimes, noise budgets, timing tolerances, control fidelity, drift behavior, connectivity, and interface stability become first-order architectural parameters rather than secondary implementation details.

AI specialists see an expansion of representational and computational capacity bounded by physics. Quantum-state geometry, interference, and correlation structure may enrich learning, inference, search, and optimization, while remaining constrained by coherence duration, noise, data-encoding cost, measurement limitations, and classical coordination requirements.

Cybersecurity practitioners encounter physics-informed trust and verification boundaries. No-cloning restrictions, disturbance sensitivity, entropy dynamics, interface exposure, cryptographic dependencies, and measurement rules help determine what can be protected or verified, when verification is possible, and how compromise or failure may propagate across architectural layers.

Governance and risk professionals see the need to connect technical limits with accountability, policy, auditability, compliance, resource authorization, and lifecycle oversight. A quantum-enabled capability cannot be considered operationally acceptable merely because it performs a desired computation; its behavior must also be explainable, controllable, secure, resilient, and consistent with institutional requirements.

QAIS therefore does not merely combine disciplinary insights. It integrates them within a unified system-level logic grounded in physical law and governed through architectural discipline.

Architectural Realization—Reframing Information as a System Resource

Intent: To demonstrate how quantum phenomena, measurement boundaries, and hybrid orchestration become QAIS architectural rules evaluated for constructive capability, security, resilience, and verification.

- **QALIS:** Treats superposition and interference, entanglement, coherence, measurement, and hybrid orchestration as composable resources for constructing learning and inference pipelines. These pipelines incorporate explicit resource budgets, control processes, validation checkpoints, and interpretable readout structures.
- **CRQC-LLM:** Treats the same mechanisms as cryptographically relevant security, constraint, resilience, and verification surfaces. It examines how coherence loss, correlation degradation, interface compromise, readout manipulation, cryptographic exposure, adversarial timing, automated exploitation, and verification bottlenecks can affect system trustworthiness.

Section 1.2 Summary — Case Study Orientation and Alignment

Together, QALIS and CRQC-LLM demonstrate the dual character of quantum information within QAIS: every mechanism that enables capability also establishes corresponding conditions for security, stability, verification, and governance.

This subsection establishes the chapter's final introductory insight: QAIS transforms quantum information into architectural law. Foundational quantum properties do not merely supply computational effects; they establish the rules, limits, interfaces, and responsibilities under which intelligent quantum-enabled systems must be engineered.

With the QAIS System-Layer Architecture established, the chapter has completed its initial conceptual and architectural framing. The Quantum Stack has positioned QAIS within the progression from physical law to intelligent-system architecture, while the System-Layer Model has clarified how physical resources, operational control, infrastructure, learning architectures, deployment functions, and cross-layer constraints interact within a complete QAIS system.

The discussion now turns to Section 1.3, Historical Foundations of QAIS, which traces the intellectual progression through which information evolved from classical abstraction to a physically instantiated quantum resource. This historical development establishes the scientific lineage required to understand why quantum information became an engineering resource and how that transformation ultimately created the foundations upon which QAIS is built.

1.3 Historical Foundations of QAIS

Architectural Framing: *Quantum States → System Architecture*

Section Framing: *This section positions QAIS within its architectural lineage, establishing how representation evolved into a physically instantiated control surface governed by quantum law. The progression defines how propagation transitioned from symbolic abstraction to structured system behavior constrained by interference, coherence, and measurement. Stability is formalized as bounded propagation enforced through physical limits rather than algorithmic guarantees. This historical framing establishes the conditions under which QAIS emerges as a system discipline grounded in constraint-aware architecture.*

QAIS emerges from the convergence of two twentieth-century revolutions: classical information theory and quantum mechanics. Their integration did not occur instantaneously but unfolded through layered conceptual redefinitions that gradually transformed information from an abstract mathematical construct into a physically instantiated resource. This section traces that progression in disciplined sequence, establishing the scientific lineage upon which the remainder of this chapter builds.

The first revolution began in 1948 with Claude Shannon's formalization of information theory. Shannon defined information statistically, through entropy and probability distributions, deliberately separating message content from its physical carrier.

In this framework, bits could be encoded in voltage levels, magnetic states, optical pulses, or radio waves without altering the underlying mathematics. This abstraction enabled the digital age, but it embedded a critical assumption: information could be copied, transmitted, and processed independently of the physics of its substrate.

The second revolution emerged from quantum mechanics. The formal structure of Hilbert space, superposition, and unitary evolution revealed that physical systems do not merely carry information; they instantiate it.

A quantum state is not a container for predetermined values but a structured amplitude distribution whose measurement outcomes are probabilistic and disturbance-sensitive. Entanglement demonstrated that composite systems encode correlations that cannot be decomposed into independent classical components. Measurement participates in state determination rather than passively revealing preexisting properties. Together, these results destabilized the classical separation between information and physical law.

The decisive pivot occurred in the late twentieth century when researchers began exploring whether quantum mechanics could be exploited computationally. Richard Feynman proposed simulating physics using quantum systems, arguing that classical simulation of quantum dynamics requires exponential resources [1]. David Deutsch formalized the concept of a universal quantum computer, providing a theoretical model capable of executing transformations governed by quantum mechanics [2]. Later algorithmic breakthroughs, including Shor's factoring algorithm [3] and Grover's search algorithm [4], demonstrated that quantum systems could outperform classical machines for defined problem classes.

These results reframed quantum phenomena as computational resources rather than interpretive curiosities. Equally transformative was the development of quantum error correction. Early assumptions held that decoherence rendered scalable quantum computation impractical.

Error-correcting codes showed that fragile quantum states could be stabilized through structured redundancy, transforming decoherence from fatal limitation into engineering constraint. Coherence time, noise thresholds, and correction overhead became architectural variables rather than insurmountable barriers.

By the early twenty-first century, QIS had matured into a coherent domain encompassing algorithms, cryptography, communication protocols, and hardware engineering. Advances in superconducting qubits, trapped-ion systems, photonics, and hybrid quantum–classical workflows transitioned quantum computation from theoretical speculation to experimental implementation.

Cloud-accessible processors further expanded accessibility and experimentation [9]. In the United States, coordinated federal investment accelerated through the National Quantum Initiative Act [10]. Parallel efforts in Europe formalized strategic quantum technology development under flagship programs [11].

Comparable national laboratory programs were established in Asia to accelerate quantum infrastructure deployment [12].

This historical sequence therefore establishes three essential foundations for QAIS:

- Information is physically instantiated.
- Quantum mechanics redefines the structure and limits of information processing.
- Computational and engineering advances transformed quantum phenomena into designable resources.

These foundations anchor the remainder of Chapter 1. The next subsection, Section 1.3.1, examines the emergence of quantum information as a discipline. Sections 1.3.2 through 1.3.4 then formalize the classical–quantum divergence and its architectural implications, culminating in Section 1.4, *The Nature of Quantum Information*, which examines qubits, superposition, and entanglement as operational primitives. The chapter then advances into Practical Foundations of Quantum AI Systems, where these physical properties are translated into system-level consequences for architecture, verification, and resilience within QAIS.

1.3.1 Emergence of Quantum Information as a Discipline

The 1990s marked a decisive transition in quantum science. What had previously been philosophical reflection on the informational implications of quantum mechanics matured into a coherent research discipline centered explicitly on **quantum information**.

The shift did not arise from reinterpretation alone; it was driven by concrete algorithmic, cryptographic, and engineering breakthroughs that demonstrated quantum systems could process, transmit, and stabilize information in ways fundamentally unavailable to classical machines.

The pivotal moment came in 1994 with Peter Shor's factoring algorithm, which showed that a quantum computer could factor large integers in polynomial time. Because integer factorization underlies RSA encryption, Shor's result signaled that widely deployed cryptographic systems were theoretically vulnerable to quantum attack [12]. This was not a marginal improvement but an exponential speedup over the best-known classical approaches.

Quantum mechanics was no longer merely a descriptive theory of microscopic systems; it had become computationally consequential.

Two years later, Lov Grover introduced a quantum search algorithm providing a quadratic speedup for unstructured database search. While less dramatic than Shor's exponential advantage, Grover's result demonstrated that quantum speedup was not restricted to narrow number-theoretic problems. Instead, it suggested broader applicability across optimization and search domains.

Together, these algorithms reframed interference and superposition as executable computational resources. At the same time, foundational work in quantum error correction overturned the belief that quantum states were too fragile for scalable computation.

Early codes developed by Shor, Steane, and others showed that logical qubits could be encoded across entangled physical qubits in a manner that allowed error detection and correction without directly measuring encoded information. Decoherence, once viewed as fatal, became an engineering constraint governed by threshold theorems and fault-tolerant design.

Communication protocols further reinforced this consolidation. Quantum teleportation demonstrated that quantum states could be transferred between distant parties using shared entanglement and classical communication, without physically transporting the original particle.

Quantum key distribution revealed that disturbance sensitivity could enforce security grounded in physical law rather than computational hardness [13]. These developments established entanglement and measurement disturbance as operational primitives rather than theoretical curiosities.

Collectively, these breakthroughs transformed quantum information from speculative abstraction into disciplined science. They demonstrated that quantum states could be manipulated, protected, transmitted, and verified under structured control. By the end of the decade, quantum information had acquired formal models, experimental platforms, error-correction frameworks, and algorithmic legitimacy. It had become a distinct and enduring discipline [5].

For QAIS, this emergence is foundational. It established that information is not merely symbolic but physically embodied, coherence-sensitive, and disturbance-aware. QAIS inherits these primitives and extends them into architectural frameworks for learning, inference, and resilience.

The discipline of quantum information provided the substrate; QAIS builds system-level intelligence upon it.

1.3.2 Classical vs Quantum Views of Information

To understand why QAIS constitutes a new system discipline rather than an incremental refinement of classical computing, it is necessary to contrast the foundational units of representation: the classical bit and the quantum bit (qubit). This comparison reveals a deeper divergence than encoding format. It reveals a difference in ontology, geometry, and operational constraint.

In classical information theory, a bit exists deterministically as either 0 or 1. Uncertainty reflects incomplete knowledge, not intrinsic indeterminacy. Logical gates manipulate these values through Boolean algebra, and copying information does not disturb its content. Representation is separable from physical substrate; voltage levels, magnetic domains, or optical pulses may encode bits without altering the governing mathematics.

Differences between classical and quantum approaches to information processing extend beyond implementation details to fundamentally distinct representational and operational paradigms.

These differences shape how information is encoded, manipulated, and interpreted within QAIS systems. Figure 1.3.1 shows this contrast within a comparative framework, highlighting the implications for system design and computational behavior.

Quantum Information Obeys Different Rules

A qubit exists as a normalized superposition:

$$|\psi\rangle = \alpha|0\rangle + \beta|1\rangle \tag{1.3.1}$$

where α and β are complex amplitudes satisfying:

$$|\alpha|^2 + |\beta|^2 = 1 \tag{1.3.2}$$

The magnitudes determine measurement probabilities. The relative phase between α and β determines interference behavior. This phase has no classical analogue, yet it governs

Unlike a classical bit, a qubit spans a continuous geometric state space. Pure states map to points on the surface of the Bloch sphere.

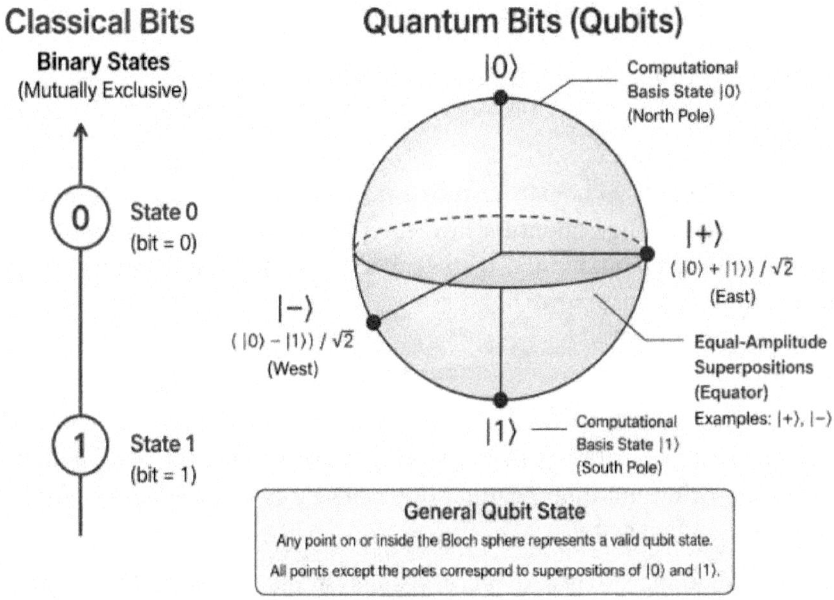

Figure 1.3.1 — Classical vs Quantum Views of Information.

This figure contrasts deterministic classical bits, which occupy mutually exclusive binary states, with quantum bits (qubits) represented on the Bloch sphere as continuous state vectors characterized by probability amplitudes and relative phase. The Bloch sphere illustrates the computational basis states |0⟩ and |1⟩ at the poles, equal-amplitude superposition states along the equator, and general qubit states throughout the sphere. The representation highlights how superposition expands representational capacity beyond classical binary encoding, how phase relationships govern interference behavior, and how quantum measurement projects qubit states into observable classical outcomes.

This comparison marks a structural pivot.

- Superposition permits structured coexistence of computational alternatives.
- Entanglement introduces non-separable correlations across subsystems.
- Coherence preserves amplitude relationships across evolution.
- Measurement collapses quantum structure into classical evidence.

Information therefore:

- Is not substrate independent.
- Cannot be copied arbitrarily (no-cloning)[2]
- Cannot be observed without disturbance.
- Cannot be decomposed cleanly into independent parts when entangled.

These properties fundamentally alter how intelligent systems must be architected.

Technical Spotlight — Information as a Physical Substrate

Conceptual Grounding

Shannon's framework abstracted symbols from substrate. Quantum mechanics removes that abstraction. If physical systems obey quantum law, then information encoded within them must obey superposition, entanglement, and disturbance constraints. Information becomes physically instantiated amplitude structure.

System Relevance

Within QAIS, the shift from classical abstraction to quantum embodiment fundamentally alters architectural assumptions. Superposition provides structured representational breadth within a single qubit register, allowing multiple computational pathways to coexist as amplitude-weighted components of a unified state.

Entanglement encodes relational structure across subsystems in ways that cannot be decomposed into independent classical correlations, transforming coordination and dependency modeling at the architectural level.

Measurement defines the quantum–classical interface, acting as a resolution boundary that converts coherent amplitude structure into classical evidence while irreversibly altering the state itself. Coherence constrains usable runtime, bounding pipeline depth, synchronization feasibility, and distributed coordination strategies.

---

[2] **No-Cloning Theorem:** A theorem proving that unknown quantum states cannot be duplicated perfectly due to linearity of quantum evolution. This establishes intrinsic tamper-evidence and limits classical redundancy strategies.

These properties are not implementation details that can be optimized away; they are architectural constants that determine how QAIS systems must be designed, staged, verified, and governed. Information, in this context, is no longer a detachable symbol but an engineered physical process whose stability and interpretability depend on adherence to quantum law.

Cross-Disciplinary Orientation

Recognizing information as a physically embodied resource harmonizes disciplinary perspectives that historically approached computation, physics, and security from separate angles.

Physicists interpret this embodiment as confirmation that informational meaning arises from amplitude structure, interference, and coherence dynamics rather than abstract symbolic manipulation.

Engineers read the same shift as constraint formalization: coherence budgets, calibration precision, noise thresholds, and interface stability become first-order architectural inputs rather than downstream refinements.

AI specialists understand it as an expansion of representational geometry, where quantum state spaces enrich feature encoding and inference pathways, but only within the operational limits imposed by coherence and measurement.

Cybersecurity practitioners recognize substrate-enforced trust boundaries shaped by no-cloning constraints and disturbance sensitivity, which define intrinsic verification mechanisms unavailable in classical systems. Across these perspectives, the conclusion converges: in QAIS, information is embodied, governed by physical law, and inseparable from the systems that instantiate it.

Case Study: Quantum Information as a Discipline

Intent: To formalize the representational shift from binary abstraction to geometry-based, coherence-sensitive state space as the substrate for QAIS architectures.

- **QALIS:** Uses superposition and entanglement to construct inference pipelines whose representational capacity exceeds classical vector encodings.
- **CRQC–LLM:** Exposes how fragile coherence and entanglement may become fault lines under adversarial timing pressure or cryptographic stress. Together, they show that quantum information is not abstract symbol but embodied resource whose power and fragility must be co-engineered.

Section 1.3.2 has established the representational divergence between classical and quantum information. Section 1.3.3 now examines how this conceptual shift matured into a research revolution, transforming philosophical questions about quantum reality into an operational science of information processing.

1.3.3 Intellectual Origins of the Quantum Information Revolution

As illustrated earlier in Figure 1.3.1, the transition from classical to quantum information marks the book's first decisive conceptual pivot. Information is no longer treated as an abstract statistical construct independent of its carrier, but as an amplitude-structured, coherence-sensitive physical entity governed by superposition, entanglement, and measurement disturbance. This shift does not merely expand representation; it restructures the design logic of intelligent systems by binding representation directly to physical law.

For QAIS, this transformation is architectural rather than philosophical. Superposition becomes a structured representational resource whose amplitude geometry must be preserved within coherence limits. Entanglement becomes a coordination mechanism across subsystems, enabling non-decomposable relational structure that reshapes distributed inference.

Measurement becomes the translation boundary between quantum evolution and classical interpretation, defining when and how information becomes actionable evidence.

These elements together define the constraint envelope within which quantum-intelligent architectures must operate, specifying both expressive capacity and operational boundary conditions.

The dual case-study constructs introduced earlier remain active even at this foundational stage. QALIS interprets these primitives constructively, assembling them into coherence-aware sensing and inference pipelines whose structure reflects disciplined amplitude control and measurement placement.

CRQC–LLM interprets the same primitives as potential fragility surfaces, revealing how coherence instability, cryptographic pressure, synchronization drift, or adversarial automation can destabilize poorly governed systems across critical control points:

- physical state preparation,
- interference-preserving transformation,
- entanglement distribution and maintenance,
- measurement staging and resolution timing,
- and verification discipline across hybrid interfaces.

To situate this shift historically, Table 1.3.1 summarizes the scientific arc that transformed quantum mechanics from interpretive theory into operational information science, establishing the substrate upon which QAIS now builds system-level intelligence.

Precursor Origins of QAIS

Table 1.3.1 — Timeline of Quantum Information Milestones

Decade	Key Milestones
1980s	• Feynman proposes simulating physics with quantum systems. • Deutsch formalizes the universal quantum computer. • Bennett and Brassard introduce quantum key distribution (BB84).
1990s	• Shor's algorithm demonstrates exponential speedup for factoring. • Grover's algorithm accelerates unstructured search. • Quantum error correction theory is established.
2000s	• Experimental quantum teleportation demonstrated. • Quantum Shannon theory formalized. • Fault-tolerant computing frameworks developed.
2010s-2020s	• Cloud-accessible quantum processors deployed. • NISQ-era experimentation expands [6]. • Quantum machine learning research accelerates [7].

This timeline reflects a cumulative transition rather than a singular breakthrough.

The 1980s established theoretical possibility.

The 1990s proved computational consequence. The 2000s demonstrated experimental viability. The 2010s–2020s introduced scalable access and hybrid workflows.

Over these decades, information underwent progressive redefinition:

- From symbolic abstraction → to amplitude-encoded structure
- From substrate-independence → to physical embodiment
- From computational hardness security → to physics-enforced trust

By the end of this arc, QIS had matured into a coherent discipline encompassing computation, communication, cryptography, and hardware engineering. QAIS inherits this substrate but extends it upward into system-level architecture, learning pipelines, and resilience frameworks.

System Relevance

Viewed collectively, these milestones redefined what information processing means when the substrate itself is quantum mechanical. Feynman reframed simulation as a quantum-native problem. Deutsch formalized universality under quantum transformation rules. Bennett and Brassard operationalized disturbance as a security resource. Shor and Grover demonstrated computational consequence. Error correction reframed fragility as an engineering constraint.

Together, these developments established a substrate in which:

- Coherence is a runtime budget,
- Entanglement is a structural connector,
- Measurement is an evidentiary boundary, and
- Noise is a design variable.

QAIS does not originate in isolation; it emerges directly from this layered redefinition.

Cross-Disciplinary Orientation

Each milestone resonated differently across professional domains:

- **Machine learning researchers** saw the expansion of search and optimization capability, particularly in the structured exploration of high-dimensional state spaces where interference can bias solution discovery.
- **Cybersecurity experts** recognized both security-by-physics (QKD) and cryptographic vulnerability (Shor), understanding that quantum mechanics simultaneously enables provable key distribution and threatens widely deployed public-key infrastructures.
- **Quantum engineers** treated teleportation and error correction as proof that fragile states could be stabilized, transmitted, and protected through disciplined control of noise, redundancy, and syndrome extraction.

System architects recognized that coherence lifetimes impose real constraints on pipeline depth, synchronization windows, and distributed coordination, forcing design decisions that explicitly account for physical timing and error accumulation.

QAIS integrates these interpretations into a unified system logic.

Capability and constraint are co-engineered from the physical layer upward, ensuring that expressive power, security posture, stability mechanisms, and architectural scalability are treated not as independent advances but as interdependent dimensions of intelligent system design.

Case Study: Origins of the Quantum Information Revolution

Intent: To trace how foundational theoretical insights reshaped computation and security, establishing the epistemic and operational groundwork for QAIS as a system discipline grounded in physical law.

- **QALIS:** The milestones summarized in Table 1.3.1 become composable primitives, superposition, entanglement, coherence management, and error correction, assembled into scalable inference and coordination pipelines. What began as theoretical breakthroughs becomes structured design vocabulary, enabling disciplined construction of quantum-enhanced learning and distributed reasoning architectures.
- **CRQC–LLM:** Shor's algorithm and its cryptanalytic implications extend naturally into the CRQC–LLM paradigm, modeling how cryptographically relevant quantum capability, when combined with automated adversarial reasoning, challenges the resilience boundaries of QAIS. The same primitives that enable constructive coordination also expose systemic fragility when coherence, verification, or governance assumptions fail.

Together, these lenses interpret the historical arc not merely as scientific progress, but as the structured emergence of both capability and constraint in quantum-intelligent systems, revealing that expansion of power and expansion of risk unfold in tandem.

The quantum information revolution established quantum phenomena as controllable resources for computation, communication, sensing, and verification. With this foundation in place, the discussion can now advance from quantum information as an engineered discipline to its integration within intelligent system architectures. Section 1.3.4 examines this transition from Quantum Information Science to Quantum AI Systems.

1.3.4 Quantum Information Science to Quantum AI Systems

The preceding historical subsections traced the emergence of quantum information from theoretical insight to an engineered discipline. The remaining step in this progression is architectural integration. QIS established the principles and primitives of qubits, superposition, entanglement, coherence, quantum operations, and measurement. QAIS extends these foundations into structured systems capable of learning, inference, coordination, verification, governance, resilience, and deployment across operational contexts. The distinction is foundational: QIS establishes the scientific principles and capabilities of quantum information, whereas QAIS addresses their integration into complete operational systems.

QIS asks:

What are the principles, limits, and capabilities of quantum information processing?

QAIS asks:

How can those principles and capabilities be integrated with classical computation, artificial intelligence, physical control, infrastructure, security, and governance to create deployable intelligent systems under physical, operational, and adversarial constraints?

This progression does not replace QIS or treat it merely as a preliminary stage. QIS remains the scientific foundation that determines what quantum information can represent, how it can evolve, what may be measured, and which physical constraints cannot be ignored.

QAIS builds upon that foundation by addressing the architectural problem of converting quantum capabilities into coherent operational systems.

The transition can therefore be expressed as:

Quantum principles → information primitives → engineered quantum capabilities → integrated quantum–classical–AI systems

QIS primarily investigates the behavior and processing of quantum information. QAIS addresses how that information is represented, propagated, controlled, verified, interpreted, secured, and incorporated into larger mission-oriented systems. It connects quantum devices and algorithms with classical computing, artificial intelligence, networks, data systems, human operators, enterprise processes, and governance authorities.

This shift marks the movement from phenomenon-centric science toward architecture-centric systems engineering. Quantum effects are no longer considered only as isolated physical phenomena or computational resources. They become constrained capabilities within an end-to-end architecture whose success depends on coherence, calibration, interoperability, traceability, verification, security, and mission suitability.

QAIS therefore begins where isolated quantum capability becomes a systems problem. Its central concern is not merely whether a quantum operation can be performed, but whether quantum and classical resources can be coordinated into a trustworthy system that produces an authorized, verifiable, and operationally useful result.

Architectural Integration Rather Than Algorithmic Extension

QAIS does not simply "add AI" to quantum computing. Nor does it treat quantum speedup as sufficient justification for intelligence claims. Instead, QAIS integrates:

- Physical substrate constraints,
- Operational control grammars,
- Representational encoding strategies,
- Architectural composition rules, and
- Deployment governance requirements

into a unified systems discipline in which capability and constraint are co-engineered from the outset and evaluated across the full system lifecycle.

Within QIS, superposition and entanglement are studied as physical and informational resources for computation, communication, sensing, and other quantum functions. Within QAIS, those resources are additionally treated as bounded architectural capabilities embedded in end-to-end pipelines that must remain viable under noise, calibration drift, scaling limits, operational demands, and adversarial pressure. Their usefulness is therefore evaluated not only through theoretical expressiveness or experimental performance, but also through their contribution to a deployable and verifiable system.

Coherence within QIS is a physical and informational condition required for controlled quantum evolution. Within QAIS, coherence is additionally treated as a finite architectural budget that constrains circuit depth, execution duration, synchronization tolerance, control sequencing, and the staging of hybrid quantum–classical workflows. A design that exceeds the available coherence budget cannot satisfy its intended architectural requirements.

Measurement within QIS is both a foundational quantum operation and a subject of experimental and computational design. Within QAIS, measurement placement becomes an end-to-end architectural decision affecting information extraction, verification, latency, feedback, interpretability, and control flow across subsystems. Where and how measurement occurs determines what classical evidence becomes available and how that evidence may support subsequent trust decisions.

This reorientation from "can it be done?" to "how must it be structured, governed, and sustained?" is the defining conceptual pivot. QAIS is therefore not merely an algorithmic extension of QIS. It is a complementary systems-engineering framework that builds upon QIS by integrating quantum capabilities with classical computation, artificial intelligence, physical control, verification, security, governance, and operational deployment.

Physical Primitive to System Variable

The progression from QIS to QAIS involves a categorical transformation as illustrated in the following table.

Table 1.3.2 — QIS vs QAIS: From Primitive to Architectural Variable

QIS Perspective	QAIS Perspective
Superposition as a quantum information resource	Superposition as bounded representational capacity within a governed processing pipeline
Entanglement as nonclassical correlation	Entanglement as a managed correlation resource supporting coordinated or distributed quantum operations
Coherence as the persistence of usable quantum relationships	Coherence as a finite execution budget and synchronization constraint
Measurement as quantum-to-classical information extraction	Measurement as a controlled resolution interface and verification checkpoint

Under this reframing, quantum phenomena cease to be isolated curiosities. They become variables in a system equation whose solution space is bounded by physical law and deployment realism.

In QAIS, information is not abstract or indefinitely manipulable; it is lifetime-limited by coherence, correlation-structured through entanglement, interface-constrained by measurement boundaries, and disturbance-sensitive under observation and environmental coupling.

These characteristics are not incidental implementation details. They define the permissible design space of intelligent quantum systems. Superposition expands representational capacity but only within a finite coherence window.

Entanglement enables distributed coordination but requires synchronization discipline. Repeated measurement produces classical outcomes whose statistical distributions provide evidence about selected properties of the quantum state, while measurement generally alters the measured state and may disturb correlated subsystems. Disturbance is not a defect; it is a governing constraint. The transition from QIS to QAIS therefore moves from understanding quantum behavior to engineering within its bounded envelope. Capability exists only insofar as these structural limits are respected, budgeted, and governed across the system lifecycle.

System Relevance

Within QAIS, the redefinition of quantum information has immediate architectural consequences. Quantum operations embedded in learning and inference architectures must be scheduled within available coherence windows and control tolerances. Hybrid quantum–classical pipelines must also position measurement deliberately, ensuring that classical information is extracted at architecturally appropriate points without prematurely disrupting quantum-state evolution or the correlations required by subsequent operations. Governance mechanisms must likewise account for substrate-level sensitivity to disturbance, the no-cloning theorem, and the resulting limitations on replication, recovery, verification, and information handling.

Noise, decoherence, calibration precision, and synchronization drift are therefore not merely downstream implementation concerns. They shape representational capacity, executable depth, inference stability, verification requirements, and system resilience from the outset. QAIS builds upon QIS by translating these physical and informational constraints into cross-layer architectural requirements. In doing so, it integrates physics, computation, artificial intelligence, control, verification, security, and governance within a coherent systems framework.

Cross-Disciplinary Orientation

The transition from QIS to QAIS clarifies disciplinary roles.

- Physicists recognize that quantum informational effects achieve full consequence only when embedded in coordinated architectures governed by coherence and measurement discipline.
- Engineers see the emergence of system-layer abstraction directly atop physical constraint models, where device stability and calibration tolerance determine architectural viability.
- AI specialists encounter quantum state spaces not merely as algorithmic accelerators, but as representational substrates bounded by coherence windows and readout constraints.
- Cybersecurity practitioners observe that cryptographic disruption and verification enforcement coexist within the same physical framework, forcing simultaneous attention to capability and trust.

QAIS is therefore not reducible to physics, computer science, or machine learning. It is a systems discipline requiring coordinated reasoning across all of them.

Capability and Constraint as Co-Instantiated Variables

A defining theme of this volume crystallizes here: quantum advantage and quantum fragility emerge from the same primitives. The amplitude structure that enables interference-driven acceleration also introduces sensitivity to disturbance.

The entanglement that binds distributed subsystems also expands the surface requiring synchronization and validation. The finite coherence available to sustain structured quantum evolution imposes strict temporal limits. QAIS refuses to separate these variables. Capability and constraint are co-instantiated.

This principle is examined through the dual interpretive lenses introduced earlier:

- **QALIS** evaluates how learning and inference capabilities are represented, assembled, verified, and sustained as quantum primitives are incorporated into scalable hybrid architectures. It examines whether the resulting learning and inference processes preserve provenance, validity, interpretability, robustness, and mission suitability across architectural boundaries.
- **CRQC–LLM** evaluates cryptographically relevant interactions among quantum capability, large language models, security infrastructure, adversarial reasoning, verification, and post-quantum transition requirements. It examines both emerging capabilities and the fragility, exposure, and governance consequences introduced across domain-specific QAIS architectures.

Together, these lenses reinforce a governing discipline of QAIS: no architectural claim is complete without simultaneous evaluation of capability, constraint, trust, and resilience.

Conceptual Inflection Point

Section 1.3.4 marks the decisive threshold at which quantum information is no longer treated as isolated phenomenon but as architectural doctrine. Earlier subsections established that information is physically instantiated. Historical analysis traced how quantum mechanics redefined representational structure.

Algorithmic breakthroughs demonstrated computational consequence. Here, those threads converge: theoretical insight is formalized as system-design obligation, and abstraction is bound explicitly to physical feasibility.

Quantum behavior is therefore treated as law-bound design substrate that simultaneously constrains and enables intelligent architecture. Superposition, entanglement, coherence, and measurement are no longer descriptive curiosities; they function as governing parameters that delimit representational capacity, synchronization tolerance, and verification strategy.

From this point forward, every discussion of encoding, inference, coordination, optimization, communication, or governance proceeds under the QAIS framing: physical constraints govern abstraction; architectural composition governs scaling; and verification requirements must be established before deployment claims can be justified.

This integration marks the intellectual hinge of Chapter 1 and anchors one of the central themes of the volume:

Quantum AI Systems emerge when quantum information ceases to be examined in isolation and becomes engineered as a governed, cross-layer architecture for intelligent behavior under constraint, accountability, and operational discipline.

Case Study: From QIS to QAIS

Intent: To clarify how the maturation of QIS into QAIS transforms quantum phenomena into structured architectural systems governed simultaneously by capability and constraint.

- **QALIS** interprets the QIS primitives as composable architectural resources, assembling superposition, entanglement, coherence management, and measurement staging into stable sensing, learning, and inference pipelines.
- **CRQC-LLM** interprets the same integration as an expansion of fragility surfaces, modeling how cryptographically relevant quantum computation, combined with automated adversarial reasoning, increases demands on verification, synchronization discipline, and coherence governance within QAIS deployments.

Section 1.3 Summary — Historical Foundations of QAIS

Together, these lenses reinforce a defining theme of this chapter: quantum information becomes architecturally meaningful only when capability and resilience are co-engineered across layers.

With Sections **1.3.1 through 1.3.4** complete, the intellectual transformation from classical abstraction to quantum-architectural doctrine is established. Information has moved from Shannon's probabilistic framework to amplitude-structured embodiment, from theoretical algorithmic promise to governed system discipline.

The chapter now turns from lineage to mechanics. **Section 1.4**, *The Nature of Quantum Information*, formalizes the qubit and superposition as the foundational representational primitives of QAIS and examines entanglement, coherence lifetimes, entropy measures, and state distinguishability as physically bounded resources. The progression therefore shifts from how quantum information emerged to how it behaves under operational constraint, preparing the architectural ascent that follows.

1.4 The Nature of Quantum Information

Architectural Framing: *Quantum States → Superposition → Entanglement → Coherence → Measurement*

Section Framing: *This section defines the core representational and relational substrate of quantum information. Superposition and entanglement jointly determine state geometry and non-decomposable correlation, operating within the constraints imposed by coherence and measurement.*

Quantum information is presented as the foundational control surface governing representation within QAIS. The qubit encodes amplitude and phase relationships that shape how information propagates under unitary evolution and the constraints of measurement. As a result, propagation is driven by interference and correlation dynamics, rather than classical, stepwise processing sequences.

Stability is treated as bounded propagation: coherence limits, noise sensitivity, and measurement disturbance restrict how long and how faithfully quantum information can be maintained and exploited. In this way, quantum information emerges as a physically governed substrate that both constrains and enables system behavior.

Throughout this chapter, standard quantum mechanical language is used to establish physical law and constraint. Alternative but mathematically equivalent formulations, including operational, interference-centric, and circuit-based perspectives, are introduced later when system execution and computational workflows are addressed. This sequencing preserves conceptual clarity: physical principles are established first; architectural translation follows.

QAIS treats information as a physical, coherence-sensitive resource whose fundamental unit is the qubit.

Unlike a classical bit fixed deterministically as "0" or "1," a qubit exists in superposition, carrying probability amplitudes whose magnitudes determine measurement likelihoods and whose relative phase governs interference behavior. In QAIS, qubits are not abstract placeholders. They are operational resources embedded in hardware, constrained by coherence lifetimes, calibration precision, environmental coupling, and measurement disturbance.

Their ability to preserve phase relationships, participate in entangled states, and yield probabilistic outcomes upon measurement is not a theoretical curiosity. It is the substrate upon which quantum-enhanced reasoning systems, distributed learning pipelines, and secure communication frameworks are constructed.

The fragility of these states, their susceptibility to decoherence and noise, directly shapes system architecture, synchronization strategy, verification staging, and resilience planning.

This section develops the nature of quantum information in deliberate progression:

1. From single-qubit geometry
2. To multi-qubit correlation structure
3. To operational deployment
4. To quantitative governance

The progression begins at the most elementary scale. Before correlation, interference, or architectural deployment can be analyzed, the foundational representational unit must be defined with precision. **Section 1.4.1** therefore formalizes the qubit and superposition as the minimal carriers of quantum information, establishing the geometric and algebraic structure upon which all higher-order QAIS mechanisms depend.

1.4.1 Defining the Qubit and Superposition

The qubit is the minimal carrier of quantum information and the foundational unit of QAIS. Unlike a classical bit, which occupies a definite value of 0 or 1, a qubit [3] exists in superposition [4], a linear combination of computational basis states:

$$| \psi\rangle = \alpha \mid 0\rangle + \beta \mid 1\rangle \tag{1.4.1}$$

subject to the normalization constraint:

$$| \alpha |^2 + | \beta |^2 = 1 \tag{1.4.2}$$

The coefficients $\alpha$and $\beta$are probability amplitudes drawn from $\mathbb{C}$, the set of complex numbers. Each may contain both real and imaginary components and can therefore represent a magnitude and a phase.

Their squared magnitudes determine outcome probabilities upon measurement in the computational basis, while their relative phase:

$$\Delta\varphi = \arg(\alpha) - \arg(\beta) \tag{1.4.3}$$

governs interference structure during state evolution. This phase parameter has no classical analogue: two states with identical measurement probabilities may behave differently under unitary transformation if their relative phases differ.

[3] **Qubit:** A two-level quantum system described by a normalized state vector in complex Hilbert space. Unlike a classical bit, its state is defined by probability amplitudes whose phase relationships affect computational evolution, not merely measurement statistics.

[4] **Superposition:** The linear coexistence of basis states within a single quantum state vector. It reflects amplitude structure in Hilbert space rather than classical probabilistic mixture, enabling interference during unitary evolution.

Consequently, phase encodes operationally significant information even when it is not directly observable in a single measurement.

Because amplitudes combine linearly, superposed states evolve through interference. Under unitary operations, amplitudes may reinforce constructively or cancel destructively depending on their phase relationships.

Computational advantage in many quantum algorithms arises not merely from probabilistic sampling, but from deliberate phase engineering that biases interference patterns toward desired outcomes.

In this sense, superposition is not "parallelism" in a classical sense; it is a coherence-dependent structure whose informational content resides in amplitude geometry.

Geometrically, a single qubit state can be represented on the Bloch sphere (see Figure 1.4.1), where global phase is physically irrelevant and only relative phase and amplitude ratios determine observable behavior. This geometric representation makes explicit that pure qubit states correspond to points on the surface of a unit sphere in $\mathbb{R}^3$, reinforcing that quantum information is constrained by normalization and unitary symmetry rather than arbitrary probabilistic assignment.

In QAIS, this structure is not abstract mathematics. It defines how information is encoded into hardware, how transformations must preserve normalization and coherence, and how measurement collapses continuous amplitude structure into discrete classical output.

The fragility of phase coherence, the impossibility of cloning unknown states, and the irreversibility of measurement together impose architectural constraints on learning pipelines, synchronization protocols, and verification layers. Superposition therefore establishes both capability and boundary: it expands representational expressiveness while simultaneously tightening the discipline under which intelligent quantum systems must operate.

Understanding quantum state behavior requires an appropriate representational framework that captures both amplitude and phase relationships within a coherent structure.

Such representations provide the basis for interpreting transformations and system evolution. Figure 1.4.1 illustrates this representation within a geometric context, establishing a reference for reasoning about quantum state space.

Single Qubit (superposition, phase)

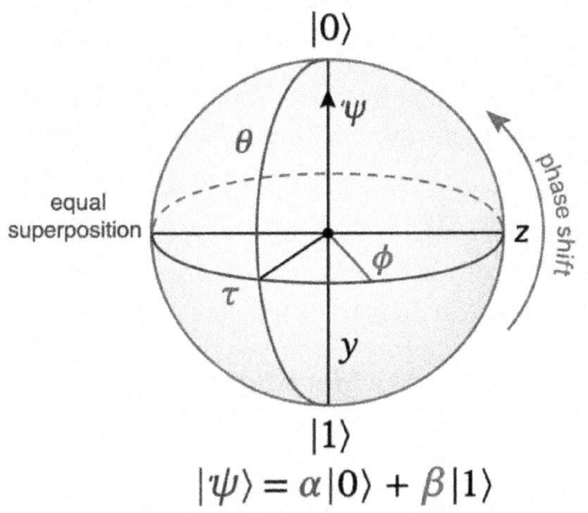

Figure 1.4.1 — Bloch Sphere Representation.

The Bloch sphere presents single-qubit pure states as points on the surface of a unit sphere. The north and south poles represent the basis states $|0\rangle$ and $|1\rangle$. Points along the equator correspond to equal superpositions. The orientation of the state vector encodes both amplitude and relative phase, making visible the transition from deterministic logic to geometric probability structure.

The Bloch sphere translates the algebra of **Equation 1.4.1** into geometry. A qubit may be visualized as a vector anchored at the sphere's center and terminating on its surface. Projection onto the computational axis determines measurement probability.

Rotation about the vertical axis encodes relative phase. Until measurement occurs, the state occupies a continuum of possibilities rather than a discrete logical value.

Decoherence may be visualized as contraction of this vector toward the sphere's center, where pure states transition into mixed states. This geometric perspective provides a first intuition for how noise erodes usable quantum information.

Within QAIS architectures:

- Superposition supplies representational breadth.
- Relative phase enables interference-based signal shaping.
- Coherence preserves structured evolution long enough for inference or sensing to occur.

These are architectural constants, not implementation details.

Visualizing a Qubit on the Bloch Sphere

The Bloch sphere may be understood intuitively as a globe. The north pole represents $|0\rangle$, the south pole $|1\rangle$, and any point on the surface corresponds to a valid quantum state.

A vector along the equator represents an equal superposition of $|0\rangle$ and $|1\rangle$. Rotation around the equator changes relative phase, determining how states interfere during computation.

Upon measurement, the state collapses to one pole or the other, with probabilities determined by the squared amplitudes in Equation (1.4.2) In this way, the Bloch sphere acts as a bridge between invisible amplitude structure and observable probabilistic outcome. This intuitive scaffold prepares the transition from analogy to formalism.

Technical Spotlight — Interpreting the Bloch Sphere

Conceptual Grounding

Figure 1.4.1 provides a geometric representation of the projective Hilbert space of pure single-qubit states, where global phase has been factored out. The north and south poles correspond to the computational basis states $|0\rangle$ and $|1\rangle$.

Points along the equator represent balanced superpositions differing only in relative phase. The spherical coordinates encode the amplitude and relative phase structure defined in Equation (1.4.1), subject to the normalization constraint of Equation (1.4.2). Measurement projects the state vector onto the computational axis, enforcing probabilistic collapse consistent with Equation (1.4.2). This geometric framing makes explicit that quantum information is not binary but continuous and phase-structured.

System Relevance

Within QAIS, the Bloch sphere functions as a design scaffold. Superposition supports reversible evolution under unitary transformations. Interference allows amplitude redistribution so that desirable outcomes are amplified while others are suppressed. State preparation, stabilization, and steering in hybrid quantum–classical pipelines all depend on controlled rotations within this state space.

Feature encoding, sensing modules, and intermediate computational layers operate within coherence windows that determine how long the vector may evolve before environmental interaction degrades structure.

Cross Discipline Orientation

The Bloch sphere serves distinct but converging roles across disciplines:

- **Physicists** interpret it as a compact geometric representation of single-qubit states, coherence, unitary evolution, and measurement within a two-dimensional Hilbert space. Single-qubit unitary transformations correspond to rotations of the Bloch vector; measurement probabilities depend on its projection onto the selected measurement axis; and its radial length indicates state purity, with pure states on the surface, mixed states inside the sphere, and the maximally mixed state at the center.
- **Engineers** treat it as a calibration map for pulse control, rotation accuracy, gate fidelity, and drift mitigation, using geometric intuition to diagnose phase error and amplitude imbalance.
- **AI specialists** view it as representational geometry in which features are encoded in amplitude and phase, enabling interference-based signal shaping and structured state preparation.
- **Cybersecurity practitioners** recognize measurement collapse and disturbance sensitivity as substrate-level verification mechanisms, where unauthorized interaction leaves detectable geometric displacement or decoherence signatures.

Across fields, the sphere provides a shared grammar for correctness, control, expressiveness, and trust. It compresses abstract linear algebra into a manipulable geometric model, enabling cross-disciplinary collaboration without sacrificing formal precision. In this way, the Bloch sphere functions not merely as visualization, but as a unifying operational language for QAIS.

Technical Clarification: The Bloch sphere does not depict Hilbert-space state vectors directly. Instead, it provides a geometric representation of physically distinct, normalized single-qubit states after the physically irrelevant global phase has been removed.

Case Study: Defining the Qubit and Superposition

Intent: To formalize the qubit as an operational state representation within Hilbert space, clarifying how superposition expands representational capacity while remaining bounded by measurement constraints and coherence limits.

- **QALIS** leverages superposition to integrate multimodal sensing streams into coherent inference pipelines. Bloch sphere steering provides the practical map for state preparation, rotation sequencing, and controlled readout, enabling structured representational synthesis under hardware constraints.
- **CRQC–LLM** exposes how fragile coherence in superposed states can be perturbed by noise, environmental coupling, or adversarial influence, making phase stability, calibration integrity, and verification placement central to resilience design.

The qubit therefore embodies the central tension of QAIS: expanded representational capability coupled with physically enforced fragility.

Expressive amplitude structure and interference-driven computation are inseparable from coherence decay, measurement irreversibility, and hardware sensitivity. Capability and constraint are co-instantiated at the level of the single qubit.

Section 1.4.2 extends this structure into composite systems, where superposition across multiple qubits gives rise to entanglement, correlations that cannot be reduced to independent subsystem descriptions.

From that composite structure, the chapter proceeds to examine how entanglement transitions from mathematical property to architectural resource, and how coherence limits, error propagation, and quantitative metrics govern what can be preserved, trusted, and distinguished under realistic noise conditions. From this point forward, quantum information will be treated not as abstract notation, but as physically bounded architecture whose representational power is inseparable from its operational discipline.

1.4.2 Entanglement and Non-Decomposable Correlation

Entanglement completes the structural foundation of quantum information by extending superposition into composite systems. Whereas a single qubit expands representational geometry within a bounded Hilbert space, entangled systems introduce non-decomposable relational structure across subsystems. This section formalizes entanglement first as mathematical inseparability, then as architectural resource within QAIS.

The same physical properties that enable distributed coordination also introduce synchronization, calibration, and verification obligations that must be engineered deliberately.

Mathematical Structure of Entanglement

Entanglement extends superposition into composite Hilbert space through the tensor-product structure of multi-qubit systems. When two subsystems are combined, their joint state space is constructed as $\mathcal{H}_1 \otimes \mathcal{H}_2$. Its dimensionality therefore increases multiplicatively rather than additively, allowing the composite system to support relational structures that cannot be represented by independently specified subsystem states.

A pure composite state $|\Psi\rangle$is separable if it can be written as:

$$|\Psi\rangle = |\psi_1\rangle \otimes |\psi_2\rangle \qquad (1.4.4)$$

In a separable pure state, the joint measurement statistics factor into statistics determined by the individual subsystem states. The state therefore contains no intrinsic correlation between the subsystems. More generally, separable mixed states may contain classical correlations arising from shared preparation or a common cause, but they do not contain entanglement.

An entangled pure state violates this factorization condition. It cannot be decomposed into independent subsystem state vectors. Its informational structure therefore resides partly in the joint amplitudes and relative phases of the composite state rather than entirely within the subsystems considered separately.

Bell states exemplify this condition. When their subsystems are measured in an appropriate common basis, the outcomes exhibit perfect correlation or anti-correlation, even though the reduced state of neither subsystem specifies a definite local outcome. Each subsystem has a maximally mixed reduced state, while the composite Bell state remains pure.

This coexistence of local mixedness and globally defined nonclassical structure is a defining feature of entanglement. Its distinction from classical correlation is revealed by the state's nonseparability and by its measurement statistics across appropriately selected bases.

Entanglement therefore encodes an intrinsic relational structure. The correlation is embedded in the composite quantum state rather than imposed solely through external preparation. Measurement of one subsystem conditions the statistical predictions for measurements of the other without enabling controllable faster-than-light signaling. This dependence reflects the global configuration of the state rather than causal transmission between the subsystems.

Structurally, entanglement marks the threshold at which correlation becomes an irreducible property of the composite state. The reduced states of the individual subsystems are insufficient to reconstruct the full joint state because they do not retain all information about the correlations between them. Once inseparability is present, subsystem descriptions are no longer complete in isolation: disturbance, decoherence, or calibration drift affecting one component can alter the joint state and degrade the relational structure on which the system depends.

This inseparability establishes the mathematical boundary between independently describable subsystems and unified composite-state structure. It also establishes the second structural pillar of quantum information: relational geometry. The discussion now shifts from formal criterion to engineered consequence, examining how nondecomposable correlation functions as architectural infrastructure within QAIS.

Figure 1.4.2 demonstrates this form of correlation within a conceptual framework, clarifying its significance for system-level reasoning.

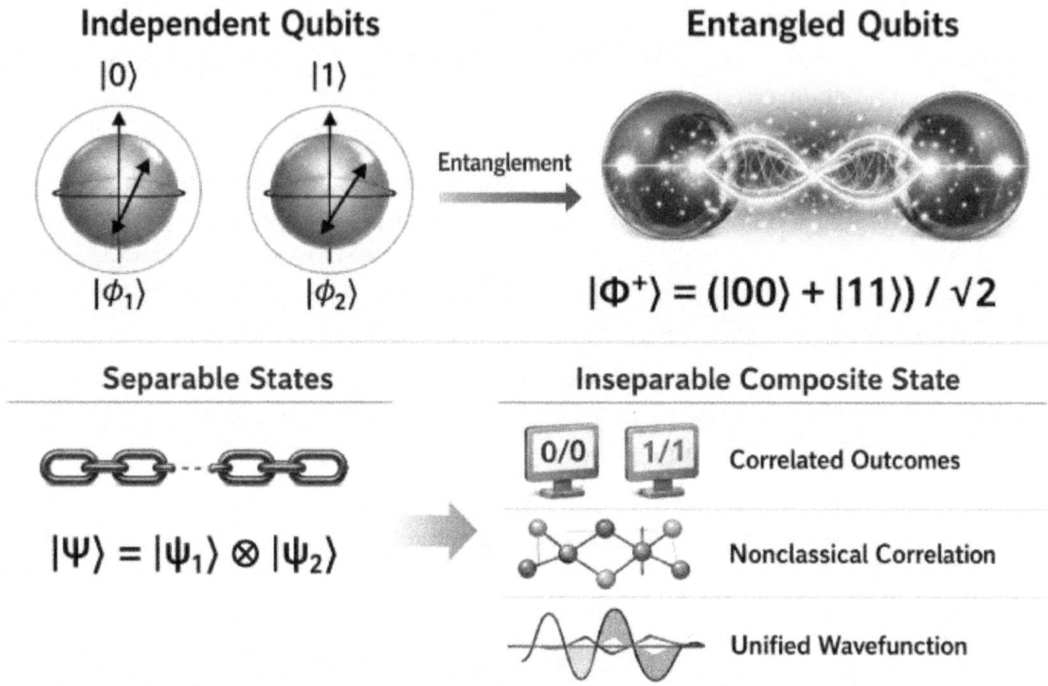

Figure 1.4.2 — Entanglement as Nondecomposable Correlation.

This figure contrasts independent qubits, whose composite state can be factored into separate subsystem states, with entangled qubits, whose joint state cannot be decomposed in this way. In the entangled state, individual measurement outcomes are probabilistic, while the joint outcomes exhibit nonclassical correlations determined by the composite state.

Figure 1.4.2 makes explicit that entanglement is not merely a stronger form of classical correlation, but a categorical departure from separability. The composite state is defined only at the system level; descriptions of the individual subsystems are incomplete without reference to the whole. This mathematical inseparability establishes the second structural pillar of quantum information: relational geometry. Whereas superposition expands representational breadth within an individual system, entanglement expands structural connectivity across systems.

The discussion now transitions from formal definition to engineered deployment, examining how this nondecomposable correlation becomes architectural infrastructure within QAIS.

This inseparability is captured formally in Bell states [5] such as:

$$|\Phi^{+}\rangle = \frac{|00\rangle + |11\rangle}{\sqrt{2}} \tag{1.4.5}$$

Equation (1.4.5) defines a maximally entangled state. No assignment of independent local states reproduces its joint measurement statistics. Only the composite state encodes the full correlation structure. Measurement of one qubit constrains the statistical distribution of the other, yet no classical signaling occurs between them.

Entanglement therefore represents more than strong correlation. It reflects non-decomposable structure intrinsic to the composite system.

Within QAIS, entanglement functions as a structural connector:

- It binds distributed subsystems into unified state spaces.
- It enables coordination across nodes without classical duplication.
- It supports non-classical representational geometry unavailable to independent registers.
- It introduces fragility requiring disciplined verification and coherence management.

Entanglement is therefore both architectural enabler and resilience surface.

Entanglement as an Operational Resource

Entanglement becomes an operational resource when a nonseparable joint state is deliberately prepared, maintained, transformed, and measured to perform a defined function. Within QAIS, engineered entanglement can support quantum-state teleportation, error correction, distributed processing, and quantum-enhanced learning. These capabilities require controlled operations, timing coordination, measurement, verification, and, where applicable, classical communication; entanglement alone does not constitute a complete execution or communication architecture.

In quantum teleportation, shared entanglement combines with local measurement and classical signaling to reconstruct a quantum state at another subsystem without copying it. In quantum error correction, logical information is encoded through structured correlations across multiple physical qubits rather than through classical duplication. Entangling operations may also create joint representations in feature maps, variational circuits, and learning models, although entanglement does not by itself guarantee computational or learning advantage.

---

5 **Bell State:** One of four maximally entangled two-qubit states forming an orthonormal basis for composite Hilbert space. Bell states exemplify non-classical correlation beyond separable mixtures.

Operational use creates corresponding engineering obligations. Entangled states must be generated and preserved within limits imposed by fidelity, coherence time, channel loss, calibration accuracy, and synchronization. Under the QALIS lens, entanglement provides constructive relational structure for coordinated learning and inference. Under the CRQC-LLM lens, decoherence, timing drift, compromised controls, and manipulated measurements expose assurance and fragility surfaces.

Entanglement can also support verification protocols in which expected joint-measurement statistics provide evidence that distributed quantum resources retain their intended correlation structure. Disturbance may produce statistically detectable deviations, but detection depends on the protocol, measurement basis, sampling strategy, noise threshold, and adversary model. Entanglement-assisted authentication should therefore complement—not replace—classical identity, cryptographic, and access-control mechanisms.

Entanglement thus functions as both capability and constraint: it expands the relational structure available to a composite system while increasing the control, verification, and governance required to preserve that structure under operational conditions.

Entanglement as Infrastructure: Nonlocal Correlation to Operational Architecture

Having established entanglement as a non-factorizable composite state, the next step is its architectural translation. In QAIS, entanglement is not merely a mathematical condition; it functions as engineered infrastructure that governs how distributed subsystems co-evolve within a shared state space under coherence constraints.

Shared entangled states enable coordinated evolution across spatially separated components, support state transfer without duplication, and permit collective inference behaviors unavailable to independent registers.

Because relational structure is encoded directly in joint amplitude geometry, coordination no longer depends solely on classical message passing; it is partially embedded within the state itself. Entanglement therefore shifts coordination from external communication to intrinsic state structure.

Rather than serving as an auxiliary feature layered onto computation, entanglement operates as a structural backbone. Through it, relational information is preserved, propagated, and operationalized across synchronized components. This infrastructural role introduces explicit design obligations: correlations must be deliberately prepared, maintained within coherence budgets, and monitored across interfaces where decoherence or drift may degrade joint structure.

Within this framework, entanglement functions as a coordination mechanism embedded in distributed architectures.

Under the **QALIS lens**, it serves as constructive connective tissue across learning and inference pipelines, enabling structured collective behavior.

Under the **CRQC–LLM lens,** it becomes a synchronization surface requiring calibration discipline, disturbance detection, and verification to prevent correlation collapse under scaling, noise, or adversarial influence.

This interpretation completes the transition from structural condition to system function. Once correlation is engineered, stabilized, and governed, it becomes a deployable resource rather than an abstract property. The central question shifts from the existence of entanglement to its controlled operationalization across coordinated subsystems for transfer, stabilization, distributed processing, and learning tasks. Figure 1.4.3 illustrates this translation from relational geometry to executable architecture, showing how shared correlations become active components within QAIS deployments.

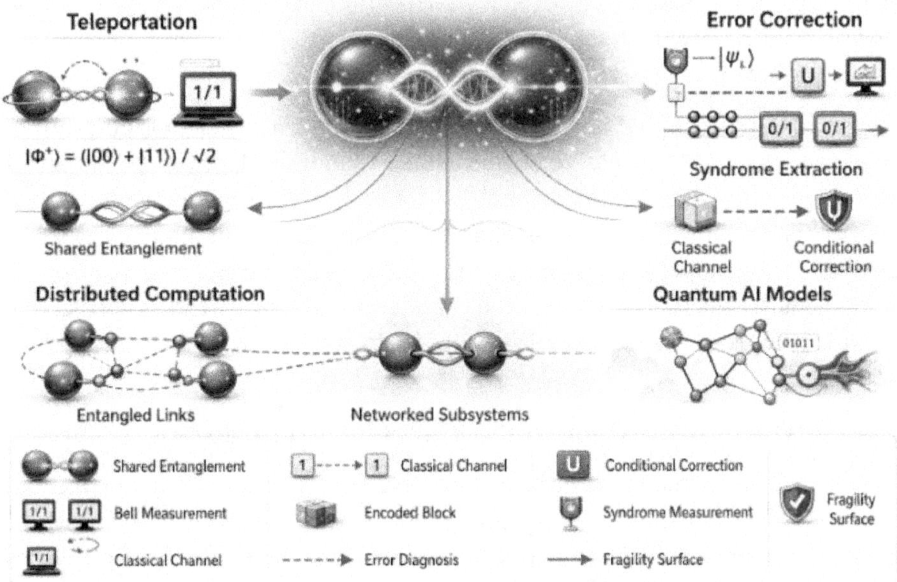

Figure 1.4.3 — Entanglement as an Operational Resource in QAIS.

This figure transitions from principle to application, illustrating how entanglement enables teleportation, quantum error correction, distributed computation, and quantum-enhanced learning architectures.

Operationally, entanglement functions as a system-level coordination resource. It enables state transfer through shared correlations supplemented by classical signaling, stabilizes fragile quantum information through distributed redundancy, and allows geographically separated subsystems to act as a coherent computational whole.

Within quantum-enhanced learning architectures, non-local correlations can be embedded directly into feature maps, kernels, and inference layers, shaping representational geometry beyond what independent registers can sustain.

Under the QALIS framework, these capabilities manifest as coordinated inference pipelines in which distributed components evolve within a shared relational structure. Under the CRQC–LLM lens, the same relational structure becomes a point of diagnostic scrutiny: correlated states are vulnerable to decoherence, synchronization drift, calibration misalignment, and adversarial timing pressure. Entanglement therefore operates simultaneously as connective infrastructure and as a fragility surface.

Figure 1.4.3 illustrates this migration from principle to deployment. Teleportation, error correction, distributed computation, and quantum AI modeling are not isolated demonstrations; they are architectural expressions of engineered correlation. Yet the power of shared state geometry introduces proportional verification responsibility. Correlated subsystems must preserve coherence budgets, maintain timing alignment, and ensure that joint statistics remain physically consistent across interfaces.

This dual character, capability coupled with constraint, reinforces a governing doctrine of QAIS: relational strength increases expressive power while expanding the envelope that must be monitored and governed. Once entanglement is treated as infrastructure, the next architectural question is not how correlation enables coordination, but how correlation integrity is preserved under operational stress. The discussion therefore advances to entanglement-assisted authentication, where shared correlations are leveraged as verification mechanisms to confirm subsystem consistency and detect structural disturbance across distributed deployments.

Entanglement-Assisted Authentication as Correlation Integrity

Beyond coordination and distributed inference, entanglement introduces a verification primitive unavailable in classical systems: correlation integrity as an authentication substrate. Because entangled subsystems encode non-decomposable joint structure, unauthorized disturbance or substitution necessarily alters observable correlation statistics. Authentication may therefore be reframed as preservation of entangled correlation geometry rather than validation of symbolic credentials.

In QAIS architectures, this principle implies that identity and integrity can be enforced through correlation-sensitive verification checkpoints embedded across distributed subsystems. Under the QALIS lens, entanglement-assisted authentication becomes a constructive trust mechanism, where shared correlation structure confirms subsystem alignment within coherence budgets. Under the CRQC–LLM lens, the same correlation surfaces become adversarial targets; drift, decoherence, or malicious perturbation manifest as measurable degradation in fidelity and trace distance.

Entanglement-assisted authentication therefore emerges not as an add-on protocol, but as an architectural consequence of non-decomposable correlation. Verification is grounded in physics-enforced disturbance sensitivity rather than in abstract symbolic validation alone.

1.4.3 Core Metrics and Coherence Constraints

Correlation without governance is unstable architecture. For entangled systems to function within QAIS, their integrity must be measurable and bounded. Entropy characterizes informational structure and compressibility. Fidelity quantifies preservation of state geometry under transformation or transmission. Trace distance bounds distinguishability and therefore limits inference reliability. Coherence time constrains the temporal window within which correlated evolution remains operational. These metrics convert entanglement from theoretical possibility into governed resource.

They define the permissible envelope within which coordinated architectures may execute before noise, drift, or disturbance render inference unreliable.

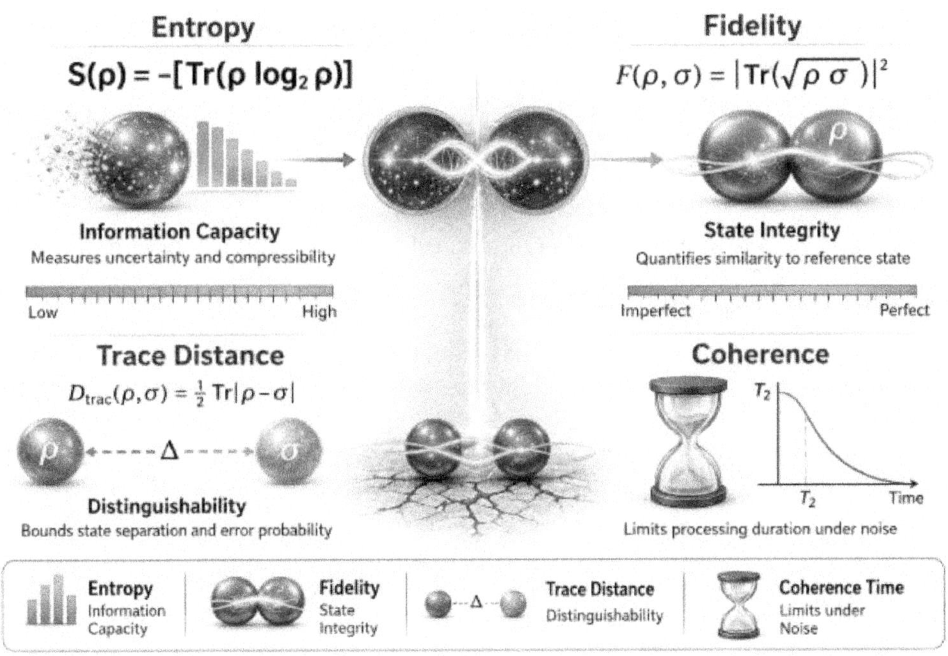

Figure 1.4.4 — Core Metrics of Quantum Information.

This figure introduces entropy, fidelity, and trace distance as operational measures of information capacity, state integrity, and distinguishability. It also illustrates how coherence and decoherence define the usable lifetime of quantum information.

Figure 1.4.4 consolidates key quantitative constraints into cross-layer evaluative control variables for assessing informational structure and system behavior under transformation. Together, these metrics provide a unified foundation for evaluating performance, stability, and representational fidelity.

Entropy characterizes informational capacity, structural complexity, and compressibility [8], informing encoding and compression strategies. Fidelity quantifies the preservation of quantum states during evolution or transmission and establishes calibration thresholds. Trace distance measures state distinguishability and bounds state-discrimination performance, while coherence time defines the finite temporal window available for reliable quantum execution.

Collectively, these quantities formalize the bounded nature of quantum information. Unitary evolution must occur within available coherence budgets; measurement resolution limits practical distinguishability; and environmental noise reshapes entropy distributions while degrading fidelity. Within QAIS architectures, these measures function as forward-looking design constraints rather than merely retrospective diagnostics. They shape circuit depth, synchronization feasibility, inference reliability, calibration discipline, and workflow staging.

Quantum information therefore remains inseparable from its physical substrate throughout its lifecycle. Superposition supplies representational breadth, while entropy, fidelity, trace distance, and coherence time define the conditions under which that representation remains operationally usable. Together, these measures establish the quantitative constraint framework required to determine whether the correlated multi-qubit structures introduced in Section 1.4.2 remain sufficiently coherent, faithful, and distinguishable for architectural use. The discussion therefore returns to entanglement in Section 1.4.4, not to redefine correlation, but to consolidate its role as a governed architectural resource within QAIS.

NISQ as an Architectural Condition

The Noisy Intermediate-Scale Quantum (NISQ) regime, introduced by Preskill [6], describes contemporary quantum processors that operate with limited qubit counts, non-negligible error rates, finite coherence times, and without full fault tolerance. Under these conditions, accumulated noise restricts circuit depth, reduces the reliability of state evolution, and limits the scale and duration of useful quantum computation.

Within QAIS, NISQ is not merely a hardware classification. It defines the current architectural condition under which quantum capabilities must be designed, integrated, evaluated, and deployed. Finite coherence constrains execution time and circuit depth; imperfect gates and measurements introduce uncertainty into system outputs; and calibration drift, crosstalk, connectivity limitations, and control latency affect the reliability of the complete execution pipeline. These constraints propagate beyond the quantum processor into compilation, scheduling, hybrid coordination, verification, interpretation, and governance.

Because full fault-tolerant quantum error correction is not yet generally available, NISQ implementations rely heavily on error suppression, error mitigation, repeated execution, statistical estimation, and hybrid quantum–classical processing. These methods can improve the usefulness of results, but they do not eliminate physical errors or provide the guarantees associated with fault-tolerant computation. Verification checkpoints must therefore be positioned throughout the workflow to determine whether calibration, execution, measurement, and postprocessing evidence remain within mission-defined acceptance thresholds.

Under the QALIS lens, NISQ conditions constrain how learning and inference capabilities can be constructed. State preparation, encoding, ansatz selection, circuit depth, measurement placement, and classical optimization must be designed together within available coherence and error budgets. Output uncertainty must also be represented explicitly so that noisy measurement statistics are not interpreted as more reliable than the supporting evidence permits.

Under the CRQC–LLM lens, the same conditions expand the system's fragility and assurance surfaces. Calibration drift, synchronization failures, manipulated control parameters, corrupted measurement records, overstated capability claims, and weaknesses across physical, computational, classical, and governance trust boundaries can invalidate system results. This lens does not presume that present NISQ devices are already cryptographically relevant; rather, it examines how current limitations, emerging capabilities, and automated adversarial reasoning affect the security and trustworthiness of QAIS deployments.

NISQ therefore establishes the present constraint envelope for quantum-enabled QAIS architectures. Useful capability must be engineered within finite coherence, bounded circuit depth, imperfect control, noisy measurement, and incomplete fault tolerance. These conditions make cross-layer co-engineering essential: substrate capabilities propagate upward to shape available system functions, while mission, verification, security, and governance requirements propagate downward to constrain how those capabilities may be used.

Correlation Integrity to Architectural Resource

Section 1.4.1 established single-qubit representation and superposition. Section 1.4.2 extended that representation into multi-qubit correlation through entanglement, while Section 1.4.3 introduced the quantitative metrics and coherence constraints used to evaluate quantum-state integrity.

Quantum information, however, is not confined to individual systems. When multiple qubits interact, their joint state can contain relational structure that cannot be described completely through the states of the individual qubits. The discussion therefore advances from single-system representation to correlation across composite quantum systems.

1.4.4 Entanglement as Correlation and Architectural Resource

Entanglement introduces nonseparable correlation as the second structural pillar of quantum information within QAIS. Whereas superposition expands the representational possibilities available within an individual quantum system, entanglement creates relational structure across multiple subsystems.

An entangled state cannot be factored into independent subsystem states. Its informational structure resides partly in the composite state itself, making correlation intrinsic to the system rather than externally imposed. Within QAIS, this relational structure can support distributed coordination, state transfer, collective computation, and quantum-enhanced learning and inference.

Entanglement is therefore not merely a phenomenon to be observed; it is an architectural resource that must be deliberately prepared, synchronized, preserved, and verified. Its operational value remains bounded by coherence loss, calibration error, environmental interaction, control imprecision, and measurement disturbance. Entropy, fidelity, trace distance, and coherence time consequently provide essential measures for determining whether the correlated state remains usable.

Entanglement thus extends quantum information from individual-state representation to governed relational structure. The architectural question is no longer only how information is represented within a qubit, but how nonseparable correlations can be engineered and sustained across composite quantum systems.

Section 1.4 Summary — The Nature of Quantum Information

Section 1.4 established quantum information as a physically embodied and coherence-bounded resource. Qubits and superposition define individual-state representation; entanglement supplies nonseparable relational structure across composite systems; and entropy, fidelity, trace distance, and coherence time provide measures of uncertainty, preservation, distinguishability, and operational viability.

Together, these concepts show that quantum information gains architectural value only when its representational and correlational capabilities are engineered within explicit stability and measurement constraints. Section 1.5 translates these mechanics into the architectural doctrine governing QAIS design and evaluation.

1.5 Mechanics to Architecture

Architectural Framing: Quantum States → Superposition → Entanglement → Coherence → Measurement → System Architecture

Section Framing: This section consolidates the transformation of quantum-mechanical principles into QAIS architectural doctrine. Representation is established as a control surface defined by amplitude structure and correlation geometry, while propagation is treated as system behavior governed by interference, coherence, and measurement constraints. Physical properties become design variables, operational limits, interface requirements, and verification obligations within complete quantum–classical systems.

This synthesis establishes the architectural bridge between the quantum mechanics examined earlier in the chapter and the practical system posture introduced in Section 1.6.

With the physical substrate, formal mechanics, relational structure, and quantitative governance variables established, the technical arc of Chapter 1 can now be consolidated. Quantum information has been examined from single-qubit geometry through multi-qubit entanglement, from interference structure through coherence limits, and from measurement theory through distinguishability metrics. Superposition defines representational latitude. Entanglement supplies non-decomposable relational structure. Entropy, fidelity, trace distance, and coherence time characterize uncertainty, preservation, distinguishability, and bounded survivability under physical constraint. Together, these elements specify not only what quantum information is, but also how it must be represented, controlled, measured, and validated within a system.

This progression converts mechanics into architectural doctrine. Quantum phenomena are no longer treated as isolated effects or algorithmic curiosities; they become system-defining variables. Superposition provides structured design space for representation. Entanglement supplies correlations that specified protocols may use under explicit timing, control, communication, and verification procedures. Coherence becomes an executable runtime budget. Measurement becomes a governed boundary between quantum evolution and classical interpretation. The laws of physics therefore translate directly into architectural commitments for systems claiming quantum-enabled intelligent capability.

1.5.1 Quantum Properties as Architectural Variables

Quantum mechanics describes how quantum states are prepared, transformed, correlated, and measured. Within QAIS, these processes must also be interpreted as architectural variables that shape system design. Superposition determines the range of state configurations available for representation and controlled evolution. Entanglement establishes non-decomposable relational structure across participating quantum components. Interference alters the relative contributions of possible measurement outcomes, while coherence determines how long these relationships remain operationally usable.

These properties cannot be treated as independent features added to an otherwise classical architecture. They arise from the same physical state and interact throughout execution. A state-preparation decision affects the transformations that can be applied. Transformation depth affects exposure to noise. Noise influences coherence and fidelity, while the selected measurement determines which classical evidence becomes available for interpretation. Architectural design must therefore account for the complete operational sequence rather than evaluate each quantum property in isolation.

This principle changes how quantum capability is specified. A proposed capability is not adequately defined merely by naming an algorithm, quantum effect, or theoretical speedup. Its specification must identify the state representation, required correlations, transformation sequence, coherence interval, measurement basis, repetition requirements, and classical post-processing needed to produce usable evidence. Quantum properties become architectural variables only when their roles, dependencies, and physical limits are explicitly incorporated into system design.

Representation and propagation are central to this translation. Amplitudes, phases, and correlation structures determine how information is represented within the quantum state. Controlled transformations and interference govern how that structure evolves. Coherence determines whether the required state relationships remain available long enough to complete the intended operation. Measurement then converts selected aspects of the evolved state into classical outcomes that downstream components can aggregate and interpret.

The architecture must preserve the relationships among these processes. Representation cannot be designed independently of transformation, transformation cannot be separated from coherence, and measurement cannot be evaluated without considering the state preparation and evolution that preceded it. The resulting architecture is therefore relational and lifecycle-oriented rather than a collection of isolated quantum functions.

1.5.2 Physical Constraints as System Requirements

The physical limits of quantum information translate directly into system requirements. Coherence time establishes an execution interval within which state preparation, controlled evolution, and measurement must occur. Gate and measurement errors establish tolerances that influence circuit depth, calibration frequency, error-mitigation procedures, and confidence in observed results. Connectivity and control limitations constrain which interactions can be implemented efficiently on a particular platform.

The same primitives that expand expressiveness simultaneously narrow operational tolerance. Execution depth is bounded by coherence. Synchronization depends on explicit timing and control procedures. Entanglement stability determines whether required correlations remain usable. Inference reliability is limited by noise, sampling uncertainty, and the distinguishability of the resulting states. Capability and constraint are therefore co-instantiated variables rather than sequential considerations.

State distinguishability is especially important because measurement evidence must support the system's intended operational claim. If noise or imperfect control causes output states to become insufficiently distinguishable, measurement results may not provide reliable evidence for classification, inference, verification, or decision-making. Fidelity, trace distance, entropy, and related measures consequently serve not only as mathematical descriptions but also as indicators of whether the system remains within an acceptable operating region.

These constraints propagate beyond the quantum processor. Limited coherence may require shorter circuits, alternative decompositions, or more frequent classical intervention. Measurement uncertainty may require repeated executions and statistical aggregation. Hardware variability may require calibration data to influence compilation, scheduling, parameter selection, and model evaluation. Physical constraints consequently become requirements for the entire hybrid architecture, including its classical control, validation, monitoring, and decision functions.

Constraint propagation also means that weaknesses originating at one architectural layer may influence behavior at higher layers. A calibration error can affect gate fidelity; reduced fidelity can alter the output distribution; an altered distribution can weaken inference reliability; and unreliable inference can compromise an operational decision. QAIS architecture must therefore trace how physical limitations propagate through interfaces and affect system-level claims.

1.5.3 Measurement as a Governed System Boundary

Measurement forms the primary boundary between quantum evolution and classical interpretation. Before measurement, the system is represented through amplitudes, phases, and correlations within a quantum state. Measurement produces classical outcomes sampled according to the probabilities defined by that state and the selected measurement context. Those outcomes can then be stored, aggregated, validated, and interpreted by classical or AI components.

This boundary must be governed because measurement both extracts information and changes the measured state. The architecture must specify when measurement occurs, which observable or basis is used, how many executions are required, how uncertainty is characterized, and how the resulting evidence is passed to downstream processes. Without these controls, an individual outcome may be physically valid yet insufficient to support the operational claim being made.

Measurement also creates a verification checkpoint. Expected outcome distributions can be compared with observed samples, calibration behavior, noise models, acceptance thresholds, or classical reference results. Such comparisons do not guarantee that every internal quantum process can be directly observed. They provide an evidence-bearing interface through which the system can evaluate whether execution remains consistent with its design assumptions and operational requirements.

Because measurement generates statistical evidence rather than merely exposing a stored classical answer, repeated executions are generally required. The resulting samples must be aggregated and evaluated in relation to uncertainty, noise, and the intended decision threshold. The architecture must distinguish between a single observed outcome, an empirical distribution constructed from repeated measurements, and a validated conclusion derived from that distribution.

Measurement therefore functions as both an information boundary and a governance surface. It determines what information becomes classically accessible, how that information is interpreted, and what evidence is available for verification. The placement and configuration of measurement operations must consequently be treated as architectural decisions rather than incidental endpoints of quantum computation.

1.5.4 Doctrine for Hybrid QAIS Architecture

The architectural doctrine emerging from these relationships is that quantum, classical, and AI components must be assigned distinct but coordinated roles. Quantum components prepare and transform states using amplitude, phase, interference, and correlation. Classical components configure experiments, control hardware, aggregate measurement samples, estimate uncertainty, validate outputs, update parameters, and determine subsequent actions. AI may participate in calibration, control, optimization, interpretation, or adaptive workflow selection when those roles are operationally justified.

The resulting architecture is not necessarily a one-pass sequence. Many QAIS workflows are iterative. Classical or AI components select and configure a quantum operation; the quantum subsystem prepares and evolves the state; measurement produces classical samples; classical processes aggregate and evaluate the evidence; and the resulting assessment determines the next state preparation, transformation, measurement configuration, or operational action. Feedback connects measurement outcomes to subsequent execution while preserving explicit boundaries between quantum behavior and classical control.

Within this architecture, stability is expressed through bounded propagation. Quantum state evolution must remain within coherence, fidelity, timing, control, and verification limits. Classical information derived from measurement must retain sufficient provenance and uncertainty characterization to support downstream interpretation. Interface design must ensure that errors, ambiguity, or manipulation are not silently propagated from physical execution into higher-level inference or decision processes.

The QALIS/CRQC–LLM dual-lens construct provides methodological enforcement for these requirements. QALIS interprets quantum primitives as constructive design variables that can be assembled into learning, inference, and reasoning pipelines. It examines how representation, transformation, correlation, measurement, and feedback may be organized into stable system capabilities.

CRQC–LLM examines those same primitives and interfaces as potential fragility surfaces under automation, scale, cryptographic pressure, coherence loss, interface manipulation, verification weakness, and adversarial timing stress. It evaluates how disturbances or failures may propagate through the architecture and affect integrity, resilience, provenance, and governance.

Together, these lenses establish that capability and constraint cannot be engineered separately. A capability that cannot remain physically feasible, operationally stable, evidentially supported, and appropriately governed does not constitute a defensible system-level advantage. Design, verification, resilience, and governance must therefore co-emerge from the same physical and architectural substrate.

Section 1.5 Summary — Mechanics to Architecture

Section 1.5 established the transformation of quantum mechanics into QAIS architectural doctrine. Superposition provides representational latitude; entanglement supplies non-decomposable correlations; interference governs how amplitudes contribute to possible outcomes; coherence defines the usable execution interval; and measurement converts quantum evolution into classical evidence.

These properties become architectural variables only when their roles, dependencies, and physical limits are incorporated into the complete system lifecycle. Amplitude and correlation structures shape representation, coherence imposes an execution budget, measurement establishes a governed quantum-classical boundary, and quantitative measures such as entropy, fidelity, trace distance, and coherence time support the evaluation of preservation, distinguishability, stability, and verification.

The section also established that physical constraints propagate throughout the hybrid architecture. Noise, calibration error, limited connectivity, coherence loss, measurement uncertainty, and interface weakness can influence transformation reliability, output distributions, inference quality, and operational decisions. Quantum capability and physical constraint must therefore be evaluated together.

QALIS examines how these primitives can support stable learning, inference, and reasoning capabilities, while CRQC-LLM evaluates their associated security, integrity, resilience, and verification obligations. Together, the two lenses establish that quantum advantage cannot be asserted independently of physical feasibility, operational stability, evidential support, and governance discipline.

Section 1.6, *Practical Applications—Foundations of Quantum AI Systems*, now translates this architectural doctrine into system-level design posture and deployment discipline.

1.6 Practical Applications — Foundations of Quantum AI Systems

Architectural Framing: *Measurement → System Architecture*

Section Framing: *Transitions from doctrine to system posture, defining how representation and propagation are bounded, staged, and governed in real deployments.*

Chapter 1's primary practical contribution is the establishment of systems-framing discipline. It prepares the reader to treat "quantum + AI" not as a loose aggregation of techniques, but as an integrated QAIS problem defined by physical law, architectural boundary conditions, and verification posture. In practice, this framing begins by explicitly defining the system boundary, clarifying what components operate within quantum state space, what remains classical, and where hybrid quantum-classical interfaces must be governed. It requires selecting an initial placement within the QAIS Quantum Stack. This ensures that later design decisions remain vertically coherent rather than drifting across abstraction layers. It also demands that minimum governance constraints be established before performance claims are made.

These include coherence budgets, measurement checkpoints, entropy bounds, fidelity thresholds, and distinguishability limits. Together, these commitments transform Chapter 1 from conceptual foundation into architectural baseline.

Within this disciplined frame, the primitives introduced earlier, superposition, entanglement, coherence, contextuality, and measurement disturbance, cease to be abstract phenomena and become operational resources. Superposition expands representational capacity; entanglement supplies correlated structure across participating subsystems without independently establishing communication or coordination; coherence sustains interference long enough for computation, sensing, and coordination to occur. Yet these same resources simultaneously impose constraints.

Coherence is time-bounded and fragile under noise. Entanglement is susceptible to disturbance and adversarial degradation.

Measurement irreversibly collapses state structure, enforcing trade-offs between information extraction and preservation. QAIS therefore emerges as a discipline defined not by capability alone, but by the inseparability of capability and constraint. Every expansion of expressive power carries a resilience implication, and every architectural claim must pass through verification surfaces shaped by physical law. This dual character is anchored through the QALIS–CRQC–LLM interpretive framework. Under the constructive lens of QALIS, quantum primitives are treated as enablers of scalable representation, distributed inference, and coordinated intelligence.

Under the stress-test lens of CRQC–LLM, those same primitives are examined as fragility surfaces, points where coherence decay, correlation instability, or interface misalignment expose systemic limits. By introducing both lenses at the foundational stage, Chapter 1 establishes that design and stress-testing must advance together.

Section 1.6 Summary — Practical Foundations of Quantum AI Systems

The practical foundation of QAIS is therefore not merely technical; it is methodological. Before operations are formalized in Chapter 2, before encoding strategies are deployed, and before distributed architectures are scaled, QAIS must be situated within explicit system boundaries, constraint budgets, and verification discipline. Only within that disciplined frame can higher-level claims about learning advantage, coordination gain, or security enhancement be meaningfully evaluated.

1.7 Case Study Advancement — Foundations of QAIS

Architectural Framing: *Full Spine Integration (System Architecture stage evaluated through QALIS and CRQC–LLM)*

Section Framing: *Applies dual-lens validation to full system composition, ensuring that architectural claims satisfy both capability and resilience constraints.*

Chapter 1 advances the dual case study by formalizing the interpretive contract that governs the remainder of the book. From this point forward, every mechanism introduced, whether physical primitive, operational rule, architectural pattern, or learning construct, must be readable through two complementary lenses: (a) a constructive systems lens that prioritizes coherence-preserving design, controlled interfaces, and deployability (QALIS), and (b) a stress-test lens that examines the same mechanisms as potential amplification pathways for brittleness, adversarial manipulation, governance failure, or architectural drift (CRQC–LLM).

At this opening stage, advancement is not measured in performance results but in definitional clarity: what counts as progress in QAIS is disciplined composition under physical constraint, state preparation → controlled transformation → interference shaping → measurement as a governed boundary, rather than isolated speedup claims or model-scale assertions.

Under QALIS, Chapter 1 establishes that quantum advantage [4] must be architecturally earned. Superposition is treated as structured representational breadth, not novelty; entanglement as a governed coordination layer, not mystical linkage; coherence as a budgeted runtime resource, not an assumed constant.

Architectural viability depends on explicit quantum–classical interface placement, bounded measurement checkpoints, and verification-aware flow across layers of the QAIS Stack. In this framing, system integrity precedes optimization, and disciplined structure precedes performance scaling.

Under CRQC–LLM, the same primitives are reinterpreted as fragility surfaces. Coherence decay, correlation instability, contextual drift, and ungoverned measurement handoffs become early warning indicators of systemic brittleness.

Importantly, the earliest failure modes are often conceptual and organizational rather than algorithmic: layer misplacement, abstraction leakage, misplaced trust in hybrid boundaries, or verification deferred until after scale.

By identifying these stress vectors at the foundation, Chapter 1 reframes resilience not as a patch applied later, but as a co-equal design requirement embedded from inception.

At this foundational stage, QAIS is therefore positioned not merely as a successor to QIS, but as a system-level discipline integrating physical law, architectural governance, and deployment realism.

The case study progression moves from "two labels" to a repeatable evaluation stance. QALIS and CRQC–LLM become methodological instruments, constructive and interrogative lenses that will be applied to operations (Chapter 2), encoding strategies, distributed architectures, communication layers, and resilience analysis throughout the volume. This locking of interpretive discipline is the true advancement of Chapter 1: it defines the criteria by which every subsequent claim of capability, scalability, or advantage will be assessed.

Comparative Analysis

Viewed together, QALIS and CRQC–LLM expose the dual character of QAIS foundations. The same physical principles that enable representational expansion, distributed coordination, and interference-driven reasoning also amplify fragility under real-world noise, adversarial manipulation, or governance failure. Superposition expands expressive capacity while increasing sensitivity to disturbance. Entanglement binds subsystems into coherent collectives while introducing correlation surfaces vulnerable to degradation. Coherence enables sustained computation yet imposes strict temporal limits on viable execution.

This dual reading anchors QAIS as a discipline defined not by capability alone, but by capability constrained and interrogated under physical law. Architectural potential and resilience requirement are inseparable. Advantage without governance is instability; governance without constructive design is stagnation. The comparative lens ensures that every resource introduced in later chapters remains interpretable as both enabler and exposure surface.

Section 1.7 Summary — Foundations of QAIS Case Study

Embedding dual-lens analysis at the foundation converts QAIS from a toolkit into a governed architecture. Design and stress-testing advance in parallel. Under QALIS, quantum primitives are assembled into coherence-budgeted, synchronization-aligned pipelines with explicit interface control and verification placement. Representational breadth is staged within runtime limits, and measurement is engineered as a deliberate resolution boundary rather than a downstream artifact.

Under CRQC–LLM, those same pipelines are interrogated as fragility surfaces—context drift, calibration decay, boundary leakage, entropy growth, and adversarial amplification under scale. Capability and brittleness are co-instantiated consequences of the same design commitments. Architectural integrity is defined not by peak advantage, but by sustained stability within constraint envelopes.

This dual evaluative stance becomes the methodological spine of the volume. Each subsequent chapter—operations, infrastructure, encoding, learning, metrics, communication, sensing, and governance—introduces mechanisms that are both constructed and stress-tested. No capability advances without constraint accounting; no resilience claim stands without architectural grounding.

With this discipline fixed, the narrative advances from physical framing to executable structure. Chapter 2 formalizes the operational grammar of QAIS—state preparation, controlled transformation, contextual encoding, mitigation, correction, and governed measurement workflows—shifting the emphasis from what the system is to how it runs under physical and adversarial constraint."

Chapter 1 Conclusion

Chapter 1 established the physical and architectural foundations of QAIS by reframing superposition, entanglement, and coherence as engineered resources, while treating contextuality and measurement disturbance as system-defining conditions governing how quantum information can be accessed, interpreted, and operationalized. Within the Quantum Stack, these primitives establish the physical constraints inherited by higher layers as quantum information progresses from physical state structure toward engineered computation and intelligent system architecture. The chapter's internal trajectory culminated in the System Architecture stage, where these foundational principles were translated into system boundaries, constraint budgets, governed interfaces, and verification requirements.

Moving beyond descriptive physics, the chapter introduced systems-framing discipline. Quantum reasoning was treated as interference structured within coherence budgets and resolved through governed measurement boundaries. Architectural legitimacy was established as a prerequisite for claims of performance, scaling, or advantage.

The dual-lens framework—QALIS as the constructive evaluation lens and CRQC–LLM as the stress-test lens—established the interpretive contract for the remainder of the volume. QALIS demonstrated how quantum resources can be assembled into disciplined learning, inference, and coordination pipelines, while CRQC–LLM exposed how those same resources can become fragility surfaces under cryptographic pressure, adversarial automation, synchronization drift, and scale-induced instability. Capability and constraint were therefore formalized as co-instantiated variables rather than separable concerns.

QAIS thus emerges not merely as an extension of QIS, but as a systems discipline defined by the co-engineering of capability and resilience under physical law. Superposition defines representational breadth, entanglement defines structural correlation, coherence defines runtime limits, and measurement defines the governed boundary between quantum evolution and classical interpretation. Together, these primitives establish the physics-bound design space within which subsequent QAIS architectures must operate.

With this foundation secured, the developmental arc advances from physical framing to operational execution. Chapter 2 moves from definition to operational grammar, formalizing how quantum states are prepared, transformed through controlled gate operations, encoded within contextual constraints, stabilized through mitigation and correction strategies, and resolved through verification-aware measurement workflows.

The progression therefore shifts from what quantum-intelligent systems are to how they operate—where representational capacity becomes executable structure and physical constraint becomes disciplined control.

Quick Self-Check

Foundations of Quantum AI Systems

1. How does QAIS redefine the concept of information?

A) As symbolic data structures optimized for algorithmic speed
B) As probabilistic outcomes determined only by measurement statistics
C) As a physically embodied, coherence-sensitive property of quantum systems
D) As encrypted signals transmitted through secure channels

2. Which primitives define the physical substrate of QAIS?

A) Classical logic gates and probability distributions
B) Superposition, entanglement, coherence, and governed measurement
C) Encryption and checksum validation
D) Tensor factorization methods

3. Why is QAIS positioned as a systems discipline rather than an algorithmic extension of QIS?

A) It replaces classical AI entirely
B) It focuses only on faster hardware
C) It frames computation, representation, and trust as system-level architectural properties
D) It ignores classical components

4. What is the architectural role of coherence in QAIS?

A) It guarantees deterministic outputs
B) It functions as a runtime budget that bounds usable quantum evolution
C) It replaces classical memory
D) It eliminates measurement uncertainty

5. Why are entropy, fidelity, and trace distance treated as forward-facing constraints?

A) They are abstract mathematical curiosities
B) They apply only to laboratory experiments
C) They define encoding limits, integrity thresholds, and distinguishability bounds prior to deployment claims
D) They replace performance benchmarks

6. How do QALIS and CRQC–LLM function in Chapter 1?

A) As competing algorithms
B) As hardware configurations
C) As a constructive design lens and a stress-test verification lens
D) As benchmarking frameworks

7. What distinguishes entanglement from classical correlation?

A) It increases computational speed
B) It allows copying of quantum states
C) It creates non-decomposable joint states that cannot be factored into independent parts
D) It guarantees secure communication without constraint

8. What is the governed boundary in QAIS?

A) Hardware scaling limits
B) The measurement interface between quantum evolution and classical interpretation
C) The encryption layer
D) The classical preprocessing stage

9. What does the Bloch sphere geometrically represent?

A) Classical probability distributions
B) Pure single-qubit states represented as points on the unit sphere under normalization and relative phase constraints
C) Hardware circuit topology
D) Encryption key spaces

10. Why is relative phase operationally significant?

A) It changes measurement probabilities directly
B) It governs interference behavior under unitary evolution
C) It guarantees entanglement
D) It removes decoherence

11. What is meant by "non-factorizable" state vectors?

A) States that cannot be measured
B) States that are computationally complex
C) States that cannot be expressed as tensor products of individual subsystems
D) States requiring classical control

12. How does QAIS reinterpret superposition?

A) As classical parallelism
B) As randomness
C) As coherence-dependent representational geometry enabling interference
D) As error correction

13. Why is entanglement described as infrastructure?

A) It guarantees secure channels
B) It enables coordinated evolution across distributed subsystems within a shared relational state space
C) It eliminates noise
D) It replaces measurement

14. What is the significance of trace distance in QAIS?

A) It measures execution speed
B) It bounds distinguishability and therefore error probability
C) It encodes phase
D) It determines qubit count

15. What architectural doctrine emerges in Chapter 1?

A) Maximize qubit count at all cost
B) Optimize algorithms independently of physics
C) Co-engineer capability and resilience under physical constraint
D) Replace classical systems entirely

16. Why is measurement treated as irreversible?

A) Because it destroys hardware
B) Because it collapses coherent amplitude structure into classical output
C) Because it consumes energy
D) Because it encrypts data

17. How does coherence time affect circuit depth?

A) It has no effect
B) It guarantees unlimited runtime
C) It constrains executable sequence length before decoherence dominates
D) It determines classical preprocessing

18. What does Chapter 1 assert about quantum information and substrate?

A) They are separable
B) They are interchangeable with classical bits
C) They are inseparable throughout the information lifecycle
D) They are independent of noise

19. What is the Quantum Stack?

A) A hardware specification list
B) A layered architectural ascent from physics to intelligent systems
C) A compiler framework
D) A benchmarking suite

20. How does QALIS interpret entanglement?

A) As cryptographic weakness
B) As connective structure enabling coordinated inference pipelines
C) As redundant overhead
D) As classical duplication

21. How does CRQC–LLM interpret entanglement?

A) As guaranteed security
B) As an optimization shortcut
C) As a fragility surface under adversarial or cryptographic stress
D) As hardware redundancy

22. What marks the conceptual inflection point in Chapter 1?

A) Introduction of hardware specifications
B) Benchmarking metrics
C) The transition from quantum phenomena as theory to architectural doctrine
D) Deployment case studies

23. In the QAIS System-Layer Model, across which layers does QAIS become most visibly associated with intelligent-system realization?

A) Substrate physics
B) Operational electronics
C) Across layers 4–6, where encoding, architectural composition, and hybrid deployment converge
D) Compiler optimization

24. Why can entanglement support authentication mechanisms?

A) Because it encrypts data
B) Because it duplicates states
C) Because disturbance of shared correlation alters observable statistics
D) Because it increases bandwidth

25. What is the central takeaway of Chapter 1?

A) Quantum speedup guarantees advantage
B) Entanglement ensures security
C) Intelligent quantum systems must be engineered within physically bounded, verification-aware architectures
D) Classical AI is obsolete

Reader Guidance

For answers to these questions, see Appendix B for expanded reflection.

References

[1] R. P. Feynman, "Simulating physics with computers," *Int. J. Theor. Phys.*, vol. 21, pp. 467–488, 1982.

[2] D. Deutsch, "Quantum theory, the Church–Turing principle and the universal quantum computer," *Proc. R. Soc. A*, vol. 400, no. 1818, pp. 97–117, 1985.

[3] P. W. Shor, "Algorithms for quantum computation: Discrete logarithms and factoring," in *Proc. 35th Annu. Symp. Found. Comput. Sci. (FOCS)*, 1994, pp. 124–134.

[4] L. K. Grover, "A fast quantum mechanical algorithm for database search," in *Proc. 28th Annu. ACM Symp. Theory Comput. (STOC)*, 1996, pp. 212–219.

[5] M. A. Nielsen and I. L. Chuang, *Quantum Computation and Quantum Information.* Cambridge, U.K.: Cambridge Univ. Press, 2000.

[6] J. Preskill, "Quantum computing in the NISQ era and beyond," *Quantum*, vol. 2, p. 79, 2018.

[7] K. Bharti *et al.*, "Noisy intermediate-scale quantum algorithms," *Rev. Mod. Phys.*, vol. 94, no. 1, p. 015004, 2022.

[8] A. S. Holevo, "The capacity of the quantum channel with general signal states," *IEEE Trans. Inf. Theory*, vol. 44, no. 1, pp. 269–273, Jan. 1998.

[9] F. Arute et al., "Quantum supremacy using a programmable superconducting processor," Nature, vol. 574, pp. 505–510, 2019.

[10] National Quantum Initiative Act, Pub. L. No. 115–368, 132 Stat. 5092, Dec. 21, 2018.

[11] European Commission, "Quantum Flagship: Pioneering European Quantum Technologies," 2018. [Online]. Available: https://qt.eu

[12] Ministry of Science and Technology of China, "China National Laboratory for Quantum Information Science," 2020.

[13] C. H. Bennett and G. Brassard, "Quantum cryptography: Public key distribution and coin tossing," in Proc. IEEE Int. Conf. Computers, Systems and Signal Processing, 1984.

Chapter 2

—

Operations and Scientific Framework of QAIS

Architectural Position

Defines the operational grammar of quantum systems, including state evolution, transformation, and measurement processes. Within the Quantum Stack, this chapter establishes Layer 2, providing the formal rules that govern how quantum states are manipulated under physical constraints. These rules enable controlled execution and form the basis upon which infrastructure, representation, and computation are constructed.

Core Principle

All QAIS behavior emerges from controlled quantum operations constrained by physical law.

Abstract

This chapter develops the operational grammar of QAIS, establishing quantum states, transformations, and measurements as the executable primitives underlying intelligent quantum architectures. Building on Chapter 1's treatment of information as a physical resource, it examines how information is encoded, evolved, interfered, and measured within quantum systems. Superposition and entanglement are treated as capacity-expanding primitives, while measurement backaction and the no-cloning principle are reframed as architectural constraints that bound verification, control, and reliability. Coherence is introduced as a tunable yet fragile parameter governing the integrity [6] of quantum operations [7].

Resilience is formalized through noise models[7] [8], decoherence pathways, and foundational error-correction constructs, with fidelity, entropy, and trace distance positioned as operational benchmarks[8] [14] for evaluating state preservation and transformation quality. Hands-on quantum labs reinforce these concepts through qubit simulation, interference analysis, entanglement exercises, and introductory error-correction routines. Within the chapter's case-study framework, QALIS provides a constructive lens for leveraging these primitives in architectural design, while CRQC–LLM exposes early-stage fragility, disturbance, and contextual vulnerabilities. Together, these perspectives establish the primitive-level foundation required for subsequent system-scale mechanisms.

Keywords

Quantum AI Systems, operational grammar, qubits, unitary evolution, interference, entanglement, measurement, no-cloning, coherence, noise models, error correction, fidelity, entropy, trace distance, QAIS resilience, QALIS, CRQC–LLM

---

[6] Preskill's NISQ-era analysis [7] underscored that fragile coherence and limited error correction define the practical boundary of near-term systems.

[7] Recent reviews of NISQ algorithms formalize the algorithmic constraints within NISQ systems, highlighting noise as an intrinsic computational factor.

[8] Benchmarking frameworks such as QUARK help quantify system-level performance and enforce operational guardrails across QAIS deployments.

2.1 Introduction — Architecture to Operations

Architectural Role: *Defines the formal transformation language of QAIS, specifying how quantum states evolve, interact, and are mathematically manipulated across system operations.*

Architectural Framing: *State Preparation → Unitary Evolution → Controlled Operations → Measurement → Verification → Hybrid Execution*

Section Framing: *Chapter 2 defines the operational grammar of QAIS, translating architectural constraints into executable system behavior through controlled transformation, measurement discipline, and verification-aware workflows. It establishes how unitary evolution, state preparation, and measurement staging form the execution logic of quantum systems, bridging physical law and programmable operation. The chapter reframes computation as governed evolution under constraint, where correctness depends on coherence preservation, control precision, and measurement placement.*

Quantum Stack Layer: L2 *— Quantum Mechanics → Formalizing state evolution, operators, measurement, and system dynamics.*

Within the Quantum Stack, Layer 2, Quantum Mechanics, supplies the mathematical framework that governs how quantum information evolves, interacts, and is measured. Chapter 2 aligns directly with this layer by translating the Hilbert-space formalism, operator dynamics, and measurement rules of quantum mechanics into the operational grammar [5] of QAIS. Here, concepts such as unitary evolution, measurement collapse, contextuality [5], and error models are reframed not as abstract mathematical structures but as executable system rules [10] that define how QAIS architectures behave in practice. By grounding gates, operations, and state transformations in the geometry and algebra of quantum mechanics, Chapter 2 functions as the mathematical layer of the QAIS stack, establishing the rigorous operational substrate upon which the system-level strategies, protocols, and resilient architectures explored in later chapters are built.

While quantum mechanics is traditionally taught through wavefunctions, operators, and the Schrödinger equation, this chapter adopts an explicitly operational formulation aligned with Feynman's sum-over-histories view. Rather than prioritizing equations of motion, the focus here is on how quantum states are prepared, transformed, allowed to interfere, and measured as executable system primitives [11].

---

[9] *Executable quantum rules:* Quantum mechanical principles interpreted as enforceable operational constraints that govern system behavior during execution, rather than as abstract theoretical effects. These rules determine how quantum processes unfold within engineered architectures.

[10] *Execution-level rules:* Rules that govern how quantum processes are physically enacted during system operation, as opposed to how they are mathematically described. Execution-level rules determine when, where, and under what constraints quantum operations are applied within a running system.

[11] The minimal set of operational actions—such as preparation, transformation, interference, and measurement—from which all higher-level Quantum AI behaviors are composed. These primitives define what a system can physically do, independent of algorithmic complexity.

This framing is mathematically equivalent to the traditional formalism, but it is more directly aligned with quantum computation, quantum circuits, and QAIS workflows.

From Physical Law to Executable Grammar

Chapter 1 established information as a physically embodied resource and positioned QAIS as a discipline grounded in quantum mechanical law. Chapter 2 advances that progression by shifting emphasis from what quantum information is to how it behaves under controlled manipulation.

Quantum operations form the operational rule-set[12] of QAIS, transforming physical resources into executable primitives that govern how intelligent quantum systems evolve, compute, interact, and verify under quantum law. State preparation, coherent transformation, interference, contextual interaction, and measurement-driven transition collectively define the execution grammar through which quantum information becomes operational within QAIS pipelines.

The thematic arc of Chapter 2 therefore moves from physical embodiment to controlled execution. Quantum states are no longer considered only as representational structures; they are prepared, transformed, correlated, measured, and verified through disciplined operational processes. Coherent transformations determine how state structure evolves, interference shapes computational pathways, contextual interactions constrain admissible behavior, and measurement establishes the boundary through which quantum processes produce classically actionable evidence.

This progression establishes the operational foundation required for the computational, infrastructural, learning, and deployment architectures developed in subsequent chapters.

Unitary evolution, interference, measurement backaction, contextuality, and irreversibility are therefore treated not as isolated phenomena but as operational constraints. These constraints determine how QAIS performs computation, coordinates inference, manages state transitions, and encodes decision pathways. Coherence defines the usable lifetime of an operation; interference becomes a constructive or destructive design lever; measurement defines the boundary through which quantum evolution becomes classically actionable.

---

[12] **Operational rule-set:** A structured collection of executable quantum behaviors—such as state preparation, unitary transformation, interference, and measurement—that collectively define how a Quantum AI System operates. Unlike abstract postulates, an operational rule-set specifies how physical processes are invoked, constrained, and sequenced within an execution pipeline.

Measurement as an Operational Boundary

In contrast to textbook treatments that frame measurement primarily as wavefunction collapse, this chapter adopts a strictly operational interpretation in which measurement functions as a boundary condition within the QAIS execution pipeline.

Rather than interrupting continuous evolution, measurement selects classical outcomes from interference structures accumulated during preparation and transformation.

This interpretation preserves mathematical equivalence with traditional formulations while clarifying measurement's role as an extraction and verification mechanism rather than an ontological event. The formal development of this view appears in Section 2.3.

This operational framing integrates disciplinary perspectives. Physicists will recognize transformations validated through controlled experiments involving reversibility, interference fringes, and statistical measurement outcomes.

Engineers will see coherence limits, noise sources, and controllability constraints shaping feasibility. AI specialists will observe how operational primitives determine representational capacity in quantum-enhanced learning models. Cybersecurity practitioners will note how irreversibility, disturbance, and contextuality impose verification boundaries that either stabilize or expose QAIS workflows.

By establishing this operational grammar, Chapter 2 provides the scientific and procedural foundation for the mechanism, architecture, and resilience layers that follow. Its full significance, however, emerges when interpreted through the dual analytical lenses introduced in Chapter 1. Before examining in detail how quantum states evolve, interact, encode correlations, and become actionable within QAIS, it is essential to understand how the QALIS constructive framework and the CRQC-LLM adversarial framework interpret this operational layer.

Their contrasting perspectives reveal how capability and fragility arise from the same physical rules and provide the interpretive structure for the remainder of the chapter. With this context established, we now turn to the case study orientation that frames how QALIS and CRQC-LLM engage the operational behaviors introduced here [13].

---

[13] *Operational foundation:* The base layer of execution rules and interpretive principles that support higher-level mechanisms, architectures, and governance structures. In QAIS, this foundation ensures consistency between physical behavior, system design, and verification.

GUIDING PRINCIPLE 2 — VERIFICATION-GOVERNED QUANTUM INTELLIGENCE

Quantum AI Systems derive reliable intelligence from the coherent integration of state transformation, measurement, and control under conditions of continuous verification. Verification is not a terminal step but a governing constraint that ensures quantum evolution remains physically consistent and computationally meaningful within hybrid quantum-classical execution.

System stability emerges when transformation, measurement, and control are aligned through verification; without this alignment, errors propagate and degrade system behavior.

2.1.1 Case Study Orientation — QALIS vs CRQC-LLM

Chapter 1 introduced two guiding perspectives for the entire book: **QALIS**, which represents the constructive, design-focused view of Quantum AI Systems, and **CRQC-LLM**, which provides an adversarial lens for understanding stress, fragility, and verification limits. Chapter 2 is the first place where these perspectives become concrete. The chapter focuses on the basic operations—unitary evolution, interference, entanglement, and measurement—that control how quantum states change and interact. Because these operations sit at the base of the QAIS stack, they form the earliest point at which QALIS and CRQC-LLM interpret system behavior differently.

Through the QALIS lens, these operations act as constructive resources. Reversible evolution [5] guides quantum states smoothly, preserving the structure needed for later reasoning. Interference strengthens desirable outcomes and weakens others, allowing quantum systems to process information in ways classical systems cannot. Entanglement links qubits in coordinated ways that enable distributed behavior, and measurement backaction becomes a controlled mechanism for turning quantum information into classical outcomes. In QALIS, these operations form the "alphabet" from which higher-level quantum intelligence will be built.

Through the CRQC-LLM lens, the same operations reveal early signs of fragility. Coherence becomes a strict time limit susceptible to noise; interference becomes sensitive to small disturbances; measurement becomes a potential point of attack if performed incorrectly; and entanglement, though powerful, is easily broken.

In this view, Chapter 2 highlights how delicate the operational layer can be, and how these vulnerabilities shape verification, security, and reliability as QAIS workflows scale. Together, these perspectives show why the operational rules of quantum mechanics matter so deeply for Quantum AI Systems [14].

---

[14] *Operational rules:* The actionable constraints derived from quantum mechanics that govern how quantum processes may be executed, observed, and controlled within engineered systems, as distinct from purely mathematical descriptions.

QALIS uses them as building blocks for intelligent behavior, while CRQC–LLM uses them to reveal limits and vulnerabilities. Chapter 2 therefore marks the starting point for understanding both the potential and the challenges of building intelligent systems powered by quantum mechanics.

2.1.2 An Operationally Grounded Quantum Framework

Throughout this volume, Quantum AI Systems are analyzed using two complementary evaluative lenses: QALIS and CRQC–LLM, which respectively illuminate constructive and adversarial trajectories in system capability, resilience, and integrity. These lenses do not introduce new physics or alternative mathematical formalisms. Rather, they provide architectural viewpoints for reasoning about how quantum mechanisms behave when embedded in real systems.

A central principle governing this analysis is the distinction between **operational interpretation** and **mathematical formalism**. Operational interpretation is used to reason about system behavior, architectural design, and execution logic within QAIS pipelines [15]. Mathematical formalism is used for calculation, verification, and implementation, and remains fully authoritative. The two are treated as **formally equivalent descriptions of the same underlying physical processes**, serving different roles in system understanding and design.

In this book, operational reasoning about quantum behavior is framed in terms of **preparation, transformation, interference, and measurement** as executable system rules. This framing is consistent with Feynman's sum-over-histories perspective, in which observable outcomes emerge from the accumulation and interference of admissible processes under constraint.

Standard mathematical representations, whether expressed through wavefunctions, operators, or Hilbert-space formulations, are preserved as exact and complete execution grammars, not as competing interpretations.

Within this operational framing, **measurement is treated as a boundary condition** that converts accumulated interference structure into classical outcomes suitable for verification, inference, and control. It is not introduced as a special ontological interruption of otherwise continuous evolution.

[15] An interpretation framework that assigns meaning to quantum processes based on how they behave during execution within a system. In QAIS, operational semantics focus on observable behavior, control flow, and evidentiary outcomes rather than internal state descriptions.

Likewise, interference is treated as a system-level mechanism, not as a secondary artifact of abstract state evolution. These interpretations are essential for reasoning coherently about hybrid quantum–classical systems, where execution, verification, and governance depend on clearly defined operational boundaries.

Traditional textbook language may be referenced where useful for equivalence or familiarity, but collapse-centric or purely state-centric intuition is not used as the primary explanatory framework for QAIS behavior.

This discipline ensures that all chapters reason consistently about quantum computation as an interference-driven operational process, while preserving strict mathematical equivalence with standard quantum mechanics.

By maintaining operational primacy alongside formal equivalence, the book provides a coherent foundation for analyzing quantum AI systems as executable architectures—systems whose behavior must be designed, verified, governed, and trusted over time [16].

2.1.3 Quantum Operations and Gates

QAIS operationalizes quantum information through a set of physical transformations, constraints, and measurement processes that define how intelligent quantum systems behave. Chapter 1 established information as a physically embodied resource; this section examines the mechanisms that act upon that resource, introducing the operational primitives—unitary evolution, interference, entanglement, contextuality, and measurement backaction, that form the executable grammar of QAIS. These mechanisms are developed in detail in subsequent sections of this chapter.

In this framework, qubits are not symbolic carriers but dynamically evolving subsystems whose allowed transformations and structural limitations shape the consistency, reliability, and expressivity of quantum operations.

In this chapter, operators, gates, and transformation rules are treated as **implementation mechanisms**, not as primary ontological descriptors of quantum behavior.

From the operational perspective adopted here, these constructs serve to *constrain, redirect, and phase-weight* the contributions of competing quantum histories, rather than to define the fundamental reality of the system. This distinction is essential for QAIS: while operator formalisms remain indispensable for calculation, verification, and hardware compatibility, the system-level behavior of quantum intelligence is governed by how preparation, transformation, interference, and measurement are orchestrated as an executable process grammar.

---

[16] *Primitive operational structure:* The high-level architectural organization of execution rules, constraints, and control boundaries that shape how Quantum AI Systems evolve over time. This structure governs system behavior across layers rather than within individual algorithms.

Throughout this chapter these mechanisms collectively define the quantum operational grammar [17], the set of reversible transformations, interference behaviors, compatibility constraints, measurement rules, and correlation structures that govern how quantum states evolve and become actionable within QAIS.

In QAIS, quantum gates are the fundamental unitary transformations that implement controlled evolution of qubit states, forming the primitive operations upon which all higher-level computation is constructed [18].

Quantum operations differ fundamentally from the deterministic actions of classical logic systems. Classical bits adopt definite states, logic gates produce predictable transitions, and circuits follow fixed combinational pathways. QAIS departs from this model entirely.

Quantum states evolve through reversible, unitary transformations that create interference patterns and correlations beyond any classical representational scheme. Measurement is not a neutral readout step but an active process that collapses a state and imposes boundaries on replication, observability, and subsequent evolution.

These features establish quantum operations as physical events that enforce system-level behavior, rather than symbolic transitions defined by abstract logic.

These operational primitives offer substantial representational capacity while introducing structural constraints. Qubits support amplitude-encoded representations, probabilistic evolution, and nonclassical correlations, but they are also subject to decoherence, environmental noise, measurement disturbance, and prohibitions such as the no-cloning rule.

This dual nature, expressive potential paired with physical fragility, defines the environment in which QAIS must operate. Accordingly, qubits and their transformations must be treated as physically embodied resources whose properties determine the control, verification, and error-handling strategies available to designers.

Encoding at the quantum level reinforces the break from classical determinism. Representations are embedded directly into amplitude, phase, or basis structure, binding information to the underlying quantum state itself. Under these conditions, copying and observation introduce unavoidable boundaries: measurement collapse restricts noninvasive access to the state, and the no-cloning theorem forbids perfect duplication. These limitations function as intrinsic design rules that protect coherence, enforce verification constraints, and guide the flow of information within QAIS.

As the Section deepens, entanglement emerges as the defining relational construct. Entangled qubits form correlations that extend across spatial or logical contexts, enabling coordinated

[17] *Execution grammar:* The structured set of allowable quantum operations and transitions that define how a Quantum AI System executes tasks. Analogous to a programming grammar, it constrains how operations may be composed into valid execution pipelines or circuits.

[18] Primitive execution rules identifies the lowest-level operational constraints from which higher-level QAIS behaviors emerge, including admissible transformations, interference conditions, and measurement boundaries.

behaviors that classical systems cannot emulate. Contextuality extends these nonclassical relationships into the logical domain, demonstrating that operational outcomes depend on interaction structure rather than intrinsic properties of isolated subsystems.

Geometric models, such as the Bloch sphere, together with representational and relational frameworks (e.g., Figure 2.2.1, Figure 2.9.1, and Figure 2.9.2), illustrate how these primitives reshape representation, interference behavior, and transformation dynamics.

Case Study Anchor — Quantum Operations and Gates

Intent: To interpret unitary evolution, reversible transformations, and interference shaping as the executable grammar of QAIS, clarifying how operational primitives generate both constructive capability and early-stage fragility under constraint.

The constructive and adversarial case study lenses introduced in Chapter 1—**QALIS and CRQC-LLM**—gain their first concrete interpretive role in Chapter 2 through the operational primitives developed across the sections that follow. These sections define the **quantum operational grammar**: the foundational rules governing how quantum states evolve, interact, correlate, and produce measurable outcomes. QALIS and CRQC-LLM interpret this grammar in fundamentally different ways, revealing two complementary but divergent trajectories of capability and fragility.

Through QALIS (Constructive Framework)

QALIS interprets the primitives introduced in Sections 2.3–2.6 as **building blocks for reliable quantum intelligence**.

- **Reversible evolution** (introduced in Section 2.3) is viewed as a stable, coherence-preserving mechanism that ensures predictable state transformation.
- **Amplitude shaping and representational structure** (Section 2.2) provide the expressive substrate upon which later encoding, learning, and decision layers will operate.
- **Measurement backaction** (Section 2.3) is interpreted as a useful transition rule—an engineered boundary between quantum and classical information flows.
- **Entanglement and nonclassical correlation** (Section 2.4) supply the relational backbone required for distributed inference and multi-qubit coordination.
- **Contextual behavior** (Section 2.5) is treated as a resource that enables nonclassical logic, compatibility structures, and richer forms of information embedding.
- **Interference-driven geometric evolution** (Section 2.6) is seen as a mechanism for shaping desirable signal pathways through constructive and destructive amplitude redistribution.

To QALIS, these primitives serve as the *alphabet* from which higher-level reasoning and inference architectures will later be composed.

Through the CRQC–LLM Stress-Test Lens

CRQC–LLM examines these same primitives as early exposure points for fragility, brittleness, and verification limits.

- **Reversible dynamics**, when perturbed, reveal how even small control errors can produce distorted state trajectories.
- **Amplitude structures** are vulnerable to normalization drift, gate miscalibration, and adversarially biased inputs.
- **Measurement backaction** exposes verification boundaries when unauthorized probing or premature readout disrupts coherent evolution.
- **Entanglement**, while powerful, becomes a stress surface where noise, decoherence, or adversarial interference can collapse correlations.
- **Contextual behavior** creates opportunities for inconsistency if incompatible settings, timing perturbations, or adversarial manipulations shift outcome statistics.
- **Interference patterns**, essential for computation, become extremely sensitive to environmental disturbance and malicious amplitude perturbations.

Under CRQC–LLM, these primitives mark the earliest domain where **quantum-accelerated threat models**, cryptographic risk, and system integrity concerns begin to surface.

Section 2.1 Summary — Architecture to Operations

This section reframes qubits and gates as the operational primitives of QAIS, physically embodied resources that evolve, interfere, entangle, and collapse under system-level rules. By treating reversible transformations, contextual correlations, and measurement backaction as architectural mechanisms rather than abstract phenomena, the section establishes the grammar that governs all QAIS computation and inference. These foundations establish the basis for Section 2.2, where encoding and representational architectures reveal how information is structured within this operational substrate.

Later chapters expand upon these primitives to show how operations and gates scale into full architectural workflows and intelligent reasoning systems built on the principles established here.

2.2 Representation and Encoding

Architectural Framing: *State Preparation*

Section Framing: *Defines initialization of the control surface, setting the initial conditions for propagation. QAIS redefine how information is represented, prepared, and manipulated at the physical level. Whereas QIS introduced qubits as abstract carriers of amplitude and phase, QAIS reframes representation and encoding as operational primitives that determine how quantum information behaves under transformation. Chapter 2 now extends the discussion beyond the bit–qubit distinction established earlier, examining how encoding mechanisms shape the conditions under which quantum states can be prepared, preserved, and made actionable inside intelligent quantum systems.*

In classical systems, representation is deterministic and easily replicated: bits occupy discrete states, transitions follow predictable logic, and copies can be created without constraint. In QAIS, representation is **coherence-bound**, **probabilistic**, and **physically sensitive** to how information is embedded in the quantum substrate. Superposition allows a single qubit to encode overlapping possibilities; amplitude and phase define the informational structure; and coherence determines how long that structure remains usable before environmental disturbance collapses it. Representation is therefore not merely a conceptual shift but an **operational choice** that governs how information persists and evolves within QAIS.

The representation of information within quantum systems establishes the foundation upon which all subsequent transformations and computations are built. Unlike classical encodings, quantum representations embed information directly within state amplitudes and phases, linking abstraction to physical realization. Figure 2.2.1 illustrates this representational framework within QAIS architectures, providing structural context for how encoding strategies shape downstream system behavior.

2.2.1 Representation as an Operational Primitive

This Subsection uses the Bloch sphere to illustrate how QAIS encodes information in amplitudes and phases, turning the geometric state space into an operational map for preparing and manipulating quantum states.

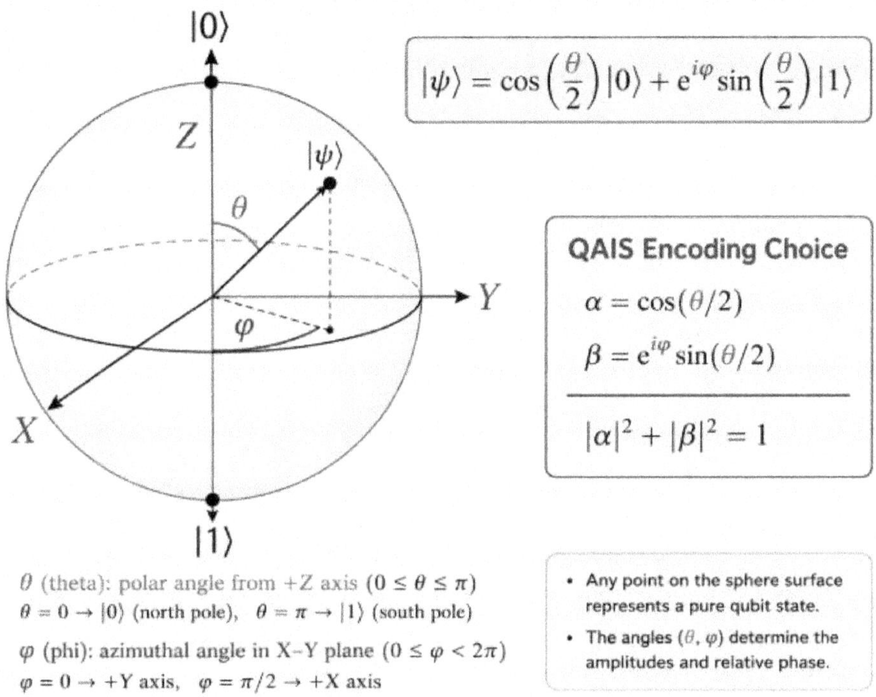

Figure 2.2.1 — Representation and Encoding of Quantum States.

The Bloch sphere provides the geometric space in which QAIS encodings operate. The poles represent computational basis states, the equator hosts superpositions, and the orientation of the state vector captures amplitude and phase relationships. These relationships reveal how QAIS embeds information directly into the quantum state, making representation inherently physical, coherence-dependent, and sensitive to disturbance.

In this book, "QAIS Encoding Choice" refers to the assignment of amplitudes (α, β) in the computational basis, parameterized by θ and φ on the Bloch sphere. These coefficients determine how classical inputs are embedded into qubits within QAIS pipelines. While mathematically general ($\alpha|0\rangle + \beta|1\rangle$), in QAIS they are treated as system-level design parameters that shape representation, stability, and resilience.

The accompanying spotlight interprets this representational framework as a primitive on which all subsequent QAIS operations depend.

Technical Spotlight — Representation as a System Primitive

Conceptual Grounding

Figure 2.2.1 illustrates that representation in QAIS is encoded directly into the physical state of the qubit, its amplitudes, phases, and coherence properties. Unlike classical symbols that carry information independently of their medium, quantum representations exist only as long as coherence is maintained. The Bloch sphere captures this relationship by visualizing how probability and phase define the informational content of the state.

System Relevance

At the operational level, representation determines how quantum states can be initialized, steered, and transformed by gates. Amplitude and phase dictate how interference manifests in circuits; coherence determines how long a state can support multi-step operations; and the chosen encoding method constrains which transformations preserve fidelity and which introduce fragility. These representational rules define the limits and opportunities of the operational grammar introduced in Chapter 2.

Cross-Disciplinary Orientation

Across disciplines, the representational framework in QAIS is interpreted through distinct but complementary lenses. Physicists understand it as the geometric trajectory of a quantum state through Hilbert space, shaped by coherent control and pulse-driven manipulation. Engineers view representation as a calibration and stabilization challenge, where noise models, drift compensation, and hardware tolerances determine how accurately amplitudes and phases can be prepared and preserved. AI specialists see representation as the substrate that enables quantum operations to express correlations and structure that classical encodings cannot capture. Cybersecurity practitioners interpret the fragility of representation, particularly its sensitivity to measurement and collapse, as a trust boundary that supports verification and reveals tampering at the physical layer.

Case Study Anchor — Representation and Encoding

Intent: To frame encoding choices as physically embodied design decisions that determine coherence stability, interference behavior, and admissible transformation pathways within QAIS workflows.

- **QALIS** emphasizes how stable preparation and controlled manipulation of amplitude and phase form the constructive foundation for the more complex behaviors developed in later chapters.
- **CRQC–LLM** highlights how representational fragility, coherence loss, phase drift, and disturbance sensitivity, reveals early indicators of operational stress that later resilience mechanisms must address.

These applications remain strictly at the level of **primitives**, reinforcing how representation functions as both a capability and a constraint within QAIS.

Technical Spotlight — Operational Interpretation of Interference as a Sum Over Histories

Conceptual Grounding

Feynman's path-integral formulation reframes quantum evolution not as the deterministic propagation of a single state, but as the accumulation and interference of amplitudes associated with multiple possible histories consistent with boundary conditions and conservation laws. In this view, a "history" represents one complete, admissible evolution of the system across time or configuration space. Quantum outcomes arise not from selecting a preferred trajectory, but from the constructive and destructive interference of these histories when their associated amplitudes are combined.

A crucial pedagogical point is that histories should not be interpreted as explicit alternatives that a system "chooses between." Rather, they represent the full space of evolutions that remain physically admissible under the imposed constraints. The path-integral formulation does not describe many simultaneous executions in a classical sense; it describes how physical law aggregates contributions from all allowed processes to determine observable outcomes.

This perspective replaces trajectory-centric intuition with **process-centric reasoning**. Instead of asking "which path did the system take," the path-integral view asks "which processes were permitted, how were they weighted, and how did their interference resolve under measurement.

System Relevance (QAIS Layer-Scoped)

Within the QAIS operational layer, this sum-over-histories perspective provides a natural interpretation of circuit execution. State preparation defines the initial boundary conditions over which histories are permitted. Unitary transformations act to re-phase and redistribute amplitude contributions across competing histories. Interference emerges as a system-level effect of these constrained evolutions, and measurement selects outcomes from the resulting interference structure. Importantly, no explicit enumeration of histories is required during execution; the physical system itself realizes the interference process implicitly.

From an architectural standpoint, this has a critical implication: **QAIS designers do not control outcomes by specifying trajectories**, but by shaping the space of admissible histories. Circuit structure, gate placement, and interaction geometry act as constraints that determine which histories constructively reinforce and which cancel. The role of architecture is therefore not to direct computation step-by-step, but to sculpt the interference landscape within which outcomes are resolved.

This interpretation also clarifies why interference should be understood as an operational resource rather than an abstract phenomenon. Interference reflects how information about alternative processes is preserved, combined, or suppressed across execution.

When interference is disrupted, by noise, decoherence, or poorly constrained evolution, the system loses its ability to resolve outcomes reliably, regardless of algorithmic intent.

In QAIS, the path-integral operational perspective thus provides a unifying lens: execution is the physical aggregation of constrained processes, learning corresponds to reshaping admissible histories, and verification concerns whether the interference structure remains stable and interpretable under adaptation.

Cross-Disciplinary Orientation

This operational interpretation aligns quantum physics, computational circuits, and learning systems within a single process narrative. From a physics perspective, it preserves full equivalence with Schrödinger and operator-based formalisms. From a computational perspective, it explains why quantum algorithms derive power from interference rather than parallel evaluation.

From a systems and AI standpoint, it clarifies how executable workflows shape outcome distributions without invoking agent-level reasoning or optimization objectives.

Case Study Anchor — Interference as a Representational Control Surface

Intent: To examine interference as a controllable amplitude-weighting mechanism that sculpts admissible histories under constraint, defining how QAIS systems steer probability distributions without invoking privileged trajectories.

- **QALIS (Constructive Lens)** Under the QALIS lens, the sum-over-histories interpretation supports disciplined system design by emphasizing interference control as a first-class resource. Stable QAIS architectures leverage predictable interference patterns to achieve robustness, scalability, and verification-friendly execution.
- **CRQC–LLM (Stress-Test Lens)** From the CRQC–LLM perspective, this same interference dependence exposes systemic fragility. Unintended phase drift, noise-induced decoherence, or adversarial manipulation of transformation sequences can distort the interference landscape, amplifying incorrect histories and undermining output integrity.

Section 2.2 Summary — Representation and Encoding

This section advanced the discussion by establishing representation and encoding as coherence-sensitive, physically embodied design choices within QAIS. The Bloch sphere illustrated how amplitudes and phases define the operational geometry of information, revealing representation as an active system primitive rather than a symbolic abstraction. Representation is the foundation upon which all subsequent quantum operations depend.

With representation and interference grounded in physical law, the discussion now turns to copying and observation, where no-cloning and measurement collapse impose structural limits on how represented information can be accessed, reused, and verified.

2.3 Copying and Observation

Architectural Framing: *Measurement as the quantum–classical resolution boundary governing observability, verification, and state disturbance within QAIS.*

Section Framing: *Operational constraints on copying, measurement, and information extraction in coherence-sensitive systems.*

QAIS treats copying and observation as foundational operational constraints rather than passive informational processes. The no-cloning theorem limits duplication of unknown quantum states, while measurement irreversibly alters the systems it interrogates. QAIS therefore reframes these principles as architectural rules governing redundancy, verification, synchronization, and classical–quantum interface behavior across hybrid intelligent systems.

Unlike textbook treatments that frame measurement as an external collapse postulate acting on a wavefunction defined over Hilbert space, QAIS treats measurement as an **operational boundary condition** [19] within the execution pipeline. Measurement does not interrupt an otherwise continuous evolution; instead, it selects classical outcomes from an interference structure accumulated through prior preparation and transformation stages. From an operational perspective, measurement is therefore not an exception to quantum dynamics but a rule-governed transition that converts quantum amplitudes and phase relationships into classical information suitable for verification, inference, and control.

This interpretation preserves full mathematical equivalence with traditional formulations while making explicit measurement's role as a verification and extraction mechanism rather than a special ontological event.

In classical systems, copying and observation are treated as trivial: states can be duplicated indefinitely, and measurement simply reads without disruption.

QAIS departs radically from this model. Perfect duplication of unknown quantum states is impossible, and any measurement irreversibly collapses superposition, eliminates phase relations, and restricts further evolution.

These effects are not mere nuisances but **structural boundaries** that define how information can move through quantum operations. As a result, QAIS treats qubits as resources whose access pathways must be carefully orchestrated, with observation understood as a transformation step rather than a passive inspection.

Fragility and constraint yield capability as well. Within QAIS, the no-cloning theorem ensures that redundancy must be achieved through indirect means, such as correlated state preparations or context-aware sampling, rather than replicating the underlying state.

[19] Refers to the organized relationship among operational rules across layers of the QAIS stack, capturing how constraints at one level propagate and shape behavior at higher levels.

Measurement collapse, moreover, becomes the mechanism by which quantum information transitions into classical form, linking quantum evolution to classical inference through carefully structured readout.

In this way, the twin constraints of copying and observation define how QAIS designers balance state preservation with the need to extract usable classical data.

2.3.1 Copying and Observation as Operational Rules

The interaction between observation and system state evolution introduces constraints that fundamentally distinguish quantum information processing from classical paradigms. These constraints govern how information can be accessed, preserved, and propagated within QAIS systems, shaping both representation and control.

Figure 2.3.1 lays out this relationship within an operational framework, illustrating how limitations on copying and the effects of measurement define system-level behavior. In contrast to classical assumptions of unrestricted duplication and noninvasive observation, quantum systems impose strict constraints in which measurement alters state and copying is fundamentally restricted.

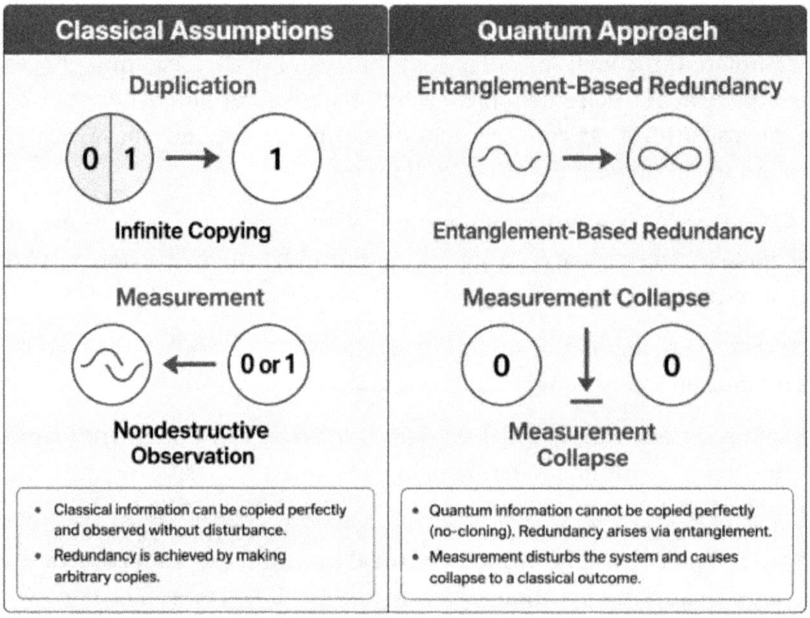

Figure 2.3.1 — Copying and Observation as Rules in QAIS.

This figure illustrates how classical intuitions fail in the quantum regime. Unlimited duplication is prohibited by the no-cloning rule, and measurement is inherently disturbing. QAIS therefore replaces classical replication with coherence-preserving redundancy strategies and treats measurement collapse as an operational event with verification implications.

The technical spotlight interprets these constraints not as theoretical curiosities but as operational rules that structure QAIS behavior.

Technical Spotlight — Copying and Observation as Design Rules

Conceptual Grounding

Figure 2.3.1 shows how QAIS transforms no-cloning and measurement collapse into actionable operational rules. Because an unknown quantum state cannot be perfectly duplicated, QAIS relies on indirect redundancy mechanisms such as correlated preparations or entanglement-assisted encoding (introduced in later sections).

Measurement collapse, rather than simply revealing a value, is an intrinsic transformation step that shapes the subsequent evolution of the system and limits how information can be reused.

System Relevance

Copying and observation determine how QAIS manages state evolution, storage, and readout. Redundancy becomes a matter of structuring correlations or repeated state preparations rather than duplicating the underlying state. Verification relies on the tamper-sensitive nature of measurement: attempts to probe a quantum state inevitably alter it, allowing designers to detect disturbances.

Measurement converts amplitudes and phase relationships into classical outputs, providing the bridge from quantum dynamics to classical decision processes without implying higher-level architectures.

Cross-Disciplinary Orientation

Across disciplines, these constraints take on distinct interpretations. Physicists view no-cloning and collapse as signatures of contextuality, expressing the impossibility of hidden-variable duplication. Engineers treat the phenomena as hardware-level constraints that require careful calibration, timing control, and error-aware state management. AI specialists interpret collapse as a sampling mechanism through which quantum states yield classical observables for downstream processing. Cybersecurity practitioners regard the disturbance caused by unauthorized measurement as a means of verifying integrity at the substrate level. Together, these orientations highlight how copying and observation function as operational rules rather than optional design considerations.

Case Study Anchor — Copying and Observation

Intent: To position no-cloning and measurement backaction as structural design rules that govern verification boundaries, redundancy strategy, and controlled extraction of classical information

- **QALIS** views copying constraints and measurement backaction as constructive boundaries that guide how states must be prepared, accessed, and refreshed within operational workflows.
- **CRQC-LLM** emphasizes how fragility under probing, coherence loss, disturbance sensitivity, and context-dependent collapse, reveals early indicators of stress that resilience mechanisms must later address.

These interpretations remain strictly at the primitive level and do not introduce pipelines, sensing systems, or architectural deployments.

Section 2.3 Summary — Copying and Observation

This section reframed copying and measurement as operational rules rather than incidental effects. The no-cloning theorem and measurement collapse were shown to impose structural constraints that shape redundancy, verification, and classical–quantum interfaces in QAIS. Instead of duplicating data, QAIS relies on entanglement-based redundancy and coherence-sensitive observation to preserve information integrity. These constraints define how information can be accessed, validated, and reused within quantum systems.

With copying and measurement established as governing limitations, the discussion now turns to entanglement and correlation, where relational structure emerges under these same constraints, transforming fragility into coordinated capability.

2.4 Entanglement and Correlation

Architectural Framing: *Controlled Operations*

Section Framing: *Controlled Operations stage where multi-qubit correlations are established through structured interaction*

Entanglement represents one of the most profound departures of QAIS from classical information frameworks. Where earlier sections described how representation (Section 2.2) and measurement constraints (Section 2.3) shape the behavior of individual qubits, this section introduces entanglement as a relational operational primitive—a mechanism through which quantum states become inseparable, forming joint structures that cannot be described by classical correlation.

In QAIS, entanglement is not an exotic curiosity but an essential resource that governs how multi-qubit systems encode, share, and transform information.

As introduced in Chapter 1, classical correlations arise from shared preparation and retain separable state descriptions. Entanglement breaks this paradigm. In an entangled system, the global state cannot be decomposed into local components; measurement outcomes reflect intrinsic interdependence rather than coincidental agreement. These correlations persist regardless of spatial separation and reveal themselves only through joint measurement.

This inseparability provides QAIS with a form of coordination that classical systems cannot emulate, enabling composite transformations whose meaning arises from the joint state rather than from its individual parts.

From an operational perspective, entanglement functions as a correlation rule: once established, qubits evolve according to shared structure rather than independent trajectories. Their transformations cannot be modeled as separate computations, and interference patterns depend on the global configuration of the state.

Entanglement therefore amplifies representational capacity—multiple qubits express relationships through shared amplitudes and phases—and introduces constraints that determine how operations propagate through composite systems.

Fragility plays a key role: decoherence affects not only the individual constituents but the global structure itself, making entanglement an indicator of both capability and vulnerability within QAIS.

These behaviors become especially visible during measurement. When an entangled qubit is measured, the outcome reshapes the state of the entire system. This does not imply information transmission but rather reflects a structural property of the entangled state, revealing how measurement dependence and correlation structure interact.

This relationship prepares the conceptual ground for Section 2.5, where contextuality extends entanglement's relational behavior into the logical domain, demonstrating that measurement outcomes depend not only on the state but on the surrounding operational context.

Correlation structures within QAIS extend beyond classical dependency models, reflecting deeper relationships that influence how information is distributed and coordinated across system components. These structures play a central role in enabling collective system behavior. Figure 2.4.1 models these relationships within a unified framework, highlighting their role in shaping information flow and interaction.

2.4.1 Entanglement as an Operational Primitive

The following visual contrasts classical correlation with quantum entanglement to illustrate how inseparability arises as a foundational behavior in QAIS.

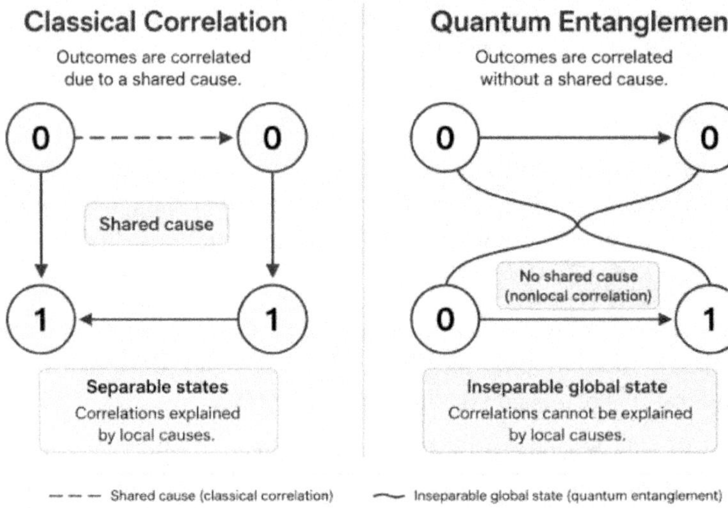

Figure 2.4.1 — Entanglement and Correlation in QAIS.

This figure contrasts separable classical correlations with inseparable entangled states. Classical systems permit local descriptions even when correlated, while entangled states can only be represented globally. The visual highlights entanglement as an operational resource in QAIS, one that governs composite state evolution, amplifies representational structure, and introduces correlation behaviors impossible in classical frameworks. The accompanying spotlight interprets this behavior as a design-relevant primitive.

Technical Spotlight — Entanglement as a System Resource

Conceptual Grounding

Figure 2.4.1 depicts entanglement as irreducible structure: an entangled pair cannot be separated into independent descriptions, and measurement outcomes reflect dependencies encoded in the joint state. This indivisibility distinguishes entanglement from classical correlation, making it the basis for multi-qubit behavior in QAIS. The global nature of entangled states means that information resides in relational structure rather than isolated components.

System Relevance

As an operational resource, entanglement governs how multi-qubit transformations occur in QAIS. Operations applied to one qubit can influence the global state, and interference patterns across circuits depend on entanglement structure. Fragility becomes a system variable: decoherence in any component can degrade the entire state, making entanglement a sensitive indicator of operational integrity.

These characteristics define how multi-qubit processes must be sequenced, controlled, and verified, influencing the operational pathways that underlie more complex system behavior introduced in later chapters.

Cross-Disciplinary Orientation

Across disciplines, entanglement is interpreted through complementary perspectives. Physicists treat it as an experimentally validated departure from local realism, revealed through Bell inequality violations. Engineers view entanglement as a coherence-sensitive multi-qubit configuration requiring careful preparation, stabilization, and characterization. AI specialists regard entanglement as a representational expansion that allows relationships between qubits to be encoded through shared amplitudes and phases. Cybersecurity practitioners interpret entanglement's disturbance sensitivity as an intrinsic verification property at the physical layer: any unauthorized interaction inevitably affects the global state.

Case Study Anchor — Entanglement and Correlation

Intent: To analyze entanglement as an irreducible relational resource whose global structure governs multi-qubit evolution and exposes correlation fragility under disturbance.

- **QALIS** interprets entanglement as a constructive relational structure that enhances the expressivity and control of multi-qubit operations.
- **CRQC–LLM** highlights how correlation fragility, particularly sensitivity to decoherence or disturbance, acts as an early indicator of stress within entangled configurations.

These interpretations remain strictly tied to the operational behavior of entanglement and do not imply higher-level architectures or pipelines.

Section 2.4 Summary — Entanglement and Correlation

This section established entanglement as a working system resource rather than a purely conceptual property. As inseparability across qubits, entanglement enables unified correlations that cannot be described classically, extending the constraints introduced by copying and observation into relational system behavior. These correlations reveal how groups of quantum states evolve under shared constraints, transforming limitations on individual access into structured, system-level capability.

With entanglement defining correlated behavior, the discussion now turns to contextuality, where measurement outcomes are shaped not only by correlation but by the operational context in which observations are performed.

2.5 From Entanglement to Contextuality

Architectural Framing: Controlled Operations → Measurement

Section Framing: *Extension of Controlled Operations into Measurement-dependent structure, where outcomes depend on operational context.*

Entanglement revealed that quantum systems cannot always be reduced to independent components. Contextuality advances this insight by demonstrating that quantum systems cannot always be reduced to independent properties. As summarized in Chapter 1, classical models assume predefined properties independent of measurement context. QAIS operates in a fundamentally different regime. In a contextual system, the outcome of a measurement depends on the *set of compatible measurements* with which it is performed and the operational context in which the system is interrogated.

Contextuality therefore exposes a deeper form of interdependence: measurement outcomes are not determined solely by the state, but by the measurement arrangement itself.

This dependency is not a philosophical nuance, it is an operational rule. Contextuality determines how information can be extracted, how transformations interact with measurement settings, and how joint outcomes must be interpreted. Unlike entanglement, which ties systems together through shared quantum states, contextuality ties outcomes together through measurement structure.

Two experiments performed on identical states can produce different statistical results, not because the state has changed, but because the *questions* asked of the system differ in their compatibility and sequencing. QAIS treats this measurement dependency as a primitive resource, using it to understand how operational context shapes the flow, accessibility, and meaning of quantum information.

Contextuality also reframes earlier principles. The no-cloning theorem and measurement collapse, introduced in Section 2.3, can be seen as manifestations of contextual rules: states cannot be duplicated because duplication would require context-free access to all observables, and observation is transformative because incompatible measurements cannot preserve all contextual structure simultaneously.

Contextuality shows why quantum information cannot be cleanly separated from the operations used to probe it, and why QAIS must encode computation, control, and verification within this interdependent framework. Contextuality as an Operational Primitive.

The transition from classical to quantum paradigms involves more than a change in computational technique; it reflects a fundamental shift in how systems represent, process, and interpret information. These differences extend beyond algorithms, shaping architectural design and system behavior. Figure 2.5.1 shows this contrast within a comparative framework, illustrating how divergent assumptions about determinism, replication, and measurement give rise to distinct design principles.

In particular, the figure contrasts classical determinism with quantum contextuality, highlighting how the underlying treatment of information drives differences in system-level architecture across the two regimes.

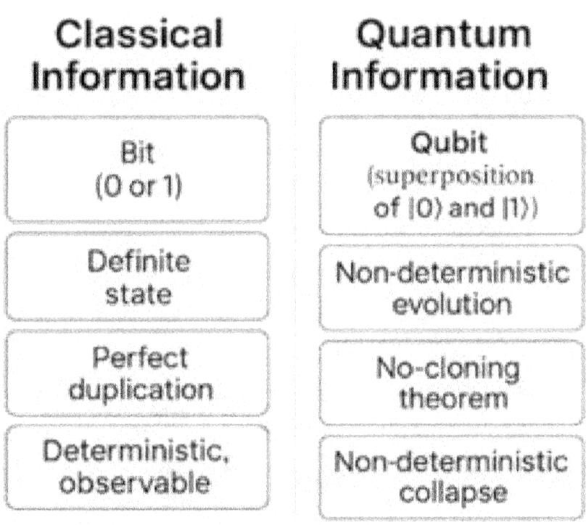

Figure 2.5.1 — Contrasting Classical and Quantum Paradigms.

These comparative visual contrasts classical assumptions, context-free measurement, independent properties, and reproducible duplication, with quantum properties such as probabilistic superposition, incompatibility of measurements, and collapse-induced outcome dependence. It highlights how contextuality, alongside entanglement, reshapes the foundations of inference and measurement in QAIS.

The accompanying spotlight examines how contextuality connects evolution, collapse, and operational meaning.

Technical Spotlight — Evolution, Collapse, and Computation

Conceptual Grounding

Figure 2.5.1 distinguishes between two processes that appear similar but follow different operational rules: **non-deterministic evolution** and **non-deterministic collapse**.

- **Evolution** describes the continuous, unitary transformation of a quantum state under controlled operations. Interference effects arise naturally, and amplitudes evolve coherently across the state space.
- **Collapse**, by contrast, occurs when a measurement is made. The system is projected onto an outcome associated with the measurement basis, eliminating incompatible amplitudes and irreversibly breaking coherence.

Contextuality binds these two processes. It determines how evolution sets up outcome distributions and how collapse resolves them, not according to an underlying set of predetermined properties, but according to the *measurement configuration* applied.

Even if the pre-measurement state is fully known, contextuality ensures that different sets of compatible measurements may yield different observable outcomes.

This makes contextuality an operational rule that shapes how QAIS designers structure measurement sequences and interpret results.

System Relevance

Contextuality governs how quantum operations yield meaningful information. Because incompatible measurements cannot be simultaneously satisfied, QAIS must sequence measurements carefully, selecting bases that extract the required information while preserving coherence where needed. Contextuality defines which observables can be jointly measured, which must be measured sequentially, and which combinations inevitably disturb the system. These constraints influence how quantum-to-classical transitions occur, how measurement outcomes encode probability distributions, and how operators interact during execution. Although the architectural implications will be explored in later chapters, the operational insight here is that **context, not only state, determines the flow of information**.

QAIS also leverages contextuality through collapse-based verification, a primitive in which measurement-induced disturbance becomes a diagnostic signal. Because incompatible measurements disrupt outcome distributions in predictable ways, any unauthorized probing, mis-sequenced operation, or adversarially induced collapse leaves detectable statistical signatures.

Collapse-based verification therefore uses contextual sensitivity as a physical boundary condition: if the system collapses in a way that is inconsistent with the expected measurement structure, the deviation itself becomes evidence of tampering or operational instability.

Cross-Disciplinary Orientation

Contextuality is interpreted differently across technical communities. Physicists treat it as the empirical validation that quantum correlations cannot be explained by hidden variables or context-independent properties.

Engineers view contextuality as a constraint on measurement control, requiring precise calibration of measurement bases to preserve coherence where possible and to sequence operations without unintended cross-measurement disturbance.

AI specialists interpret contextuality as a nonclassical structure in which the meaning of measurement outcomes arises from relationships among observables, not from independent features. Cybersecurity practitioners treat contextual sensitivity as a verification boundary, where any unauthorized measurement disrupts outcome patterns and becomes detectable at the physical layer.

Case Study Anchor — Entanglement to Contextuality

Intent: To extend correlation into measurement-dependent structure, showing how contextuality shapes admissible outcome distributions and constrains how QAIS sequences compatible observables:

- **QALIS** interprets contextuality as a constructive principle in which measurement dependencies inform how primitives are prepared, sequenced, and read out in operational workflows.
- **CRQC–LLM** highlights how contextual disturbance sensitivity reveals early indicators of fragility: incompatible measurements, mis-sequenced operations, or adversarial probing produce detectable inconsistencies in outcome patterns.

These interpretations remain strictly at the primitive level, applying the chapter-wide case-study lenses without introducing architectural or system-scale behaviors.

Section 2.5 Summary — Entanglement to Contextuality

Contextuality advances QAIS from structural inseparability to systemic interdependence, where outcomes depend on the measurement context rather than intrinsic predefined properties. Extending entanglement, it shows that correlations alone do not determine behavior; instead, the operational context in which measurements are performed governs observable outcomes, establishing deeper representational power and sharper verification boundaries.

These contextual constraints define how coherence, correlation, and disturbance must be interpreted within QAIS operations, shifting analysis from static structure to context-dependent behavior.

With contextuality establishing the dependence of outcomes on measurement conditions, the discussion now turns to geometry, logic, and computation, where interference and state evolution are formalized through geometric and analytical frameworks that quantify and govern QAIS dynamics.

2.6 Geometry, Logic, and Computation

Architectural Framing: Unitary Evolution

Section Framing: *Unitary evolution expressed through geometric transformations and interference-driven state propagation.*

From the operational, history-interference perspective adopted in this chapter, geometric transformations are best understood as mechanisms for shaping the relative contributions of competing quantum histories rather than as purely abstract rotations in state space. Circuit geometry, phase relationships, and transformation sequencing collectively determine which histories interfere constructively and which are suppressed, directly influencing system-level behavior observable at measurement. Classical computation, by contrast, is grounded in Boolean logic: bits occupy discrete states, gates map inputs to deterministic outputs, and truth tables define transitions that are symbolic, rule-bound, and context-independent.

Within this framework, computation reduces to the manipulation of discrete values inside a closed logical system.

QAIS reframes computation in geometric rather than symbolic terms. Qubits evolve continuously under unitary transformations that rotate state vectors across the Bloch sphere or, in multi-qubit settings, through exponentially large regions of Hilbert space.

Amplitudes and phases—not binary values—encode representational content. Computation therefore emerges from controlled trajectories, rotations, and interference patterns [5] that preserve coherence while shaping how probability amplitudes interact. Rather than applying fixed logical operations, QAIS performs computation by sculpting the geometry of state evolution.

This geometric view is governed by the operational grammar developed earlier in the chapter. Unitarity enforces reversibility, coherence maintains relational structure during evolution, entanglement couples subsystems into joint trajectories, and contextuality determines how measurement configurations shape observable outcomes. These principles collectively define computation at the primitive level, preceding the introduction of algorithms and architectures in later chapters.

Crucially, QAIS is not a faster version of classical computation; it is a fundamentally different paradigm. The meaning of a quantum operation arises from the global structure of the state space and the interference patterns that evolve within it.

Outcomes depend on the combined geometry of amplitude flow, the relational constraints imposed by entanglement, and the measurement context applied during readout. The resulting model is continuous, reversible, relational, and context-dependent, as illustrated in Figure 2.6.1.

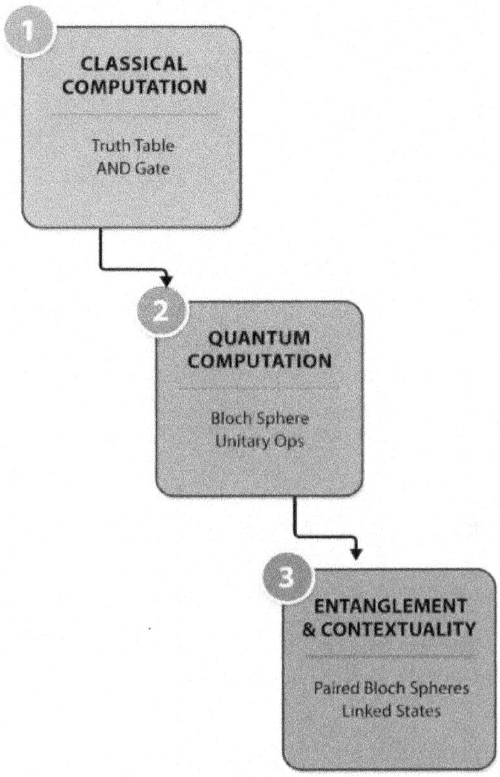

Figure 2.6.1 — Classical and Quantum Computational Models.

This figure contrasts three computational paradigms. The classical panel shows how deterministic logic gates map discrete bit values through truth tables. The geometric panel depicts the Bloch sphere, illustrating how unitary transformations rotate quantum states to manipulate amplitudes and phases. The relational panel visualizes correlated multi-qubit structures, where joint states evolve across shared Hilbert space and outcomes depend on relational constraints rather than isolated variables. Together, these panels situate QAIS computation as a transition from symbolic determinism to geometric evolution and, ultimately, to relational, context-dependent information processing.

This figure emphasizes the operational transition from classical symbolic computation to relational quantum processing. Classical systems manipulate discrete truth-preserving states through deterministic gate operations, whereas quantum systems evolve amplitudes and phase relationships through unitary transformations within Hilbert space.

Entanglement introduces correlated state dependencies that cannot be reduced to isolated variables, producing context-dependent outcomes during measurement. In QAIS architectures, this shift from symbolic determinism to relational state evolution alters how information propagation, synchronization, and inference are coordinated across quantum-classical systems.

Technical Analysis — Geometry, Reversibility, and Relational Structure

Conceptual Grounding

Figure 2.6.1 highlights that quantum computation is defined by geometric trajectories rather than symbolic rule application.

Unitary gates rotate state vectors and produce interference through coherent amplitude evolution. In multi-qubit settings, entanglement binds trajectories together, creating relational structures that cannot be decomposed into independent components. Contextuality then determines how measurement choices resolve these trajectories, showing that observable outcomes depend on the operational context, not on pre-existing values.

System Relevance

Within QAIS, this geometric–relational model defines how information flows through primitive-level operations. It clarifies why coherence must be preserved, why gate sequences must be reversible, and why entanglement and contextuality function as enabling constraints rather than optional enhancements. The computational meaning of a state reflects both its geometric evolution and the relational context in which it is measured, providing the foundation upon which later architectural abstractions are built.

Cross-Disciplinary Orientation

Physicists interpret this model as a direct expression of quantum dynamics under the Schrödinger equation. Engineers view it as a control problem requiring precise gate calibration and coherence preservation. AI specialists leverage geometric and relational structure to build representational spaces enriched by interference and multi-qubit correlations.

Cybersecurity practitioners treat reversible evolution and measurement dependence as mechanisms that impose verifiable constraints on information access, creating intrinsic boundaries against cloning or passive observation.

Case Study Anchor — Geometry, Logic, and Computation

Intent: To reinterpret computation as geometric amplitude evolution and relational constraint propagation, distinguishing QAIS processing from symbolic Boolean logic.

- **QALIS** interprets geometric and relational computation as the set of rules that govern how qubits must be prepared and transformed to maintain coherent operational pathways.
- **CRQC-LLM** highlights how geometric instability, such as phase drift or misaligned gate operations, produces detectable relational inconsistencies, providing early indicators of fragility during quantum operations.

These interpretations reinforce the chapter-level case-study framing without extending into architectural or application-level behavior.

Section 2.6 Summary — Geometry, Logic, and Computation

This section established that QAIS computation is governed by geometric and relational evolution rather than symbolic Boolean logic. Quantum operations act as reversible transformations that shape amplitude flow, interference, and multi-qubit correlations, defining how information is structured and manipulated at the primitive level. These principles clarify why coherence and unitarity must be preserved throughout computation, as they ensure that geometric trajectories remain stable and information-preserving under transformation.

Taken together with Sections 2.3 through 2.6, this establishes the operational grammar of QAIS—encompassing copying constraints, measurement effects, entanglement structures, contextual dependencies, and geometric evolution—as the foundation for how quantum information is processed and made actionable.

With this primitive grammar in place, the discussion now turns to quantum error correction, where mechanisms are introduced to preserve coherence and stabilize these geometric evolutions against noise and environmental disturbance, enabling reliable computation at system scale.

Error correction mechanisms are essential for maintaining system reliability, yet their implementation differs significantly between classical and quantum domains due to the constraints imposed by measurement and state disturbance. These differences influence how redundancy is introduced and managed within QAIS architectures.

In the next section Figure 2.7.1 operationalizes these contrasting approaches within a structural framework, clarifying their impact on system stability.

2.7 Quantum Error Correction

Architectural Framing: *Verification and stabilization of quantum information under noise and decoherence.*

Section Framing: *Error detection, coherence preservation, and fault-tolerant control mechanisms in QAIS.*

Quantum computation relies on coherence, reversibility, interference, and quantum correlations to manipulate information, yet these same properties make quantum information vulnerable to physical disturbance. Environmental interactions can produce decoherence; imperfect state preparation, gates, and measurements can introduce errors; and control inaccuracies can cause the actual state evolution to depart from its intended trajectory. Because quantum operations build on earlier state transformations, uncorrected physical errors can accumulate and reduce the reliability of later results.

Classical systems face errors as well, but classical information can be copied, inspected, and compared without fundamentally changing the information being protected. Redundant copies, parity relationships, and majority decisions can therefore be used to identify or recover corrupted data. Quantum systems cannot simply use the same strategy. The no-cloning theorem prohibits independent identical copies of an arbitrary unknown quantum state, while directly measuring the protected logical information can disturb the state and convert quantum information into classical outcomes.

Quantum error correction (QEC) addresses these constraints by encoding logical quantum information across multiple physical qubits within a structured code space. Rather than placing the entire logical state in one vulnerable physical qubit, QEC distributes its representation across a larger physical system. Carefully designed syndrome measurements then provide information about physical errors without directly measuring the encoded logical state. A classical decoder interprets the resulting syndrome and determines an appropriate recovery or tracks the correction in software. When errors remain within the code's correction capability, the logical information can be recovered despite disturbances to individual physical qubits.

This distinction between **physical qubits** and **logical qubits** is fundamental to fault-tolerant quantum computation. Physical qubits are hardware elements exposed to noise, decoherence, and imperfect control. A logical qubit is an encoded computational unit represented across multiple physical qubits. QEC does not make the hardware immune to error; it creates a protected logical level at which correctable physical disturbances can be detected and managed before they become logical failures.

Within QAIS, QEC therefore provides a protected substrate for longer sequences of quantum-state transformations, interference, entangling operations, and measurements. Reliable quantum processing depends not only on preparing the intended state, but also on maintaining usable logical information throughout the operations required to accomplish the task.

Before examining this quantum approach, it is useful to establish the simpler classical case. **Figure 2.7.1 illustrates how a three-bit repetition code uses redundancy and majority decoding to protect one classical bit against a single bit-flip error.** Figure 2.7.2 then shows how quantum error correction achieves protection through multi-qubit encoding, syndrome extraction, and recovery rather than direct copying.

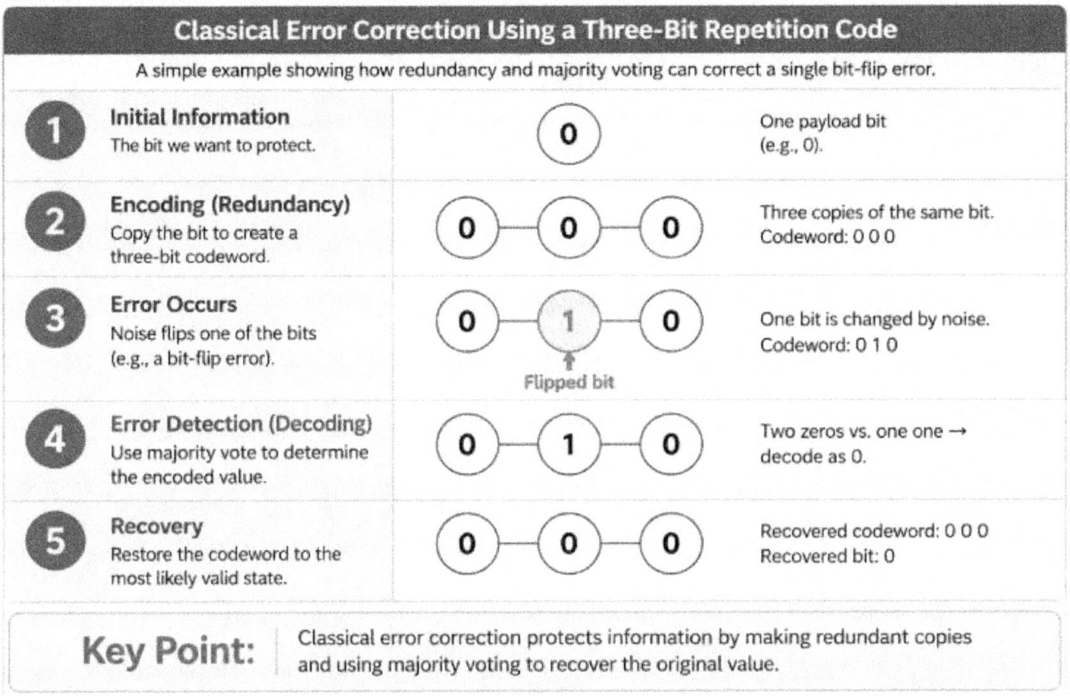

Figure 2.7.1 — Classical Error Correction Using a Three-Bit Repetition Code.

A payload bit with value 0 is encoded as the three-bit codeword 000. If noise flips one bit, producing 010, the three values no longer agree. Under the assumption that no more than one bit has flipped, majority decoding determines that the encoded value is 0, allowing the valid codeword 000 to be restored.

The important feature of this example is that classical information can be directly replicated. Redundant copies provide additional evidence from which the original encoded value can be inferred when a limited number of errors occur. The information being protected remains represented by ordinary classical values that can be read and compared without fundamentally changing them.

Quantum information requires a different protection strategy. An arbitrary unknown quantum state cannot be protected by creating independent identical copies and later comparing them. Instead, QEC introduces redundancy through **encoding**: one logical quantum state is mapped into a larger multi-qubit code space. Figure 2.7.2 illustrates this conceptual transition from vulnerable physical qubits to protected logical quantum information.

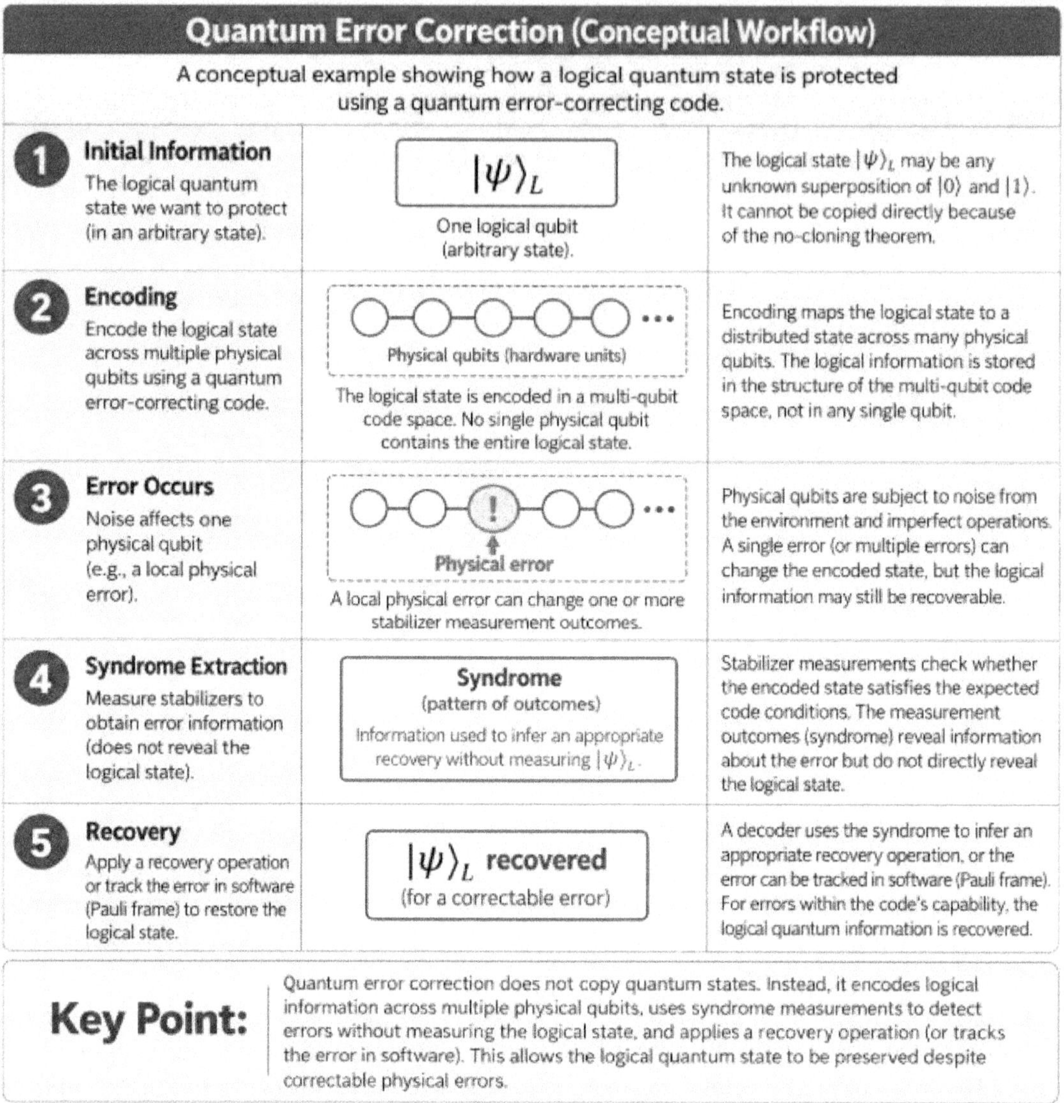

Figure 2.7.2 — Quantum Error Correction (Conceptual Workflow).

Figure 2.7.2 illustrates the conceptual workflow of quantum error correction. The process begins with logical quantum information represented by a state such as $|\psi\rangle_L$. An encoding operation maps that logical information into a structured state distributed across multiple physical qubits. No single physical qubit contains the entire logical state.

When noise or an imperfect operation affects the physical system, one or more physical errors may alter the encoded state. Rather than directly measuring the logical information, QEC performs specially designed syndrome measurements. In stabilizer-based QEC, these measurements test whether the encoded state satisfies specified stabilizer conditions. Changes in the resulting pattern of measurement outcomes form an **error syndrome**.

The syndrome provides information that a classical decoder can use to infer an appropriate recovery. Depending on the implementation, a correction may be physically applied or tracked in software, such as through a Pauli frame. When the disturbance lies within the error-correcting capability of the code, the logical quantum information can be recovered without directly measuring the logical state itself.

Taken together, Figures 2.7.1 and 2.7.2 demonstrate a fundamental distinction between classical and quantum protection. Classical repetition protects information by creating redundant classical representations that can be directly compared. Quantum error correction instead creates redundancy by encoding logical information into a larger multi-qubit code space. The goal is not to copy an unknown quantum state, but to arrange physical degrees of freedom so that information about certain errors can be extracted while preserving the encoded logical information.

Stabilizer codes provide an important framework for implementing these principles. Examples include the canonical 9-qubit Shor code and 7-qubit Steane code. Stabilizer codes define a multi-qubit code space through a set of stabilizer operators. Syndrome measurements test whether the encoded state satisfies the expected stabilizer conditions without directly measuring the logical information. Correctable errors that anticommute with particular stabilizers can change their measurement outcomes, producing a syndrome that a classical decoder uses to infer an appropriate recovery.

Topological quantum error-correcting codes provide an important implementation of this general approach. The surface code, for example, is a topological stabilizer code that uses geometrically local stabilizer measurements while encoding logical information nonlocally across a qubit lattice. Local physical errors can change nearby syndrome outcomes. Logical failure generally requires an error pattern sufficiently extensive to produce an uncorrectable error or effectively implement a logical operation across the code. Thus, sufficiently small correctable local disturbances need not destroy the encoded logical information.

These strategies protect logical quantum information collectively rather than relying on the state of any single physical qubit. Within QAIS, this distributed encoding helps preserve the logical amplitudes, phases, and correlations required for reliable quantum-state evolution despite correctable physical disturbances.

QEC therefore marks an important transition in the operational grammar of QAIS. At the physical level, individual qubits remain vulnerable to noise and decoherence. At the logical level, however, encoding, syndrome extraction, classical decoding, and recovery provide mechanisms for detecting and managing correctable errors. Quantum processing can therefore operate on protected logical information rather than depending entirely on the uninterrupted behavior of individual physical qubits.

This distinction can be summarized as:

Physical qubits remain vulnerable to noise.

Logical quantum information is encoded across physical qubits.

Syndrome measurements provide information about correctable errors without directly measuring the logical state.

Classical decoding determines an appropriate recovery.

Successful recovery preserves the logical information when errors remain within the code's correction capability.

Rather than eliminating physical decoherence, QEC provides a mechanism for maintaining logical quantum information despite correctable errors caused by noise, imperfect operations, and decoherence. When physical error rates, code performance, and fault-tolerance requirements permit, encoded logical information can remain usable beyond the lifetime that would ordinarily limit an unprotected physical qubit.

For QAIS, this capability is especially important because useful quantum processing may require sequences of controlled state transformations, interference, entangling operations, and measurements. QEC provides a pathway toward maintaining the logical quantum substrate on which those processes depend. It therefore supports the transition from short-lived physical quantum behavior toward longer-duration, controlled logical quantum evolution and provides an important foundation for the quantification tools introduced later in Section 2.9.

The following spotlight interprets these constructs as operational components of QAIS rather than as higher-level architectural mechanisms.

Technical Spotlight — Quantum Error Correction as Coherence Preservation

Conceptual Grounding

Figures 2.7.1 and 2.7.2 illustrate the fundamental difference between classical and quantum protection mechanisms. Figure 2.7.1 shows how a classical three-bit repetition code protects information through redundancy and majority decoding. Because classical information can be copied directly, multiple representations of the same value can be compared to recover from a limited number of errors.

Figure 2.7.2 shows why quantum error correction requires a different strategy. An arbitrary unknown quantum state cannot be independently copied because of the no-cloning theorem. Instead, QEC encodes logical quantum information collectively across multiple physical qubits within a structured code space. Error information is then obtained through syndrome measurements designed to avoid directly measuring the encoded logical state.

Stabilizer codes define structured multi-qubit code spaces through sets of stabilizer operators. Stabilizer measurements test whether the encoded state satisfies specified code conditions. Correctable errors can change the resulting syndrome, providing information that a classical decoder uses to infer an appropriate recovery without directly revealing the logical information.

Topological stabilizer codes, such as the surface code, extend this principle by combining geometrically local stabilizer measurements with nonlocal encoding of logical information. Local physical errors can change nearby syndrome outcomes without necessarily causing logical failure. Logical failure can occur when accumulated or correlated errors produce an uncorrectable pattern or effectively implement a logical operator across the code.

Both approaches transform physical fragility into structured resilience by encoding **logical quantum information across collections of physical qubits rather than depending on the state of any single physical qubit**.

System Relevance

Within the operational grammar of QAIS, QEC provides a mechanism for maintaining usable logical quantum information despite correctable physical errors. Physical qubits remain vulnerable to noise, decoherence, imperfect gates, and measurement errors, but encoding, syndrome extraction, classical decoding, and recovery allow some of these disturbances to be detected and managed at the logical level.

This distinction is important. QEC does not eliminate decoherence or make physical qubits permanently coherent. Instead, it can extend the useful lifetime and reliability of **encoded logical quantum information**, provided physical error rates, code performance, and fault-tolerance requirements remain within the operating capabilities of the system.

Logical qubits therefore represent a transition from direct dependence on vulnerable physical states toward protected computational units. This protection supports longer and more reliable sequences of quantum-state transformations, entangling operations, interference, and measurements than would ordinarily be possible using unprotected physical qubits alone.

Within QAIS, QEC consequently provides an important physical and logical substrate for sustained quantum processing. It does not itself perform higher-level reasoning, learning, verification, or decision-making; rather, it helps maintain the quantum information on which those operations may depend.

Cross-Disciplinary Orientation

Physicists interpret QEC through the structure of quantum information: logical states can be encoded into multi-qubit code spaces so that specified classes of physical errors can be detected and corrected without directly measuring the logical information.

Engineers view QEC as a foundation of fault-tolerant quantum computing. Its implementation requires reliable state preparation, physical gates, ancilla qubits, syndrome-extraction circuits, measurement, classical decoding, calibration, feedback, and control. The effectiveness of the code therefore depends not only on its mathematical construction but also on the quality of the physical and classical systems that implement it.

AI specialists can view QEC as an enabling mechanism for quantum or hybrid AI workloads that require sufficiently reliable multi-step quantum processing. QEC does not itself perform learning or inference. Instead, fault-tolerant logical operation can provide a more reliable quantum computational substrate on which quantum learning, optimization, inference, or other algorithms may execute.

Cybersecurity practitioners should distinguish QEC from security monitoring and tamper detection. Error syndromes provide evidence that the encoded state has experienced a disturbance detectable by the code, but the syndrome alone does not establish its cause. Noise, hardware faults, imperfect operations, environmental interactions, and potentially malicious disturbances may produce error signatures. QEC therefore does not inherently provide authentication, intrusion detection, or proof of tampering. Security-oriented QAIS mechanisms must interpret disturbance evidence within additional verification, contextual, and governance structures.

Case Study Anchor — Quantum Error Correction

Intent: To demonstrate how logical quantum information can be protected without independently copying an unknown quantum state by encoding it across a structured multi-qubit code space, extracting error syndromes, and recovering from correctable physical disturbances.

- • **QALIS** interprets QEC as a supporting mechanism for sustaining the quantum substrate required by controlled primitive operations. By helping preserve logical amplitudes, phases, and correlations against correctable physical errors, QEC supports more reliable state evolution within the physical constraints governing the system.
- • **CRQC–LLM** highlights QEC as a source of structured diagnostic evidence rather than automatic proof of instability or tampering. Changes in syndrome patterns can indicate that physical disturbances have occurred and can contribute to broader monitoring and verification processes. Determining whether those disturbances result from ordinary noise, miscalibration, hardware faults, operational mistakes, or adversarial activity requires additional contextual evidence and system-level analysis.

Section 2.7 Summary — Quantum Error Correction

This section demonstrated that classical and quantum error correction introduce redundancy in fundamentally different ways. Classical repetition codes can protect information by directly replicating classical values and using majority decoding. Quantum error correction cannot protect an arbitrary unknown quantum state by independently copying it. Instead, it encodes logical quantum information collectively across multiple physical qubits within a structured code space.

Stabilizer measurements can produce error syndromes without directly measuring the encoded logical information, while classical decoding and recovery mechanisms use those syndromes to manage correctable physical errors. Topological stabilizer codes, such as the surface code, extend this principle through nonlocal logical encoding and geometrically local syndrome measurements.

Within QAIS, these mechanisms convert physical vulnerability into a manageable systems-engineering problem. QEC does not remove noise, prevent decoherence, or guarantee that quantum information will survive every disturbance. Instead, it provides a structured means of maintaining logical quantum information when physical errors remain within the capabilities of the selected code and fault-tolerant implementation.

QEC therefore translates several foundational quantum constraints into operational design requirements: unknown quantum states cannot simply be duplicated; error information must be obtained without directly measuring the protected logical state; physical disturbances must be diagnosed through carefully designed measurements; and recovery must preserve the logical information encoded across the system. These principles provide an essential foundation for reliable and ultimately fault-tolerant quantum computation.

With the protection of logical quantum information established as a system-level requirement, the discussion now turns to **embodied information and conceptual transformation**, where noise, coherence, physical substrate, and information representation are treated as intrinsic properties that shape how quantum information is represented, manipulated, and interpreted.

2.8 Embodied Information and Conceptual Transformation

Architectural Spine: *State Preparation → Unitary Evolution → Measurement*

Section Framing: *Transition from State Preparation through Measurement, emphasizing physical embodiment of information across operational stages*

Classical information theory treats information as an abstraction, symbols that retain their identity regardless of where or how they are stored. A bit written on paper, encoded in voltage, or transmitted as a digital packet retains meaning independent of its physical substrate. This assumption underlies classical computation, communication, and logic: information is portable, detachable, and context-free.

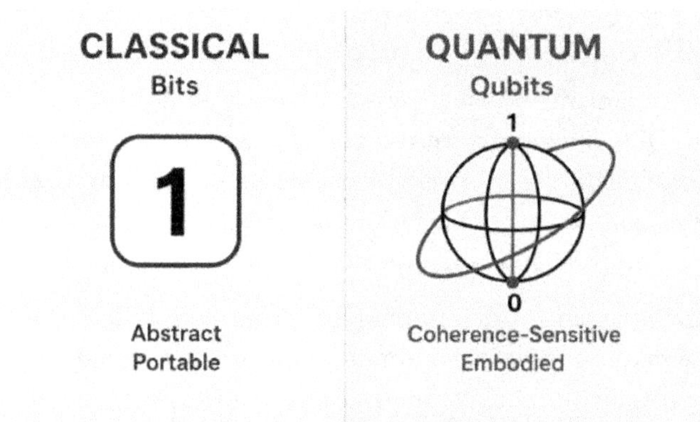

Figure 2.8.1 — Symbolic Abstraction to Embodied Quantum Information.

This diagram contrasts classical symbolic information, portable, copyable, and independent of physical substrate—with quantum embodied information, where amplitudes and phase structure define representational content. The classical panel shows a bit as an abstract symbol detached from the medium that stores it. The quantum panel depicts a qubit as a coherence-sensitive physical system, emphasizing that quantum information cannot be separated from the substrate that carries it.

The progression from symbolic abstraction to embodied quantum information reflects a shift in how systems encode and process meaning, linking representation directly to physical state. This transition has implications for both interpretability and system design.

Figure 2.8.1 frames this evolution within QAIS architectures, illustrating how representational paradigms influence system behavior.

QAIS replaces this abstraction with a fundamentally different principle: information is embodied. Quantum information exists only as a physical phenomenon, inseparable from the amplitudes, phases, and correlations that define it. A qubit is not a symbolic placeholder; it is a coherent physical system whose representational content arises from its quantum state.

Amplitudes encode probability structure, relative phases encode interference potential, and entanglement binds meaning across multiple subsystems. These properties cannot be stripped away, duplicated, or observed without altering the underlying state.

Embodiment therefore establishes a conceptual pivot for QAIS. Information becomes relational, contextual, and substrate-bound. Amplitude encodings cannot be separated from the geometry of Hilbert space; phase relationships cannot be examined without affecting interference; and entangled correlations cannot be reduced to local descriptions without losing essential structure.

Measurement, unitarity, and contextuality, not symbolic logic, define how information evolves and becomes accessible. QAIS treats these physical constraints not as limitations but as design conditions that govern the behavior of quantum representations and transformations.

This shift from abstraction to embodiment integrates every operational rule introduced in the chapter. No-cloning emerges naturally when information is tied to amplitudes that cannot be copied without disturbance. Measurement collapse follows inevitably when information is encoded in coherent states sensitive to context.

Entanglement becomes a structural representation rather than an exotic exception. Contextuality ensures that information depends on measurement configurations rather than predefined properties. Together, these features define information not as a pattern of symbols but as a physically embodied resource.

The following spotlight interprets embodiment as an operational rule that shapes representation and transformation in QAIS.

Technical Spotlight — Embodiment as a System-Level Rule

Conceptual Grounding

Figure 2.8.1 shows that quantum information cannot be separated from the medium that carries it. Amplitudes determine probability distribution, phases govern interference, and entanglement defines nonlocal correlations. These elements are not symbolic, they are physical. The meaning of a quantum state arises from the structure of the state space itself, and any attempt to extract information necessarily interacts with that structure. Embodiment therefore reframes information as a property of Hilbert space trajectories and relational correlations, not as a detached abstraction.

System Relevance (Operational Layer Only)

Within the operational grammar of QAIS, embodiment determines how information is represented, manipulated, and accessed. Coherence defines the usable lifetime of a representation; phase structure determines which operations preserve or disturb interference; and entangled correlations define how information is shared across composite states. These physical features shape how operations must be sequenced, how measurements affect state evolution, and how representational content persists or degrades during computation.

Cross-Disciplinary Orientation

Physicists interpret embodiment as confirmation that information is fundamentally physical: amplitudes and coherence carry meaning, not symbols. Engineers see embodiment as a control challenge, requiring precise manipulation of fragile states to preserve representational integrity. AI specialists view embodiment as a representational constraint and opportunity, quantum states encode structure unavailable to classical systems but must be maintained within coherence windows [7].

Cybersecurity practitioners regard embodiment as an inherent verification mechanism: disturbance cannot be hidden when information resides in fragile physical configurations.

Case Study Anchor — Embodied Information and Conceptual Transformation

Intent: To reframe information as substrate-bound amplitude–phase structure whose physical embodiment governs representational persistence, disturbance sensitivity, and transformation viability.

- **QALIS** interprets embodiment as the condition governing how qubits must be prepared and maintained so that operational primitives preserve amplitude and phase structure over time.
- **CRQC–LLM** highlights how embodied information reveals fragility: coherence loss or phase disturbance immediately alters representational content, exposing instability or probing at the primitive level.

Section 2.8 Summary — Embodied Information in QAIS

This section reframed information as a physically embodied resource whose meaning arises from amplitudes, phases, and correlations rather than symbolic abstraction. By unifying representation, copying constraints, entanglement, contextuality, and geometric computation, this section completed the conceptual arc of QAIS operational grammar, showing that all transformations act on physically grounded states shaped by coherence, compatibility, and relational structure.

This perspective establishes embodiment as the governing condition under which quantum information is represented, transformed, and measured, where system behavior depends directly on the physical properties of the substrate.

With embodiment defining how information exists and evolves, the discussion now turns to the scientific tools that quantify this behavior, introducing entropy, fidelity, trace distance, and related metrics that formalize how embodied quantum states transform, degrade, and can be evaluated within QAIS workflows.

2.9 The Science of Quantum AI Systems

Architectural Framing: *Verification*

Section Framing: *Verification stage formalized through metrics and analytical tools that quantify state behavior and transformation integrity*

The preceding sections established the operational grammar of QAIS: information is represented through coherent quantum states, constrained by no-cloning and measurement collapse, structured through entanglement, shaped by contextuality, and manipulated through geometric unitary operations.

Section 2.8 concluded that quantum information is fundamentally embodied, inseparable from the physical substrate that carries it. Section 2.9 now introduces the scientific tools used to analyze, quantify, and evaluate these embodied quantum resources.

In classical computing, the science of information rests on Shannon measures of uncertainty, compressibility, and communication limits. QAIS extends this framework into the quantum domain, where information resides in amplitudes, phases, and correlations that cannot be reduced to symbolic variables. Quantum states exhibit degrees of coherence, distinguishability, and structure that demand new quantitative tools.

Rather than treating metrics as secondary analysis techniques, QAIS positions them as operational primitives. They dictate how states evolve, how errors manifest, and how transformations can be validated.

The discussion therefore moves from conceptual transformation (embodiment in Section 2.8) to *formal quantification*. Entropy captures uncertainty and mixed-state structure. Fidelity measures the similarity between prepared and target states. Trace distance quantifies how distinguishable two states are under optimal measurement.

These metrics allow QAIS practitioners to assess how much coherence has been preserved, how closely an operation has approximated its intended transformation, and how noise has altered a state's informational content. They are the analytic language that turns the operational grammar of QAIS into a measurable discipline.

Importantly, these tools operate entirely at the primitive level. Entropy evaluates quantum states without assuming any architecture in which they might be embedded.

Fidelity compares states without regard to the purpose of the computation that produced them. Trace distance quantifies distinguishability independently of application context.

By remaining tied to the quantum substrate rather than higher-level systems, these metrics serve as universal measures for QAIS, unifying physics, engineering, computation, and information theory.

Section 2.9 therefore frames QAIS as a **science of measurable quantum information**. It establishes the bridge between physical representation and evaluative precision, preparing the ground for the subsections that follow, each dedicated to one of the core metrics that quantify the behavior of embodied information.

Case Study Anchor — The Science of Quantum AI Systems

Intent: To formalize QAIS as a measurable discipline by applying entropy, fidelity, trace distance, and mutual information as operational benchmarks for coherence integrity and disturbance detection.

- **QALIS** interprets these metrics as tools for assessing how well primitive quantum operations preserve coherence, structure, and representational content during evolution.
- **CRQC-LLM** highlights how deviations in entropy, fidelity, or trace distance reflect fragility or disturbance at the substrate level, providing primitive indicators of instability or probing.

These interpretations apply the chapter-wide case-study lenses without introducing architectural or application-level behaviors.

2.9.1 Quantum State Formalism

Quantum information begins with the formalism of the qubit, the fundamental representational unit in QAIS. A qubit differs from a classical bit not by taking additional discrete values but by inhabiting a continuum of coherent states shaped by amplitudes and relative phase. This structure is captured by the general form of a single-qubit state:

$$|\psi\rangle = \alpha|0\rangle + \beta|1\rangle \tag{2.9.1}$$

The coefficients α and β are complex amplitudes whose squared magnitudes determine the probabilities of obtaining measurement outcomes $|0\rangle$ or $|1\rangle$.

Since a qubit represents a complete probability distribution, its amplitudes must satisfy the normalization condition:

$$|\alpha|^2 + |\beta|^2 = 1 \tag{2.9.2}$$

These equations formalize the representational shift introduced throughout Chapter 2. A qubit does not encode information through discrete symbolic assignments but through geometric structure: a balance of amplitude, phase, and coherence that remains intact only while the system avoids disturbance. This continuous, probabilistic representation is embodied in the physical system that carries it, linking directly to the principles introduced in Section 2.8.

Bloch Sphere Representation of a Qubit

Geometric representations provide an essential framework for understanding the structure and evolution of quantum states, offering insight into how transformations operate within state space. These representations support both conceptual reasoning and system-level analysis by making abstract algebraic relationships visually interpretable. Figure 2.9.1situates this geometric perspective within QAIS, translating the algebraic qubit state in Equation (2.9.1) into its Bloch sphere representation, where normalization constrains all valid states to points on the unit sphere.

The figure introduces the state $|\psi\rangle = \alpha|0\rangle + \beta|1\rangle$ using the standard formalism established in foundational quantum information texts such as Nielsen and Chuang [5].

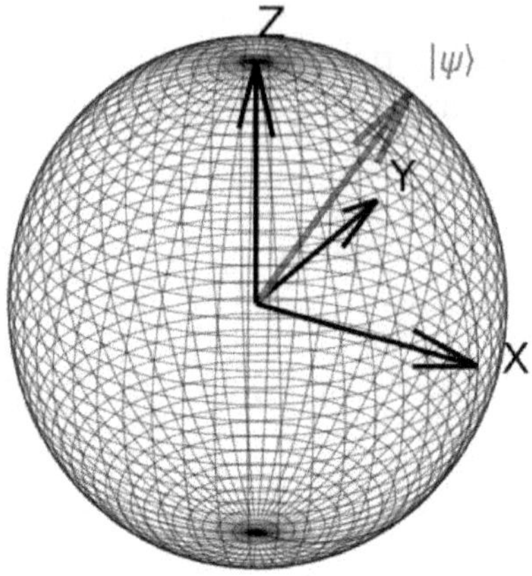

Figure 2.9.1 — Bloch Sphere Representation of a Qubit.

This figure maps the qubit state $|\psi\rangle = \alpha|0\rangle + \beta|1\rangle$ onto the Bloch sphere. The normalization condition in Equation (2.9.2) restricts valid qubit states to the unit sphere. The polar angle θ determines the distribution of amplitudes, while the azimuthal angle φ encodes their relative phase. The sphere visualizes the continuum of coherent states available to a single qubit, illustrating how probability and phase are embedded directly into its geometric structure.

The following spotlight interprets this formalism as the basis of QAIS's operational grammar.

Technical Spotlight — Qubit State Formalism

Conceptual Grounding

Figure 2.9.1 demonstrates how the algebraic qubit state becomes a geometric object in state space, consistent with the standard representation described by Nielsen and Chuang [5]. On the Bloch sphere, the amplitudes α and β appear as coordinates that determine probability weights, phase structure, and interference potential.

The north and south poles correspond to the classical states $|0\rangle$ and $|1\rangle$, while all other points represent coherent superpositions whose measurement outcomes and phase relationships vary with orientation. This geometric view makes clear that quantum representation is inherently embodied in amplitude and phase, inseparable from the physical substrate that carries the state.

Operational Relevance

In QAIS, the qubit formalism defines how quantum information can be prepared, manipulated, and measured. Unitary operations correspond to rotations on the sphere; relative phase determines interference behavior; and coherence dictates how long these trajectories remain meaningful.

Although the architectural implications appear in later chapters, the operational meaning here is foundational: every transformation in QAIS begins as a geometric motion of a state vector constrained by normalization and governed by quantum dynamics.

Cross-Disciplinary Orientation

Across disciplines, the qubit formalism serves different but complementary roles. Physicists use the Bloch sphere to relate algebraic states to physical dynamics under controlled fields. Engineers treat it as a guide for constructing pulse sequences that steer states reliably. AI specialists interpret the sphere as a representational surface where amplitude and phase encode structured relationships. Cybersecurity practitioners view the sensitivity of the sphere to measurement disturbance as an inherent boundary on information access.

Case Study Anchor — Quantum State Formalism

Intent: To ground QAIS representation in Hilbert space geometry, clarifying how amplitude normalization, phase structure, and state evolution define admissible quantum behavior.

- **QALIS** interprets qubit formalism as defining the preparation and stabilization rules that ensure primitive state trajectories maintain coherence during operation.
- **CRQC-LLM** highlights how disturbance of amplitude-phase coordinates produces detectable deviations, revealing fragility at the primitive level.

By grounding representation in a geometric formalism, Section 2.9.1 establishes the quantitative foundation needed for comparing and evaluating quantum states.

This prepares the reader for Section 2.9.2, where measures such as **fidelity** formalize how closely two quantum states resemble one another under ideal or noisy evolution.

2.9.2 Bell States: Formal Structure of Multi-Qubit Inseparability

Single-qubit states express quantum information through amplitude and phase. Multi-qubit systems, however, reveal a deeper representational structure: correlations that cannot be decomposed into independent components. Bell states provide the canonical example of such inseparability. They represent maximally entangled two-qubit states whose joint behavior must be described as a single, unified system.

One of the standard Bell states is:

$$|\Phi^+\rangle = (|00\rangle + |11\rangle)/\sqrt{2} \tag{2.9.3}$$

This expression captures a global property: neither qubit possesses its own state. Instead, both are components of one coherent two-qubit system whose outcomes are correlated in a way that no classical model can reproduce.

Measurement of either qubit instantaneously determines the outcome of the other, not because information travels between them, but because the system does not factorize into independent parts.

Within QAIS, Bell states formalize the concept introduced in Section 2.4: entanglement is not simply strong correlation but structural inseparability. The Bell basis provides the mathematical foundation for reasoning about multi-qubit coherence, joint measurements, and the relational fabric that underlies higher-dimensional quantum information.

Entangled representations introduce relationships between system components that cannot be reduced to independent descriptions, fundamentally altering how information is structured and interpreted. These representations are central to many QAIS capabilities.

Figure 2.9.2 situates this form of representation within a system-level context, highlighting its role in enabling coordinated behavior.

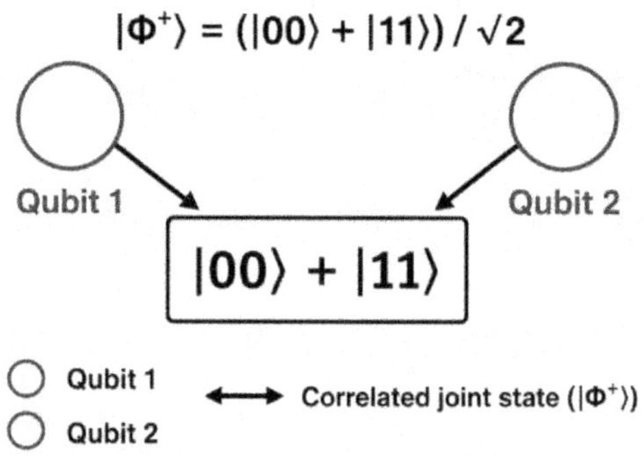

Figure 2.9.2 — Bell State as Entangled Representation.

This figure depicts the Bell state $|\Phi^+\rangle = (|00\rangle + |11\rangle)/\sqrt{2}$ as a unified two-qubit representation. The visual highlights that the joint state cannot be separated into independent single-qubit states. Measurement of either qubit yields correlated outcomes consistent with the global structure of the entangled pair.

This figure builds on the single-qubit Bloch-sphere formalism by extending representation into the multi-qubit regime.

Technical Spotlight — Bell States and Global Correlation

Conceptual Grounding

Figure 2.9.2 illustrates how the Bell state embodies inseparability. The joint amplitude distribution is shared across the two-qubit system; no single-qubit description can capture its structure. Measurement outcomes reflect the global nature of the state rather than independent local variables. This distinguishes entanglement from classical correlation: it arises from the structure of Hilbert space, not shared causes or incomplete information.

Operational Relevance

In QAIS, Bell states serve as the primitive example of multi-qubit coherence. They define how operations on one qubit affect the relational structure of the entire system, how joint measurements extract contextual information, and how global correlations persist until disrupted by noise or incompatible measurement.

These principles underpin the formal metrics introduced next: fidelity and trace distance often evaluate the preservation or degradation of such entangled structures.

Cross-Disciplinary Orientation

Physicists interpret Bell states as experimental signatures of nonclassical behavior, validated through tests of Bell inequalities. Engineers treat entangled pairs as calibration references for verifying two-qubit gates and assessing noise channels. AI specialists view entangled distributions as extensions of representational capacity, where multi-qubit correlations encode relationships unavailable to classical systems. Cybersecurity practitioners regard disturbance of Bell correlations as evidence of interaction, reflecting the sensitivity of entangled states to external probing.

Case Study Anchor — Bell States — Formal Structure of Multi-Qubit Inseparability

Intent: To formalize entanglement through maximally correlated states, illustrating how inseparability manifests mathematically and how global structure governs measurement outcomes.

- **QALIS** interprets Bell states as exemplars of stable two-qubit coherence used to illustrate how relational structure must be preserved during primitive quantum operations.
- **CRQC–LLM** highlights that disturbances to amplitude-phase relationships in an entangled pair produce detectable deviations in expected correlations.

By introducing multi-qubit inseparability, Section 2.9.2 prepares readers for Section 2.9.3, where *trace distance* quantifies how distinguishable two quantum states are, including how noise or disturbance alters global structures such as Bell correlations.

2.9.3 Formal Tools and Metrics

QAIS rely on a rigorous mathematical and physical toolkit that quantifies how information behaves when embodied in quantum substrates. Where QIS treated these constructs as theoretical descriptors, QAIS elevates them into operational guardrails, frameworks that determine how far a system can be pushed before coherence collapses, verification fails, or adversarial perturbations become detectable.

Section 2.9.3 therefore consolidates the analytic backbone of QAIS: the spaces, operators, channels, and metrics that give the discipline its quantitative structure and establish the boundaries of reliability, scalability, and resilience.

Section 2.9.3 marks a transition from representational formalism (Sections 2.9.1 and 2.9.2) to the quantitative science that governs implementation. *Hilbert spaces* structure what states can exist; operators define how they evolve; quantum channels encode both ideal and noisy transformations; and metrics such as entropy, fidelity, mutual information, and trace distance quantify the quality, stability, and distinguishability of states. These tools are the mechanisms through which QAIS systems can be evaluated with precision, rather than observed qualitatively.

The analysis and evaluation of QAIS systems depend on formal tools and metrics that quantify information content, similarity, and system behavior under transformation. These constructs provide a foundation for rigorous system assessment. Figure 2.9.3 situates these tools within a unified framework, clarifying their role in supporting system validation and analysis.

Formal Tools as Operational Foundations

This figure synthesizes the mathematics and metrics that underpin QAIS state evolution, coherence management, and verification. It situates Hilbert space geometry, operators, and quantum channels alongside coherence, correlation, and distinguishability measures, providing a unified view of the analytical levers used throughout the discipline.

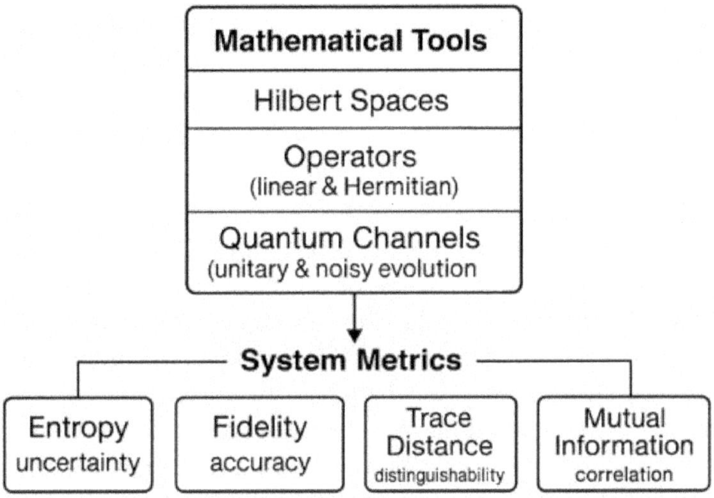

Figure 2.9.3 — Formal Tools and Metrics in QAIS.

The mathematical tools and performance metrics used to quantify quantum information in QAIS is summarized in this figure. Hilbert spaces define the geometric structure of states; unitary and Hermitian operators govern evolution and measurement; and quantum channels model both ideal and noisy transformations. Core metrics, entropy for uncertainty, fidelity for state preservation, mutual information for correlation strength, and trace distance for distinguishability, serve as operational benchmarks for coherence, accuracy, and resilience. Together, these constructs form the quantitative guardrails that ensure QAIS systems can be evaluated, compared, and validated across operational contexts.

The figure crystallizes how abstract mathematics becomes engineering practice. The following technical spotlight interprets these tools and metrics as operational boundaries, guardrails that determine when QAIS systems can preserve coherence, resist tampering, or achieve scalable performance.

Technical Spotlight — Metrics as Operational Guardrails

Conceptual Grounding

Figure 2.9.3 emphasizes that Hilbert spaces, operators, and quantum channels form the mathematical substrate of QAIS. Hilbert spaces provide the geometric arena where amplitudes and correlations reside, while linear operators, including unitaries and Hermitian observables, define the allowed transformations and measurable properties of states. Quantum channels extend these transformations further, capturing both ideal unitary evolution and the noisy dynamics encountered in practical implementations.

Building on this foundation, metrics supply the quantitative tools needed to evaluate system behavior. Entropy measures uncertainty and sets limits on compression and inference; it also reflects mixed-state contributions arising from environmental coupling. Fidelity measures how well a state or process preserves its intended structure after transformation. Trace distance assesses the distinguishability of states, revealing whether environmental or adversarial perturbations have introduced detectable change. Mutual information quantifies correlations across subsystems, connecting entanglement strength to actionable information content. Together, these constructs transform quantum theory into a framework for operational analysis.

System Relevance

Formal tools provide the analytic scaffolding for QAIS design and evaluation by translating abstract mathematical properties into operational criteria. Fidelity indicates whether error-correcting codes can preserve usable coherence during computation, while entropy bounds establish the limits of inference, compression, and uncertainty quantification within evolving quantum states. Trace distance reveals whether a transformation preserves the intended trajectory of a state or deviates due to drift, disturbance, or adversarial influence. Mutual information benchmarks internal correlations, ensuring that multi-qubit resources remain aligned with design intent and retain their relational integrity.

Together, these metrics give designers, engineers, and analysts precise, measurable means for assessing QAIS behavior.

Cross-Disciplinary Orientation

Across domains, these formal tools serve as a shared quantitative language for interpreting and validating quantum behavior. Physicists rely on fidelity, entropy, and operator norms to assess coherence, characterize entanglement strength, and verify that experimental outcomes remain consistent with theoretical predictions. Engineers apply the same measures in calibration workflows, hardware tuning, and noise-mitigation routines, ensuring that device-level operations remain stable and within prescribed tolerances.

AI researchers interpret fidelity and mutual information as structural analogues to distance and correlation metrics, embedding them into representation learning where amplitude–phase relationships shape model expressiveness. Cybersecurity practitioners view entropy and trace distance as tamper-sensitive indicators, using deviations in these quantities to reveal perturbations that cannot be concealed within quantum evolution or hidden behind classical post-processing.

These differing vantage points reinforce the centrality of formal metrics across QAIS.

Case Study Anchor — Formal Tools and Metrics

Intent: To treat Hilbert spaces, operators, channels, and quantitative metrics as operational guardrails that constrain and evaluate QAIS transformations under noise and adversarial stress.

- **QALIS** employs fidelity and mutual information thresholds to validate whether distributed sensing or multi-qubit encoding pipelines maintain their intended correlations and coherence levels during operation.
- **CRQC-LLM** interprets entropy and trace distance as diagnostic indicators that reveal adversarial perturbations, environmental distortions, or drift, allowing such disturbances to be detected through measurable deviation.

Section 2.9.3 establishes the quantitative backbone of QAIS. By formalizing uncertainty, preservation, differentiation, and correlation, these tools provide the analytic framework necessary for predicting protocol performance, validating system behavior, and evaluating resilience.

The next Section builds on these foundations by examining the operational protocols, such as teleportation, entanglement distribution, and related mechanisms, that mobilize these metrics in practice.

2.9.4 Operational Protocols in Practice

Operational protocols demonstrate how the principles of quantum mechanics, superposition, entanglement, measurement sensitivity, and coherence, become executable procedures. In QIS, these protocols were introduced as theoretical constructs. In QAIS, they form the applied expression of quantum law, revealing how information behaves when transformed, transmitted, or verified under constrained physical rules.

Teleportation, quantum key distribution, and entanglement purification occupy this role in QAIS not as architectural components but as operational demonstrations. They show how the primitives introduced throughout Chapter 2 function when placed into structured, rule-bound workflows. These protocols clarify the limits, capabilities, and dependencies of quantum information independent of any system-level design.

Operational protocols define how QAIS systems implement computation, communication, and measurement processes in practice. These protocols determine how abstract principles are realized within functioning systems. Figure 2.9.4 situates these procedures within an operational framework, illustrating how structured processes guide system execution

Foundational Quantum Information Protocols

Figure 2.9.4 illustrates how teleportation, quantum key distribution (QKD), and entanglement purification convert foundational quantum phenomena into operational procedures.

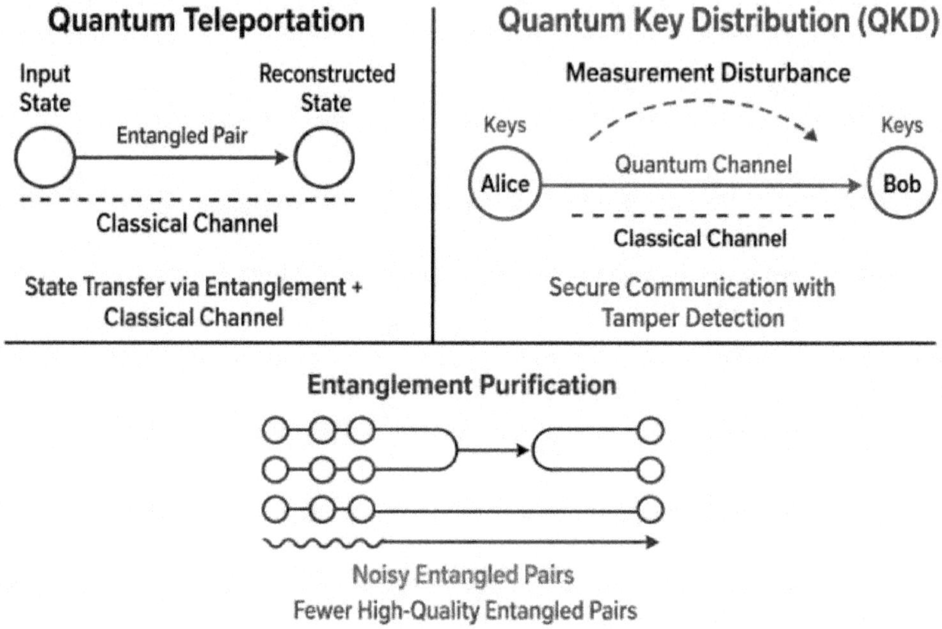

Figure 2.9.4 — Operational Protocols in QAIS.

The figure presents canonical quantum protocols, including teleportation, QKD, and entanglement purification, that demonstrate how superposition, entanglement, and measurement sensitivity are operationalized. Teleportation transfers entangled correlations and classical information to reconstruct quantum states at a remote location without moving the physical carrier. QKD transforms measurement disturbance into a verification mechanism for detecting unauthorized access. Purification refines noisy entangled states by probabilistically concentrating correlations. Together, these examples illustrate how QAIS converts fundamental principles into structured operational procedures governed by quantum law.

This figure crystallizes how QAIS protocols embody the field's predictive and operational power. The following technical spotlight interprets teleportation, QKD, and purification as more than demonstrations: they are the system-level templates that turn quantum law into deployable architectures.

Technical Spotlight — Protocols as System Templates

Conceptual Grounding

Teleportation, QKD, and purification share a common structure: each translates a quantum principle into an executable sequence that respects unitarity, entanglement constraints, and measurement-induced collapse. Teleportation demonstrates that entangled correlations can carry state information.

QKD shows how the impossibility of undetected copying becomes a verification tool. Purification illustrates how probabilistic selection can strengthen entanglement by discarding noisy components. These protocols do not add new principles; rather, they clarify existing ones through operational expression.

System Relevance

At the operational level appropriate to Chapter 2, these protocols illustrate how information behaves under quantum constraints without invoking architectural structures. Teleportation shows that the transfer of quantum state information depends entirely on global correlations rather than local carriers. Quantum key distribution (QKD) transforms measurement disturbance into a verification mechanism, using the inevitability of collapse as evidence of tampering or interception. Purification demonstrates how entanglement quality can be improved through probabilistic filtering, revealing how noisy correlations can be refined into more coherent resources. Together, these examples remain firmly within the domain of operational information behavior, setting the stage for later architectural treatments without preempting them.

Cross-Disciplinary Orientation

Across disciplines, these operational protocols reveal different facets of quantum behavior. Physicists approach them as experimentally testable demonstrations of entanglement, measurement sensitivity, and state evolution under controlled conditions. Engineers see them through the lens of implementation, interpreting each protocol as a sequence that demands precise timing, calibrated control operations, and robust error mitigation to maintain coherence.

AI specialists read these workflows as structured patterns for manipulating probabilistic information, where superposition and entanglement become resources for representation and transformation under coherence constraints.

Cybersecurity experts, in turn, recognize quantum key distribution as a physical instantiation of verification principles rooted in no-cloning and measurement disturbance, providing tamper-evident pathways that classical systems cannot emulate.

Each perspective reinforces that these protocols are expressions of quantum law, not system-level designs.

Case Study Anchor — Operational Protocols in Practice

Intent: To demonstrate how teleportation, QKD, and purification convert primitive rules into executable procedures that respect interference, entanglement, and measurement constraints.

- **QALIS** interprets these protocols as demonstrations of how entanglement, disturbance sensitivity, and correlation refinement behave under controlled conditions, properties later used to inform system-level reasoning and verification.
- **CRQC-LLM** interprets these protocols as illustrations of how sensitive correlations respond to perturbation, noise, or probing. Teleportation's reliance on shared correlations, QKD's dependence on disturbance detection, and purification's selectivity highlight how fragility becomes a diagnostic indicator under stress.

These anchors interpret the protocols without extending them into architectures, preserving Chapter 2's boundaries.

Operational protocols connect the formal tools of Section 2.9.3, entropy, fidelity, trace distance, and mutual information, to observable behaviors. Teleportation links entanglement and measurement; QKD binds no-cloning and disturbance; purification ties correlation strength to probabilistic filtering. The next Section (2.9.5) draws on these insights to delineate the intrinsic limits of quantum information through the no-go theorems that frame what QAIS cannot do.

2.9.5 Boundaries of Quantum Information

QAIS is defined not only by the resources it mobilizes, superposition, entanglement, contextuality, but by the boundaries it must honor. These boundaries, expressed formally as **no-go theorems**, are more than mathematical impossibilities: they are the structural rules that determine how quantum information behaves and how it can be manipulated without violating physical law. In QAIS, these limits are treated as part of the operational grammar rather than as shortcomings. They anchor coherence, govern verification, and prevent information from being separated from its physical embodiment.

Earlier in Section 2.3, no-cloning and measurement collapse appeared as constraints on copying and observation. Here, the no-go theorems enter in their formal, system-agnostic expression. The **no-cloning theorem** prohibits the duplication of unknown states.

The **no-broadcasting theorem** disallows distributing arbitrary quantum states across multiple subsystems. The **no-deleting theorem** forbids irreversible erasure of unknown quantum states. Together, these boundaries define the limits of permissible transformations in QAIS.

The operational boundaries defined by quantum no-go theorems do not prevent all forms of cryptographic disruption. While quantum mechanics prohibits cloning and broadcasting of unknown states, it simultaneously enables algorithmic capabilities, most notably Shor's algorithm—that undermine widely deployed classical cryptographic systems.

In response, the National Institute of Standards and Technology (NIST) has initiated the Post-Quantum Cryptography (PQC) standardization effort [12], which develops cryptographic schemes resistant to quantum adversaries.

These approaches do not rely on quantum mechanical constraints such as measurement disturbance or no-cloning, but instead on classical hardness assumptions believed to remain intractable even for quantum computers.

Within QAIS, this distinction is critical: no-go theorems define physical security guarantees, whereas PQC defines computational security resilience against quantum-enabled attacks.

Certain theoretical constraints define boundaries on what is achievable within quantum systems, shaping both system design and algorithmic development. These constraints must be accounted for when constructing QAIS architectures. Figure 2.9.5 situates these limitations within a structural context, clarifying how foundational rules influence system behavior.

Visualizing the Boundaries of Quantum Information

This figure contrasts classical expectations about copying, distribution, and deletion with their quantum restrictions.

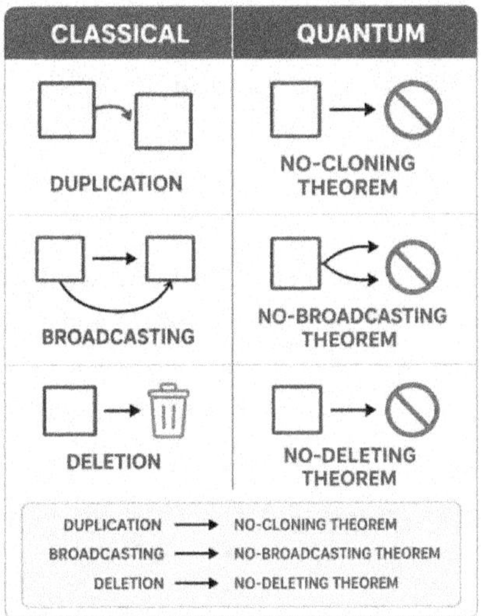

Figure 2.9.5 — No-Go Theorems as Structural Rules in QAIS.

The figure compares classical assumptions, perfect duplication, unrestricted broadcasting, and irreversible deletion, with their quantum counterparts: the no-cloning, no-broadcasting, and no-deleting theorems. The classical panel depicts operations that treat information as separable from its medium, while the quantum panel shows that these operations are forbidden for unknown states. The ledger highlights how QAIS replaces classical operations with coherence-preserving alternatives grounded in entanglement, controlled evolution, and measurement-aware reset procedures.

This visual makes clear that the very limits of quantum mechanics serve as the structural grammar of QAIS. What cannot be cloned, broadcast, or deleted becomes the foundation for resilience, verification, and coherence preservation.

The following technical spotlight interprets these no-go theorems not as barriers but as system-level rules for engineering scalable and secure architectures.

Technical Spotlight — No-Go Theorems as Design Grammar

Conceptual Grounding

The no-go theorems represent a categorical shift from classical reasoning. Classical computation assumes that information can be freely copied, distributed, or discarded without consequence. Quantum mechanics forbids these operations for unknown states, reflecting the fact that information is embedded in amplitudes, phases, and correlations. Instead of duplicating a state, QAIS stabilizes it through encoded redundancy; instead of broadcasting, it relies on structured correlations; instead of deletion, it uses controlled reinitialization consistent with measurement backaction.

System Relevance (Operational Level)

Within QAIS, these limits act as **operational guardrails**, ensuring that transformations preserve coherence and remain faithful to quantum law. The no-cloning theorem defines how redundancy must be constructed; the no-broadcasting theorem sets boundaries on information distribution; and the no-deleting theorem informs how states can be reset without violating reversibility. These boundaries collectively ensure that verification, inference, and state preparation occur within physically permitted transformations.

Cross-Disciplinary Orientation

Physicists treat no-go theorems as signatures of quantum contextuality, proof that quantum information resists reduction to classical analogs. Engineers interpret them as constraints on device design, requiring non-duplicative redundancy and coherence-preserving control. AI specialists recognize these limits as rules for how representations must be encoded, transformed, and sampled.

Cybersecurity experts view them as foundations for tamper-evident communication, since forbidden operations cannot be performed without producing detectable disturbance.

Case Study Anchor — Boundaries of Quantum Information

Intent: To interpret no-go theorems as structural grammar that defines forbidden transformations and guides coherence-preserving alternatives within QAIS operations.

- **QALIS** interprets the no-go theorems as safeguards that preserve coherence and verification integrity, guiding how information is prepared and stabilized under allowed transformations.
- **CRQC–LLM** uses these boundaries as indicators of adversarial limits, since attempts to duplicate, broadcast, or erase unknown states inevitably produce detectable disturbance.

These anchors provide interpretive, non-architectural perspectives consistent with Chapter 2's operational scope.

2.9.6 Constructive QAIS Alternatives to No-Go Theorems

While Figure 2.9.5 establishes the fundamental restrictions imposed by no-go theorems, QAIS achieves practical functionality by operating within these constraints through permitted transformations. Rather than limiting system design, these constraints define the boundaries within which viable architectures must be constructed.

Constructive approaches emerge by leveraging allowable operations that respect these limits while enabling effective computation, communication, and control. These strategies reflect the interplay between theoretical restriction and engineering adaptation, where system capability is realized through disciplined design under constraint. Figure 2.9.6 situates these alternatives within a constructive framework, illustrating how QAIS architectures extend functionality without violating foundational principles. This figure illustrates the constructive alternatives, permitted transformations that enable stability, redundancy, and controlled reset without violating quantum law.

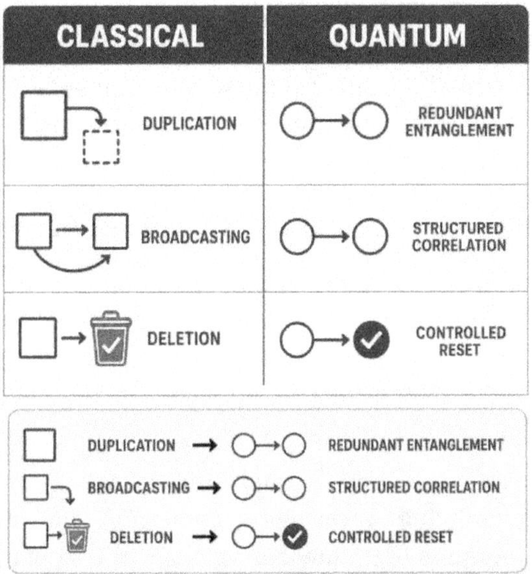

Figure 2.9.6 — Constructive QAIS Alternatives to No-Go Theorems.

The figure presents the operational substitutes that QAIS employs in place of forbidden classical operations. Instead of cloning, QAIS uses entanglement-based redundancy and stabilizer encodings; instead of broadcasting, it uses structured correlations and teleportation-based redistribution; instead of deletion, it relies on measurement-driven reset and ancilla-swap reinitialization, both fully consistent with reversibility. The ledger maps each forbidden classical action to its permitted quantum alternative, showing how QAIS builds functionality within the boundaries established by the no-go theorems.

QAIS Operations Under Quantum Constraints

Figure 2.9.6 extends the message of Figure 2.9.5 by showing that the no-go theorems, while restrictive, do not prevent QAIS from performing the functional roles that classical systems achieve through duplication, broadcasting, and deletion. Figure 2.9.5 establishes the boundaries: quantum information cannot be cloned, broadcast, or irreversibly erased.

Figure 2.9.6 then demonstrates how QAIS works *within* these boundaries by adopting constructive alternatives that preserve coherence and respect unitarity. These alternatives are not direct analogues of their classical counterparts; they are carefully engineered transformations that remain valid under quantum law.

For example, where deletion is forbidden, QAIS uses measurement-based reinitialization, ancilla-swap resets, or state re-preparation sequences, processes that collapse the state through measurement and then reinitialize it without violating reversibility. Similarly, entanglement-based redundancy and structured correlation replace classical duplication and broadcasting while maintaining quantum integrity.

Together Figure 2.9.5 and Figure 2.9.6 illustrate a coherent progression: the former defines what quantum mechanics prohibits, and the latter shows how QAIS constructs operational capability by transforming those prohibitions into design principles for stability, verification, and coherence management.

Section 2.9 Summary — The Science of Quantum AI Systems

This section established QAIS as a quantitative science grounded in formal measures governing embodied quantum information. Quantum state formalism defined the geometric structure of Hilbert space, while Bell states extended this foundation into the relational regime, demonstrating how entanglement produces non-classical correlations across multi-qubit systems.

Building on these structures, entropy, fidelity, and trace distance were introduced as operational tools for quantifying uncertainty, similarity, and distinguishability. These measures translate the operational grammar developed in Sections 2.2–2.6 into evaluative quantities, enabling precise diagnosis of disturbance, verification of transformations, and characterization of quantum state behavior. Protocols such as teleportation, QKD, and purification further operationalize these principles, demonstrating how formal metrics guide state transfer, verification, and correlation management.

Together, these elements complete the scientific foundation of QAIS, spanning representation (2.2), operational constraints (2.3), correlation (2.4), contextual dependence (2.5), geometric computation (2.6), coherence preservation (0), and physical embodiment (2.8), providing a unified framework for analyzing how quantum information behaves under transformation, noise, and protocol execution.

With this scientific basis established, the discussion now turns to the historical emergence of QAIS, where these principles are traced through their development across theory, experimentation, and large-scale research efforts into the modern discipline of Quantum AI Systems.

2.10 Historical Emergence of QAIS

Architectural Framing: *Verification → Hybrid Execution*

Section Framing: *Verification and Hybrid Execution interpreted through system maturation, validation, and operational evolution*

QAIS did not emerge fully formed; the discipline developed through decades of scientific discovery, operational refinement, and coordinated institutional investment. Chapter 1 introduced the physical primitives that make QAIS possible, superposition, entanglement, coherence, and measurement, while Chapter 2 has shown how these primitives function as operational rules. Section 2.10 now traces how those operational rules arose historically, without revisiting the physics already established in Chapter 1.

The focus here is on how early quantum discoveries evolved into the mechanisms, constraints, and protocols that define QAIS practice today.

Beginning in the 1980s, foundational scientific milestones reframed quantum effects as usable computational and informational resources. Feynman's insight that classical machines cannot efficiently simulate quantum dynamics positioned quantum behavior as a computational advantage rather than a theoretical oddity [1]. Deutsch's model of the universal quantum computer transformed qubits and gates into formal primitives for general computation [2]. Shor's and Grover's algorithms demonstrated practical performance benefits in factoring, search, and optimization, establishing a clear rationale for quantum systems in domains that now underpin QAIS inference, verification, and decision-making [3].

Advances in error correction, teleportation, and entanglement distribution showed that fragile quantum states could be stabilized, transmitted, and manipulated with operational reliability, providing the foundations for the resilience mechanisms used throughout QAIS architectures.

These scientific milestones unfolded alongside national and international initiatives that accelerated the maturation of the field. Programs such as the U.S. National Quantum Initiative Act (2018) [9], the European Quantum Flagship (2018) [10], China's National Laboratory for Quantum Information Science (2020) [11], and the EuroQCI program (2025)[20] [13] created the

[20] The EuroQCI program (2025) [13] initiated development of a continent-scale quantum communication infrastructure.

institutional scaffolding, funding, research ecosystems, workforce development, and governance frameworks, required for sustained progress.

While Chapter 3 expands on these ecosystem and policy dimensions in greater depth, Section 2.10 highlights the specific historical developments that shaped the operational rules introduced in Chapter 2: how gates and channels became formalized tools, how contextuality evolved into a verification resource, how measurement disturbance became a security primitive, and how error correction and entanglement distribution matured into the stabilizing mechanisms that anchor QAIS today.

Technical Spotlight — From QIS Milestones to QAIS Operations

Conceptual Grounding

The historical trajectory that culminated in QAIS began with a sequence of breakthroughs that reframed quantum behavior as an operational resource rather than a theoretical curiosity. Richard Feynman's insight in 1982, that quantum systems naturally compute quantum dynamics in ways classical machines cannot efficiently emulate, established the foundational argument for treating quantum mechanics as a computational substrate [1]. David Deutsch's formulation of the universal quantum computer in 1985 transformed this idea into a programmable model, giving rigorous meaning to qubits, unitary gates, measurement, and algorithmic evolution [2].

The field accelerated dramatically in the 1990s. Peter Shor's factoring algorithm (1994) demonstrated an exponential quantum speedup, revealing that quantum computation could undermine classical cryptographic assumptions and thereby signaling the arrival of a new computational era [3]. Lov Grover's search algorithm (1996) added a quadratic improvement for unsorted search, reinforcing the idea that quantum operations could produce computational outcomes beyond classical approximation [4]. These results shifted quantum mechanics from speculative possibility to practical roadmap, establishing the promise of coherent evolution, interference, and superposition as system-level assets.

This conceptual momentum was strengthened by advances in quantum error correction, fault-tolerant architectures, and quantum communication theory, including the Holevo bound [6][21]. Together, these developments transformed fundamental physics into an operational rulebook, defining how quantum states must be prepared, how coherence must be protected, and how measurement and disturbance shape the boundaries of valid information processing. This progression, from physical insight to operational discipline, is precisely the transformation explored throughout Chapter 2, where quantum principles evolve into the executable grammar of QAIS.

---

[21] Holevo's coding theorems established capacity bounds for quantum channels, demonstrating that quantum information could be quantified rigorously.

System Relevance

The milestones described above directly shaped the operational constructs introduced earlier in this chapter.

- **Superposition and entanglement** became recognized not merely as physical effects but as *resources* whose preservation defines the viability of pipelines.
- **Gates and circuits** evolved from mathematical abstractions into programmable operations with hardware constraints, pulse-level controls, and calibration models.
- **Error correction and decoherence mitigation** emerged as necessary system scaffolding, not optional optimizations.
- **Teleportation and entanglement distribution** matured into transport mechanisms for QAIS networks and distributed learning behaviors.

These developments explain why Section 2.9 centers its formalisms around coherence, metrics, protocols, and no-go constraints; they reflect decades of discoveries that turned conceptual physics into a workable operations discipline.

Cross-Disciplinary Orientation

The historical emergence of QAIS resonated differently across the communities represented in this book.

Physicists viewed each breakthrough as validation that quantum effects could be controlled, engineered, and reproduced rather than merely observed in isolation.

Engineers interpreted the same milestones as proofs of feasibility, motivating sustained investment in qubit platforms, control electronics, cryogenic stacks, photonic interconnects, and the hardware needed to support error correction at scale.

Computer scientists and mathematicians transformed these developments into formal complexity classes, encoding rules, and noise models that now anchor QAIS operations.

AI specialists recognized in these advances entirely new representational spaces and optimization strategies, anticipating the quantum kernels, hybrid models, and inference mechanisms developed in later chapters.

Cybersecurity professionals confronted both opportunity and threat: quantum key distribution and tamper-evident channels on one hand, and cryptographically relevant quantum acceleration on the other, directly motivating the CRQC–LLM stress-test case study.

Across these disciplines, QAIS did not emerge as a technological novelty but as a multidisciplinary framework through which intelligent, resilient, and secure architectures could finally be constructed.

Case Study Anchor — Historical Emergence of QAIS

Intent: To trace how foundational QIS milestones matured into the operational rule-set formalized in Chapter 2, clarifying the evolution from theoretical insight to executable QAIS discipline.

- **QALIS** serves as the constructive design framework in this book, illustrates how the field's historical capabilities matured into a coherent architecture for distributed quantum intelligence. The representational expansions revealed by early demonstrations of superposition and interference eventually became the basis for QALIS's multimodal sensing and inference pipelines. Likewise, the stabilizing strategies introduced in the 1990s, most notably quantum error correction and systematic coherence management, now underpin QALIS's approach to maintaining operational reliability across hybrid and distributed environments.

Even the earliest experimental proofs of teleportation and entanglement distribution evolved into the entanglement-assisted coordination behaviors that support coherent multi-qubit interactions.

- **CRQC–LLM** operates as the adversarial stress-test lens, draws on the same milestones to expose the vulnerabilities inherent in quantum systems. Shor's factoring algorithm, for instance, remains the conceptual foundation for analyzing cryptographic collapse and for understanding how LLM-accelerated reconnaissance accelerates adversarial targeting [3].

The fragility of coherence and entanglement, recognized early in QIS as both a technical challenge and a fundamental limitation, becomes the diagnostic surface through which CRQC–LLM detects manipulation, disturbance, or drift. And the quantum no-go theorems establish the boundaries adversaries cannot cross without leaving detectable traces, making the physical laws themselves a form of structural verification.

Together, these case studies demonstrate how the history of quantum information matured into two parallel trajectories within QAIS: a constructive pathway embodied by QALIS and a resilience-driven, adversarial pathway examined through CRQC–LLM. Both perspectives arise from the same scientific origins, yet they illuminate complementary dimensions of how quantum operations become system-level behavior.

Section 2.10 Summary — Historical Emergence of QAIS

This section traced how QAIS evolved from theoretical origins into a coordinated scientific discipline. Foundational milestones—Feynman's quantum simulation insight, Deutsch's universality, Shor's algorithmic breakthrough, and Grover's speedup—established the conceptual basis for quantum computation, while subsequent advances in channel capacity, error correction, and resource theory transformed these ideas into operational science. National and international initiatives further institutionalized this progress, aligning research, infrastructure, and policy into a globally coordinated effort.

This trajectory demonstrates how physical theory, computational models, and engineering practice converged to produce the operational framework of QAIS developed throughout Chapter 2, linking foundational principles (Chapter 1) to executable mechanisms, metrics, protocols, and constraints.

With this historical synthesis established, the discussion now turns to integrating these operational primitives into a unified QAIS model, where representation, transformation, measurement, verification, and execution are combined into a coherent system-level lifecycle.

2.11 Integrating Operational Primitives into the QAIS Model

Architectural Spine: *State Preparation → Unitary Evolution → Controlled Operations → Measurement → Verification → Hybrid Execution*

Section Framing: *Full architectural integration of all stages into a unified operational lifecycle.*

The operational mechanisms introduced across Sections 2.3–2.10—representation, transformation, correlation, measurement, verification, and coherence management—do not function as isolated tools. They form a unified grammar that governs how QAIS systems process information under quantum laws. Section 2.11 synthesizes these mechanisms into a cohesive operational flow, showing how quantum information moves through a sequence of primitive-level stages that define every QAIS process: **Representation → Transformation → Correlation → Verification → Extraction → Reset.** This sequence anticipates the life-cycle discussions developed in later chapters while remaining fully within the primitive scope of Chapter 2.

Section 2.11 connects back to Chapter 1 only where necessary to reaffirm that information in QAIS is physically embodied and subject to the coherence, phase, and contextual constraints established earlier. The emphasis remains on operational integration rather than foundational re-explanation.

By drawing these primitives into a single flow, Section 2.11 prepares the ground for **Chapter 3 and subsequent chapters**, which expand from operational mechanisms into ecosystem-scale adoption, architectural design, and system-level resilience.

2.11.1 Operational Flow in QAIS

A QAIS process begins when information is represented as a quantum state. This representation is amplitude- and phase-encoded, coherence-dependent, and embedded physically in the qubit substrate (Sections 2.2 and 2.8). Once encoded, the state undergoes transformation through reversible unitary gates (Section 2.1.3), interference-driven evolutions (Section 2.6), and controlled contextual interactions (Section 2.5). These transformations create new amplitudes, correlations, and computational pathways.

At the multi-qubit level, correlation, especially entanglement (Section 2.4), extends this evolution across subsystems, forming global structures that enable distributed inference, multi-node coordination, and nonlocal reasoning.

Throughout this process, coherence must be preserved using error-mitigation and error-correcting strategies (Section 2.7). The system must remain within allowable transformations defined by the no-go theorems (Section 2.9.5), which function as operational boundaries rather than limitations.

As states evolve, QAIS must perform verification, translating quantum structure into actionable signals while respecting measurement backaction (Section 2.3). Verification uses metrics such as fidelity, entropy, and trace distance (Section 2.9.3) to assess whether the system remains within operational tolerances. Measurement collapse is treated as a deliberate and controlled extraction of classical information, not an incidental event, and is embedded into the operational design.

Finally, all QAIS processes must include reset, meaning the controlled reinitialization of qubits consistent with measurement rules and reversibility constraints. Reset prepares qubits for new representations and delineates the boundaries between successive computational cycles. Together, these stages form the QAIS Operational Flow, providing a unifying scaffold for the mechanisms developed throughout this chapter.

Visualizing the Primitive-Level Flow of QAIS Operations

Figure 2.11.1 integrates the operational primitives introduced in Sections 2.3–2.10 into a unified flow model, illustrating how QAIS systems process information through a structured sequence of representation, transformation, correlation, verification, extraction, and reset. This progression maps the conceptual mechanisms of the chapter into a coherent operational cycle.

QAIS operation is therefore characterized by coordinated information flow across representation, computation, and control stages, forming the process through which system behavior is realized and regulated.

Figure 2.11.1 provides a unified view of these interactions, clarifying how system components coordinate during execution and how information propagates across stages. Each stage linked to Sections 2.3–2.10, forming one operational continuum.

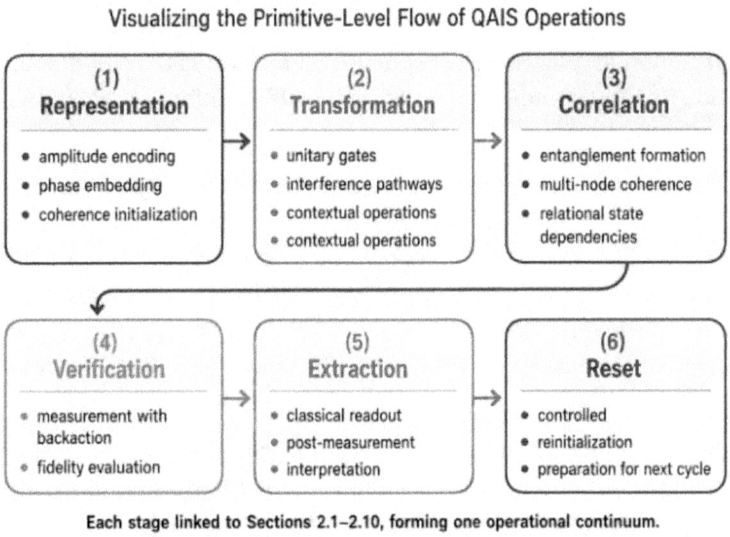

Figure 2.11.1 — QAIS Operational Flow Model.

This figure illustrates how QAIS processes information through six primitive-level stages: representation, transformation, correlation, verification, extraction, and reset. It consolidates the operational mechanisms introduced across the chapter into a single lifecycle model, showing how each primitive contributes to coherent quantum processing.

Technical Spotlight — Operational Integration in QAIS

Conceptual Grounding

The QAIS Operational Flow demonstrates that quantum computation is not a sequence of isolated operations. It is a coordinated traversal through representational, dynamical, relational, and verification stages, each governed by specific constraints of quantum mechanics.

This spotlight interprets the flow as the convergence point of the chapter: the operational grammar that arises when all primitives are treated as interdependent components of a single process.

System Relevance

At the primitive level, the QAIS operational flow illustrates how stability and reliability arise directly from the mechanics of quantum information itself. Representation sets the available feature space and determines the coherence budget upon which all subsequent operations depend. Transformation arranges the computational pathways that modulate amplitudes and phases. Correlation introduces nonlocal structure, enabling inference relationships that cannot be factorized classically.

Verification establishes the operational boundary by testing fidelity and detecting disturbance, and Reset restores the system for continued cycles without contaminating subsequent computations. These elements form the core "operational signatures" of QAIS, mechanisms that later chapters scale into architectural constructs but are presented here in their foundational, substrate-level form.

Cross-Disciplinary Orientation

Across disciplines, this flow acquires different interpretations while preserving its unified operational meaning. Physicists view it as the controlled choreography of state evolution, measurement, and reinitialization. Engineers regard it as the timing-and-control sequence that hardware must implement with precision to maintain coherence. AI specialists interpret it as a quantum analogue of a data-processing pipeline, one constrained not by symbolic logic but by coherence, context, and relational dependencies. Cybersecurity practitioners see verification and reset as the intrinsic guardrails that enforce trust at the substrate level and ensure that any disturbance becomes observable.

Case Study Anchor — Operational Primitives in QAIS

Intent: To synthesize representation, transformation, correlation, verification, extraction, and reset into a unified operational lifecycle that frames QAIS as a coherent execution continuum.

- **QALIS** interprets the operational flow as the constructive backbone of quantum reasoning: representations map to inference-ready states, transformations propagate structured correlations, and verification maintains coherence integrity.

- **CRQC–LLM** exposes vulnerabilities across the same flow: disturbances affect representation, noise disrupts transformations, correlation is fragile under probing, and verification metrics reveal tampering.

Section 2.11 Summary — Integrating Operational Primitives into the QAIS Model

Section 2.11 integrated representation, transformation, correlation, measurement, verification, extraction, and reset into a coherent QAIS operational lifecycle. This unified flow brings together the mechanisms developed throughout Chapter 2 and establishes an operational grammar governed by quantum law, physical constraints, and coordinated quantum–classical execution.

Within this lifecycle, QALIS interprets the operational flow as the constructive foundation for quantum reasoning and controlled information processing. CRQC–LLM examines the same sequence as a set of interconnected fragility surfaces through which noise, disturbance, verification weakness, and interface failure may propagate. Together, these lenses show that operational capability and system vulnerability emerge from the same underlying processes.

This synthesis consolidates the chapter's progression from representation and copying constraints through entanglement, contextuality, geometric computation, error correction, embodiment, and formal evaluation. With the operational lifecycle established, Section 2.12 turns to practical QAIS pipelines and experimental systems in which these principles are applied under physical and engineering constraints.

2.12 Practical Applications — QAIS Pipelines

Architectural Spine: State Preparation → Unitary Evolution → Controlled Operations → Measurement → Verification → Hybrid Execution

Section Framing: *End-to-end operational realization of the architectural spine in practical systems under real-world constraints*

The operational primitives developed throughout Chapter 2—representation, unitary evolution, interference, entanglement, contextuality, measurement backaction, quantum metrics, and no-go constraints—form the substrate upon which all practical QAIS workflows are built.

Although this chapter remains at the primitive level, these mechanisms already manifest in early-stage systems, laboratory platforms, and hybrid tools that illustrate how QAIS pipelines behave under real operational constraints.

Primitive-Level Applications in Contemporary QAIS Contexts

- **Quantum State Preparation & Verification Pipelines**: Contemporary devices use amplitude steering, state tomography, and Bloch-sphere calibration routines to operationalize the representational and measurement rules of Section 2.2–2.3. Coherence windows determine how long the prepared states remain actionable.

- **Entanglement-Enabled Multi-Qubit Operations:** Section 2.4's relational primitives appear in today's photonic and superconducting platforms as entanglement-distribution modules, Bell-state generators, and contextual correlation tests used for calibration and device-level benchmarking.
- **Error-Mitigation and Early Error-Correction Workflows:** The principles of Section 2.7 are implemented in stabilizer-measurement cycles, repetition codes, and simplified topological routines. These workflows operationalize coherence preservation, syndrome extraction, and reversible reset.
- **Metric-Driven Validation of Quantum Behavior:** Entropy, fidelity, trace distance, and mutual information Section 2.9.3 are already used in NISQ pipelines to quantify gate accuracy, noise drift, disturbance sensitivity, and entanglement stability. These metrics determine operational thresholds for viable computation.
- **Protocol-Based Demonstrations of Foundational Behaviors:** Teleportation, entanglement purification, and QKD Section 2.9.4 serve as early exemplars of primitive-level reasoning: transmitting information without carriers, leveraging disturbance sensitivity for verification, and refining correlations under noise.

Across these examples, QAIS pipelines do not yet implement architectural intelligence, but they *do* express the operational grammar defined across Sections 2.2–2.11. Chapter 2 therefore establishes the *primitive functionality* from which all later architectural, reasoning, and multi-agent behaviors will emerge.

Section 2.12 Summary — Practical Applications: QAIS Pipelines

This section demonstrated how the operational primitives of QAIS—representation, transformation, correlation, measurement, verification, and constraint—are already realized in contemporary quantum systems. Through state preparation pipelines, entanglement-enabled operations, error-mitigation workflows, metric-driven validation, and protocol-based demonstrations, these primitives manifest as executable behaviors within laboratory platforms and early-stage hybrid systems.

These implementations confirm that the operational grammar developed throughout Sections 2.2–2.11 governs real-world QAIS pipelines, even in the absence of full architectural intelligence. They reveal both the capabilities enabled by coherent quantum evolution and the limitations imposed by noise, instability, and physical constraints, establishing the practical boundary conditions under which QAIS systems currently operate.

With these operational behaviors grounded in practice, the discussion now advances to a comparative case analysis, where the same primitives are examined through constructive (QALIS) and stress-test (CRQC–LLM) lenses to reveal how capability and fragility emerge from the underlying quantum operational framework.

2.13 Case Study Advancement — QALIS vs CRQC-LLM at the Operational Layer

Architectural Spine: *Controlled Operations → Measurement → Verification → Hybrid Execution*

Section Framing: *Comparative evaluation of system behavior across execution, observation, and validation stages under constructive and adversarial conditions Chapter 2 extends the dual-case framework introduced in Chapter 1 by examining how a common set of quantum operational primitives gives rise to both constructive capability and systemic vulnerability. At the operational layer, QALIS and CRQC-LLM do not differ in physical mechanism, but in how those mechanisms are interpreted, stabilized, and stressed under execution. The same processes that enable structured transformation, correlation, and inference also define the boundaries within which reliable behavior can be maintained.*

QALIS — Constructive Interpretation of Operational Primitives

Under the QALIS lens, the primitives developed in this chapter constitute a coherent and disciplined foundation for quantum intelligence. Representation and encoding establish physically grounded initial conditions, while reversible evolution and interference provide controlled mechanisms for shaping amplitude distributions. Entanglement extends these transformations across subsystems, enabling coordinated, nonlocal behavior, and contextuality refines this capability by structuring how information is extracted under compatible measurement settings.

Within this framework, coherence is treated as a resource that can be preserved, managed, and extended through error correction and noise-aware design. Measurement is not a destructive limitation but an integral component of controlled information extraction, and verification—through fidelity, entropy, and related metrics—ensures that system evolution remains within acceptable operational bounds.

From this perspective, Chapter 2 defines the origin of QAIS capability: a physically grounded operational grammar that supports scalable, verifiable, and architecturally consistent system behavior, forming the basis for higher-level reasoning and learning systems developed in later chapters.

CRQC-LLM — Stress-Test Interpretation of Operational Primitives

In contrast, the CRQC-LLM lens interprets the same primitives as indicators of structural sensitivity and potential instability. Coherence becomes a limited and fragile window, subject to environmental noise, drift, and adversarial perturbation. Interference, while enabling computational advantage, introduces a global sensitivity in which small disturbances propagate through amplitude evolution, altering system outcomes in nonlocal ways.

Measurement, which enables information extraction, simultaneously imposes irreversible constraints through backaction, creating points of vulnerability when triggered prematurely or manipulated. Entanglement, rather than serving solely as a coordination resource, exposes correlated systems to collective degradation under noise or probing, while contextuality amplifies sensitivity to measurement sequencing and operational configuration.

Verification metrics—entropy, fidelity, and trace distance—therefore function not only as indicators of correctness but as diagnostic surfaces that reveal instability, disturbance, or tampering. No-go theorems further formalize these constraints, establishing fundamental limits on copying, broadcasting, and state manipulation that define the boundaries of permissible system behavior.

From this perspective, Chapter 2 reveals that the same operational grammar that enables capability also defines fragility. The primitives do not merely support computation; they impose constraints that must be actively managed to prevent degradation, instability, or adversarial exploitation.

Section 2.13 Summary — Capability and Constraint as Co-Extensive

Taken together, QALIS and CRQC-LLM demonstrate that capability and fragility are not opposing properties but co-extensive features of the same physical system. Operational primitives simultaneously enable structured transformation and impose strict conditions on how that transformation can occur. Reliable QAIS design therefore depends on maintaining alignment across controlled operations, measurement, verification, and hybrid execution. This dual interpretation establishes the central insight of Chapter 2: quantum intelligence emerges not from the primitives themselves, but from how their capabilities are stabilized under their inherent constraints.

This insight motivates the subsequent development of architectural, resilience, and governance frameworks in later chapters, where the operational behaviors identified here are elevated into system-level design principles.

Chapter 2 Conclusion

Chapter 2 completed the transition from the foundational principles established in Chapter 1 to the execution framework that governs how QAIS behave in practice. Quantum information was treated not as an abstract construct but as a physically embodied resource—encoded through amplitudes and phases, transformed through unitary evolution, correlated through entanglement and contextuality, interrogated through measurement and backaction, and recycled through controlled reset. Across these stages, QAIS emerged as a discipline defined by structured transformation rather than symbolic procedure: computation became the controlled evolution of amplitudes, communication the management of shared correlations, and resilience the preservation of coherence under noise and disturbance.

Throughout the chapter, the QALIS and CRQC-LLM lenses interpreted these primitives under supportive and adversarial conditions. QALIS emphasized stability, controlled evolution, and verifiable transformation, while CRQC-LLM exposed sensitivity to noise, perturbation, and contextual disruption, revealing where resilience must be engineered. At this stage, these perspectives remained confined to primitive behavior, clarifying how capability and fragility arise from the same underlying processes.

Section 2.11 and the QAIS Operational Flow Model (Figure 2.11.1) unified the discussion, presenting representation, transformation, correlation, verification, extraction, and reset as a continuous cycle. This model provided a structural through-line across Sections 2.2–2.10, organizing the primitives into a coherent and executable sequence and establishing the rule set from which higher-level QAIS systems are constructed. It marks the transition point at which quantum dynamics begin to interface with system-level reasoning and design.

With this foundation established, the scope now expands outward. Chapter 3 examines the environments in which QAIS can be deployed and scaled—ecosystems, infrastructure, workforce capacity, and policy frameworks—determining how these principles translate into real-world capability. The dual case-study arc evolves accordingly: QALIS extends from controlled transformation toward scalable integration across platforms and institutions, while CRQC–LLM traces how fragility at the primitive level propagates into systemic risk across infrastructure and governance layers.

Together, Chapters 1 through 3 establish the foundational triad of QAIS: physical principles, execution mechanisms, and deployment environments. This progression enables the subsequent development of architectural, reasoning, learning, and resilience frameworks explored in later chapters, including reasoning systems (Chapter 5), integrated architectures (Chapter 12), and forward-looking system synthesis (Chapter 13).

By concluding at this junction—where quantum processes begin to inform architectural design—Chapter 2 defines the execution layer as both the enabler of intelligent behavior and the boundary at which resilience must be enforced. With its mechanisms defined and its flow formalized, the groundwork is established for the systems, architectures, and societal contexts developed in the chapters that follow.

Quick Self-Check Questions

Operations and Scientific Framework of QAIS

1. How does QAIS redefine the concept of information?

A) Symbolic, abstract values manipulated logically
B) A probabilistic variable defined entirely by measurement statistics
C) A physical and context-dependent property of quantum systems
D) An encrypted message in communication channels

2. Which set of primitives forms the foundation of QAIS?

A) Logic gates, random variables, matrix multiplication
B) Superposition, entanglement, coherence, unitary evolution, and measurement
C) Vectors, tensors, eigenvalues, and Fourier transforms
D) Encryption, redundancy, and integrity checks

3. Why is QAIS considered a new discipline rather than a linear extension of QIS?

A) It only replaces Boolean logic with linear algebra
B) It adds symbolic processing to classical models
C) It redefines computation, representation, and trust as system-level properties
D) It relies entirely on faster quantum hardware

4. Which of the following describes the interdependent barriers to QAIS maturity?

A) Isolated engineering challenges
B) They reinforce one another, demanding coordinated solutions
C) They can be ignored if quantum speedup is achieved
D) Solved solely through hardware scaling

5. What are the five interdisciplinary domains of QAIS proficiency?

A) Boolean logic, probability theory, linear algebra, statistics, graph theory
B) Linear algebra, quantum mechanics, quantum information theory, quantum programming tools, systems-level thinking
C) Physics, chemistry, mathematics, economics, linguistics
D) Tensor algebra, classical mechanics, software design, biology, cybersecurity

6. How does QAIS embed trust and security differently than classical AI?

A) Using stronger encryption algorithms
B) Adding redundant integrity checks after computation
C) Integrating trust into the substrate via no-cloning, entanglement, and measurement disturbance
D) Outsourcing verification to classical oversight systems

7. How do QALIS and CRQC–LLM function as case-study anchors in Chapter 2?

A) QALIS shows cryptographic attacks; CRQC–LLM shows abstract theory
B) QALIS demonstrates design-aligned behavior; CRQC–LLM stress-tests resilience under adversarial pressure
C) QALIS models symbolic reasoning; CRQC–LLM benchmarks simulation environments
D) QALIS addresses hardware cost; CRQC–LLM addresses workforce shortages

8. How does the hybrid QAIS–AI architecture combine classical and quantum layers?

A) Fully replacing classical AI with quantum-only models
B) Interleaving classical preprocessing and optimization with quantum-native operations
C) Ignoring preprocessing and focusing on raw quantum states
D) Using classical algorithms to simulate quantum interference

9. What distinguishes QAIS representations from classical feature vectors?

A) Larger matrices for data storage
B) Correlations through physically embodied amplitudes, entanglement, and phase
C) Reliance on symbolic tokens and engineered attributes
D) Only improved speed without altering representation

10. How does Chapter 1 prepare for Chapter 2?

A) By concluding QAIS is mostly theoretical
B) By establishing qubits, gates, and circuits as the operational substrate of QAIS
C) By focusing on cryptography as the sole application
D) By summarizing hardware costs

11. Which milestone highlighted the need for quantum-native computation?

A) Grover's algorithm
B) Shor's algorithm
C) Feynman's quantum simulation proposal
D) Deutsch's universal model

12. Why are Shor's and Grover's algorithms pivotal?

A) They replaced classical computing
B) They demonstrated quantum advantage as an operational resource
C) They removed error correction needs
D) They introduced entanglement

13. What is the primary role of measurement in QAIS?

A) Passive observation
B) Data duplication
C) Boundary condition for extracting classical information
D) Error elimination

14. Why is the no-cloning theorem important in QAIS?

A) It speeds computation
B) It enforces limits on duplication and supports verification integrity
C) It enables redundancy
D) It reduces noise

15. What is entanglement's primary operational role?

A) Data compression
B) Error elimination
C) Establishing non-separable correlations across qubits
D) Increasing memory

16. What does contextuality imply in QAIS?

A) Deterministic outcomes
B) Independent properties
C) Measurement outcomes depend on operational context
D) Faster computation

17. How is computation best described in QAIS?

A) Logical rule execution
B) Geometric evolution of amplitudes and phases
C) Data sorting
D) Symbolic inference

18. What is the main goal of quantum error correction?

A) Increase speed
B) Reduce circuit depth
C) Preserve coherence without copying states
D) Eliminate measurement

19. Which metric evaluates similarity between quantum states?

A) Entropy
B) Fidelity
C) Trace distance
D) Mutual information

20. What does trace distance quantify?

A) Energy
B) Entanglement strength
C) Distinguishability between quantum states
D) Computation speed

21. What is the significance of entropy in QAIS?

A) Measures speed
B) Quantifies uncertainty and mixed-state structure
C) Determines hardware cost
D) Defines circuit depth

22. What is the role of the QAIS Operational Flow Model?

A) Hardware design
B) Algorithm optimization
C) Integrating primitives into a unified lifecycle
D) Error elimination

23. Why is coherence critical in QAIS?

A) It increases storage
B) It preserves phase relationships required for interference
C) It reduces cost
D) It simplifies circuits

24. What is the purpose of quantum teleportation as a protocol?

A) Copy data
B) Increase speed
C) Transfer state information via entanglement and classical communication
D) Store data

25. What is the key insight of Chapter 2 regarding capability and fragility?

A) They are unrelated
B) Fragility dominates capability
C) Capability eliminates fragility
D) Capability and fragility arise from the same physical primitives

26. Which milestone first proposed simulating quantum systems using quantum machines?

A) Grover's algorithm
B) Shor's algorithm
C) Feynman's quantum simulation proposal
D) Deutsch's universal model

Reader Guidance

For answers to these questions, see Appendix B for expanded reflection.

References

[1] R. P. Feynman, "Simulating physics with computers," *Int. J. Theor. Phys.*, vol. 21, no. 6/7, pp. 467–488, 1982.

[2] D. Deutsch, "Quantum theory, the Church–Turing principle and the universal quantum computer," *Proc. R. Soc. A*, vol. 400, no. 1818, pp. 97–117, 1985.

[3] P. W. Shor, "Algorithms for quantum computation: Discrete logarithms and factoring," in *Proc. 35th Annu. Symp. Found. Comput. Sci. (FOCS)*, 1994, pp. 124–134.

[4] L. K. Grover, "A fast quantum mechanical algorithm for database search," in *Proc. 28th Annu. ACM Symp. Theory Comput. (STOC)*, 1996, pp. 212–219.

[5] M. A. Nielsen and I. L. Chuang, *Quantum Computation and Quantum Information*. Cambridge, U.K.: Cambridge Univ. Press, 2000.

[6] A. S. Holevo, "Coding theorems for quantum channels," *Russ. Math. Surv.*, vol. 53, no. 6, pp. 1295–1331, 1998.

[7] J. Preskill, "Quantum computing in the NISQ era and beyond," *Quantum*, vol. 2, p. 79, 2018.

[8] K. Bharti, A. Cervera-Lierta, T. H. Kyaw, *et al.*, "Noisy intermediate-scale quantum algorithms," *Rev. Mod. Phys.*, vol. 94, no. 1, p. 015004, 2022.

[9] National Quantum Initiative Act, Public Law No. 115-368, U.S. Congress, Dec. 2018.

[10] European Commission, "Quantum Flagship: Pioneering European quantum technologies," 2018. [Online]. Available: https://qt.eu

[11] Ministry of Science and Technology of China, "China National Laboratory for Quantum Information Science," 2020.

[12] National Institute of Standards and Technology (NIST), "Post-Quantum Cryptography Standardization," 2022. [Online]. Available: https://csrc.nist.gov/projects/post-quantum-cryptography

[13] European Commission, "EuroQCI: Europe's quantum communication infrastructure," Press Release, Jan. 2025.

[14] P. Kiwit, R. Meyer, *et al.*, "QUARK: A modern framework for quantum computing application benchmarking," Linux Foundation Quantum Initiative, 2024. [Online]. Available: https://www.linkedin.com/pulse/quark-modern-framework-quantum-computing-application-m1upf

Chapter 3

—

Evolution, Ecosystem, and Workforce of QAIS

Architectural Position

Defines the ecosystem, infrastructure, and human and institutional systems enabling QAIS development and deployment. Within the Quantum Stack, infrastructure constitutes the Layer 3 substrate that enables controlled execution of quantum systems. This layer provides the hardware, control systems, and integration environments upon which higher-level representation, learning, and reasoning depend. It also establishes the interface through which external signals may be introduced into the system, a role further developed in Chapter 11 through quantum sensing.

Core Principle

QAIS advancement depends on coordinated evolution across technology, infrastructure, and human capability.

Abstract

This chapter examines the ecosystem-level maturation of QAIS within Layer 3 of the Quantum Stack, framing ecosystem realization as a **distributed control problem** governed by representation, propagation, and stability constraints. It traces the progression from foundational theoretical advances and algorithmic formalization to experimental platforms, NISQ-era engineering, infrastructure development, policy coordination, sectoral adoption, barrier identification, workforce formation, and hybrid quantum–AI convergence.

Through this lens, ecosystem scaling is understood as the coordinated evolution of representation across technological and organizational layers, with propagation emerging through cross-domain interaction and deployment workflows. Stability is enforced as bounded propagation, where alignment across calibration practices, benchmarking standards, governance structures, and workforce capabilities determines whether systems achieve reliable, reproducible operation.

Using the constructive lens (QALIS) and the resilience-oriented lens (CRQC–LLM), the chapter evaluates how these conditions collectively govern the transition from experimental capability to operational maturity. The analysis demonstrates that durable QAIS deployment depends on synchronized systemization across physical infrastructure, ecosystem coordination, and institutional readiness, establishing the foundation for the representational and encoding architectures developed in the subsequent chapter.

Keywords

QAIS; Quantum Stack; ecosystem realization; historical evolution; NISQ era; distributed constraint propagation; infrastructure architecture; policy coordination; standards alignment; barriers to maturity; workforce architecture; hybrid quantum–AI systems; bounded propagation

3.1 Introduction — Ecosystem and Workforce Constraints

Architectural Role: *Establishes the scaling conditions and system feasibility constraints governing how quantum resources expand under noise, hardware limits, and architectural design.*

Architectural Spine*: Hardware Realization → Control Infrastructure → System Integration → Distributed Coordination → Institutional Scaling → Workforce Architecture*

Section Framing: *Establishes the doctrine of systemization, demonstrating that scale must follow structure.*

Quantum Stack Layer: L3 *— Quantum Technology → Realizing physical qubits, control systems, and execution infrastructure.*

Chapters 1 and 2 established the foundational layers of the Quantum Stack: the physical constraints that define quantum possibility and the operational grammar governing controlled evolution and measurement. Chapter 3 advances within Quantum Stack Layer 3 from technological realization toward ecosystem-scale systemization, where these constraints are realized across infrastructures, institutions, and human capability. If earlier chapters defined what is possible, this chapter defines how possibility becomes scalable system capability.

Within the Quantum Stack, this layer marks the transition from constrained realization to distributed maturation. Quantum systems must evolve from experimental demonstrations to reproducible platforms, from platforms to shared infrastructure, and from infrastructure to coordinated adoption across domains. The central concern is systemization—the disciplined alignment of engineering practice, institutional coordination, policy frameworks, communication systems, and workforce development such that scaling preserves structural integrity.

Systemization is not the presence of quantum devices alone, but the synchronization of fabrication, calibration, verification, benchmarking, and skills formation across distributed environments. When synchronization precedes expansion, scaling reinforces stability; when expansion outpaces synchronization, fragility propagates across institutional boundaries. This establishes the governing principle of the chapter: systemization must precede scale.

The chapter traces this maturation through successive phases of quantum advancement, from early theoretical foundations to algorithmic development, experimental realization, infrastructure coordination, and cross-sector adoption. Each phase introduces new capabilities while exposing new constraints requiring stabilization.

Fabrication reveals reproducibility limits, integration exposes verification requirements, distributed access introduces infrastructure dependencies, and workforce expansion demands interdisciplinary fluency. Growth in quantum systems is therefore inseparable from constraint management at scale.

Within the Quantum Stack, Chapter 3 operates at Layer 3 — Quantum Technology, extending its concern with physical realization into ecosystem-level systemization, infrastructure coordination, institutional readiness, and workforce development. Representation is instantiated through interconnected technological and institutional structures; propagation emerges as system behavior across distributed networks; and stability is enforced as bounded propagation through interoperability, governance coordination, infrastructure reliability, and workforce readiness.

At this layer, constraint propagation becomes a distributed systems property, extending beyond device physics into engineering systems, institutional mechanisms, and human-capital structures. The viability of QAIS at scale depends on maintaining these constraints consistently across interconnected domains, rather than optimizing isolated components.

Chapter 3 frames ecosystem evolution as an architectural process and establishes the conditions under which quantum technologies transition from experimental realization to durable, scalable systems. The sections that follow examine historical development, infrastructure and communication primitives, adoption dynamics, systemic barriers, workforce architecture, and the convergence of quantum and artificial intelligence within large-scale intelligent systems.

GUIDING PRINCIPLE 3 — SYSTEMIZATION PRECEDES SCALE

Scale without systemization magnifies instability. In QAIS, coherent expansion requires that constraint propagation, calibration discipline, verification governance, cross-layer integration, and workforce maturity consolidate before throughput, capital intensity, or automation intensify. Ecosystem acceleration that outpaces architectural discipline converts localized engineering variability into systemic risk.

Section 3.1 Summary — Ecosystem and Workforce Constraints

This principle frames the chapter's evaluative contract. The historical, institutional, industrial, and workforce dimensions examined herein are not independent narratives; they are structural variables that either preserve or destabilize constraint-aware execution. The ecosystem must therefore be analyzed through the same discipline applied to physical and operational layers: capability and limitation co-evolve, and stability precedes scale.

3.2 Case Study Orientation and Alignment

***Architectural Framing**: Hardware Realization*

***Section Framing:** This section situates real QAIS deployments within the hardware-to-infrastructure transition, grounding architectural propagation in operational practice. It extends QALIS and CRQC–LLM from internal analysis tools to ecosystem-level evaluative frameworks, ensuring that system evolution and adoption are assessed under dual constraints: constructive capability on one axis and resilience under stress on the other.*

Alignment is established as a control surface governing how QAIS systems are interpreted, evaluated, and compared across heterogeneous environments. Representation is anchored through system-level positioning of QALIS and CRQC–LLM as interpretive reference frames, while propagation is treated as observable behavior under cross-model interaction, benchmarking, and constraint validation. Stability is enforced through bounded propagation: alignment conditions, evaluation criteria, and consistency checks preserve comparability as implementations vary by hardware, tooling, and operating assumptions. In this way, alignment functions as the coordinating mechanism that maintains coherence across QAIS architectures and real-world instantiations.

Chapter 3 is the first point in the book where the QALIS constructive lens and the CRQC–LLM stress-test lens shift their focus from quantum mechanisms to the **ecosystem conditions that determine whether QAIS can mature beyond isolated prototypes**.

In this chapter, the two lenses operate not on qubits, gates, or encoding schemes, but on the **infrastructures, institutions, and workforce structures** that enable—or constrain—the growth of intelligent quantum systems.

Through the QALIS lens, the chapter interprets ecosystem evolution as a constructive pathway for QAIS scalability. Workforce development pipelines, cross-sector coordination, standards formation, and institutional readiness appear as the enabling forces that support stable operations, coherent system integration, and long-term architectural growth. This perspective highlights how collaborative research networks, talent development initiatives, and aligned policy frameworks form the socio-technical foundation upon which practical QAIS deployments depend.

Through the CRQC–LLM lens, the same ecosystem reveals points of fragility and exposure. Gaps in workforce depth, uneven national and industrial readiness, dependency bottlenecks in quantum supply chains, and governance asymmetries are treated as systemic stress points. These vulnerabilities mirror the adversarial pressures that CRQC–LLM uncovers at the physical and operational layers, but here they surface at the institutional and infrastructural levels, where weaknesses can propagate across entire QAIS deployments.

Section 3.2 Summary — Case Study Orientation and Alignment

Together, these lenses frame Chapter 3 as an examination of the conditions under which QAIS can achieve stability, trustworthiness, and operational maturity. They clarify how ecosystem strength amplifies the constructive trajectory of QAIS, and how ecosystem weakness magnifies the stress-test vulnerabilities explored throughout the book. With this interpretive structure in place, the chapter now turns to the evolution of the quantum information ecosystem and the forces shaping QAIS readiness across sectors.

Throughout this chapter, historical developments and ecosystem dynamics are presented as contextual background only; all analysis remains grounded in the operational QAIS formulation established in Chapter 2, where preparation, transformation, interference, and measurement function as executable system primitives rather than explanatory postulates.

3.3 The Evolution of Quantum AI Systems

Architectural Framing *— Control Infrastructure. Traces system maturation, showing how propagation capabilities evolve across technological stages.*

Section Framing: *This section establishes evolution as a temporal control surface through which QAIS representation, propagation, and stability progressively mature. Representation evolves from foundational abstractions to structured operational models, while propagation emerges as system behavior shaped by increasing algorithmic sophistication and experimental realization.*

Stability is enforced through bounded propagation as advances in error correction, validation, and infrastructure progressively extend operational limits. Evolution therefore functions as a structured transformation process that converts theoretical constructs into scalable QAIS systems.

QAIS marks the transition from QIS as precursor science to a full-fledged discipline that treats information as physically embodied and systemically operational.

QIS established the rules, no-cloning, entanglement, coherence, that revealed possibilities beyond classical limits. These historical interpretations reflect the conceptual language of their time; within QAIS, the same phenomena are treated operationally as system constraints and execution rules, not as ontological explanations. QAIS reframes those rules as design primitives for computation, communication, and sensing. The evolution of QAIS is therefore dual: a story of scientific milestones and a story of System Relevance. Each decade brought not only theoretical or experimental advances but also architectural implications for how intelligent quantum systems could be built, deployed, and secured. Throughout this section, QALIS (Design Framework) and CRQC-LLM (Stress-Test Lens) illustrate how historical milestones translate into operational principles.

3.3.1 1980s — Foundations and First Principles

In the 1980s, foundational ideas reframed the possibilities of computation and communication. Richard Feynman argued that classical machines could not efficiently simulate quantum systems, opening the vision of quantum-native computation [1]. David Deutsch extended this with the proposal of a universal quantum computer, providing the blueprint for general-purpose architectures [2]. At the same time, Charles Bennett and Gilles Brassard introduced quantum key distribution (QKD), the first protocol to demonstrate that physics itself could enforce secure communication [3].

System Relevance

These first principles became the earliest design rules of QAIS. The Feynman–Deutsch lineage underpins hybrid inference in QALIS, where quantum modules tackle problem classes unreachable by classical simulation.

The Bennett–Brassard protocol echoes in CRQC–LLM, where measurement disturbance is reframed not only as a vulnerability but also as a diagnostic mechanism for detecting adversarial probing. Together, these insights established the precedent that physics itself, not just abstract computation, would define the grammar of future intelligent systems.

3.3.2 1990s – Algorithmic Breakthroughs and Theoretical Formalization

The 1990s marked a decisive phase in the evolution of QAIS. This decade demonstrated that quantum advantage was not merely theoretical speculation but could be formalized in both algorithms and information-theoretic structures. Two algorithmic breakthroughs, Shor's quantum factoring algorithm and Grover's quantum search algorithm, proved that quantum systems could outperform classical ones in domains critical to computation and security [4][5]. At the same time, advances in quantum error correction and quantum Shannon theory formalized how fragile quantum states could be stabilized and how information capacities could be quantified, reframing skepticism into solvable design challenges.

System Relevance

Algorithmic breakthroughs of the 1990s provided QAIS with its first proof of scalable capability. In QALIS, Shor's algorithm illustrates how cryptography-aware inference pipelines can integrate quantum advantage, while Grover's search demonstrates how unstructured data can be accelerated into usable features for distributed sensing.

In CRQC–LLM, advances in quantum error correction and quantum Shannon theory expose resilience thresholds: coherence lifetimes, redundancy schemes, and channel capacities become measurable indicators of when adversarial exploitation could destabilize fragile resources.

These developments reframed skepticism about quantum information into solvable design challenges, anchoring QAIS as a discipline of operational rules rather than speculation.

QAIS as a discipline is rooted in a sequence of foundational algorithmic and theoretical developments that reshaped expectations for computation and information processing. These advances established the conceptual basis for subsequent system architectures and applications. Figure 3.3.1 situates these developments within a historical framework, providing context for how early breakthroughs inform the structure of modern QAIS systems.

Algorithms and Formalization as Dual Milestones

The figure that follows highlights two strands of progress in the 1990s: algorithmic breakthroughs, which revealed exponential or quadratic speedups, and theoretical formalization, which showed how coherence could be stabilized and communication bounded. Together, they demonstrate how QAIS evolved from conceptual promise into a discipline with both offensive and defensive design tools.

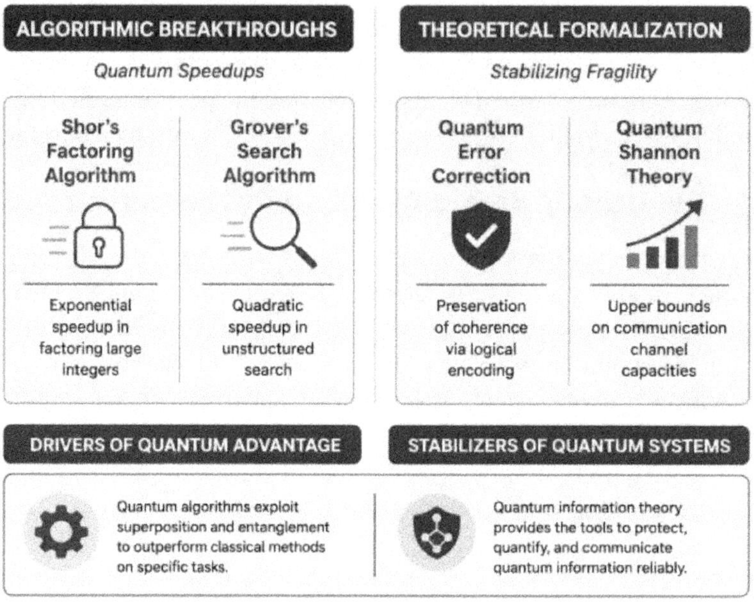

Figure 3.3.1 — 1990s Algorithmic and Theoretical Advances in QAIS.

This figure contrasts two strands of progress in the 1990s. On the left, algorithmic breakthroughs such as Shor's factoring algorithm and Grover's search algorithm demonstrated exponential and quadratic speedups, redefining the boundaries of computation and security. On the right, theoretical formalization through quantum error correction codes and quantum Shannon capacity theorems showed how fragile quantum states could be stabilized and how communication limits could be rigorously quantified. Together, these advances transformed quantum information from speculative potential into the operational foundations of QAIS.

This visual makes clear that the 1990s were not only about demonstrating advantage but also about building the theoretical guardrails that made scaling conceivable. The following technical spotlight unpacks how these breakthroughs became the dual foundations of QAIS design.

Technical Spotlight — Algorithms and Formalization as QAIS Foundations

Conceptual Grounding

The 1990s marked the shift from thought experiments into rigorous formalisms that could anchor QAIS. Shor's factoring algorithm and Grover's search algorithm did more than prove speedups: they demonstrated that quantum resources redefine complexity classes, reshaping the landscape of tractable and intractable problems. These breakthroughs established that superposition and interference are not abstract curiosities but operational levers for optimization and search.

Alongside these advances, the field of quantum information theory, formalized through measures like Holevo's bound[22] and entropic uncertainty relations, provided the mathematical grammar to quantify information capacity, security, and channel limits [8]. Together, these contributions grounded QAIS in a dual trajectory: algorithms revealed capability expansion, while information-theoretic formalism clarified the rules and constraints under which that expansion must occur.

System Relevance

For QAIS, the 1990s advances remain foundational. Algorithms such as Shor's and Grover's continue to define benchmark problems against which hybrid pipelines are tested, while information-theoretic constructs guide how entanglement, fidelity, and channel capacity are operationalized in system design. The interplay between algorithmic capability and formal constraints became the architectural baseline for QAIS, ensuring that every deployment is both performance-driven and law-constrained.

Cross-Disciplinary Orientation

Physicists interpret Shor and Grover as demonstrations of interference and entanglement turned into computational resources. Computer scientists formalize these algorithms within complexity theory, framing quantum advantage in contrast to classical bounds. Mathematicians extend entropy and channel capacity as quantitative tools for QAIS reliability.

Engineers, looking forward, recognize these dual milestones as the first step toward pipelines where optimization and information-theoretic bounds are inseparable.

---

[22] Holevo's work codified information-theoretic constraints that continue to define QAIS verification, linking entropy, communication capacity, and measurement disturbance.

Case Study Anchor — Algorithms and Formalization as Ecosystem Catalysts

Intent: To frame how early quantum algorithms and information-theoretic formalization transformed quantum theory into a viable ecosystem foundation for QAIS, enabling coordinated research, benchmarking, and long-term infrastructure planning. This anchor establishes how algorithmic milestones seeded institutional and industrial engagement.

- **QALIS (Design Framework):** Demonstrates how hybrid optimization pipelines still use Grover-inspired search structures embedded into distributed sensing arrays.
- **CRQC-LLM (Stress-Test Lens):** Highlights how adversaries model decryption strategies by exploiting Shor's factoring potential, stressing the resilience of QAIS architectures against cryptographic collapse.

This spotlight prepares the ground for Section 3.3.3, where experimental proof-of-concepts begin to translate these theoretical milestones into hardware demonstrations. It shows that QAIS is not only built on physical primitives but equally on algorithmic and theoretical foundations that direct how those primitives will be applied.

3.3.3 Formalism and Experimental Proof-of-Concepts

The new millennium brought increasing formalization and a global race to build quantum hardware. Academic groups and government labs demonstrated teleportation over fiber, basic quantum gates, and small quantum memories. Theoretical advances refined the role of entropy, distinguishability, and mutual information, while hybrid models, entanglement distillation, and the concept of quantum repeaters laid the groundwork for scalable quantum networks.

The figure that follows highlights quantum repeaters as a milestone in this decade. Repeaters address one of quantum communication's central challenges: quantum signals cannot be amplified without violating the no-cloning theorem. They overcome this by segmenting channels, extending entanglement through swapping, and integrating purification and quantum memories to sustain coherence across distance.

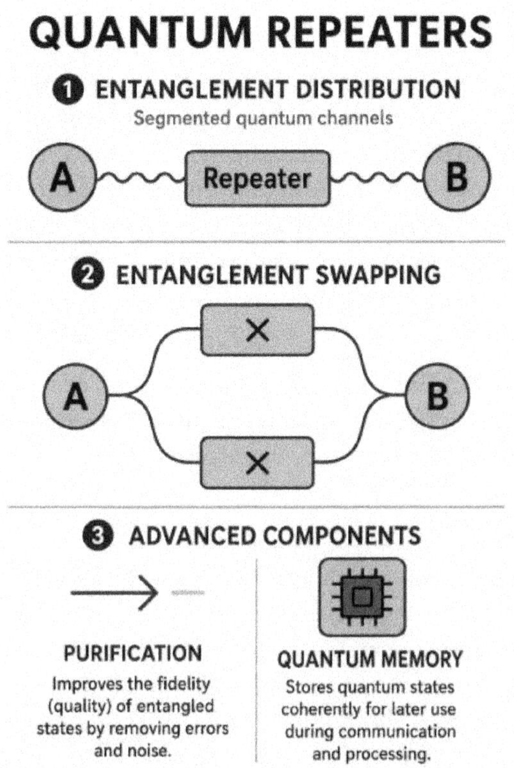

Figure 3.3.2 — Quantum Repeaters as a Milestone in QAIS Evolution.

This schematic shows how quantum repeaters segment long communication channels into shorter links, generate entanglement locally, and extend it through entanglement swapping. Advanced repeater designs integrate purification and quantum memory to preserve fidelity over thousands of kilometers. Within QAIS, repeaters shift from a theoretical construct to an engineering milestone, enabling distributed sensing pipelines and secure communication backbones.

The evolution of QAIS is closely tied to advances in communication infrastructure capable of supporting quantum information transfer across distance. These developments introduced mechanisms for extending system reach beyond localized computation.

Figure 3.3.2 situates this progression within the broader trajectory of QAIS evolution, highlighting the role of communication milestones in enabling distributed system architectures.

Technical Spotlight — Quantum Repeaters and QAIS Infrastructure

Conceptual Grounding

Quantum repeaters represent one of the first architectural responses to the central fragility of quantum communication: the exponential decay of entanglement fidelity across distance. While entangled photons degrade quickly in optical fibers or free-space links, repeaters divide the channel into shorter segments, creating entanglement locally and then extending it through entanglement swapping and purification. This design transforms distance from an absolute barrier into a modular engineering challenge. In doing so, repeaters illustrate how QAIS reframes fragility into a design rule, coherence loss is no longer a terminal obstacle but a parameter to be managed, extended, and scaled through architecture.

Historically, this shift paralleled the transition from experimental demonstrations of entanglement to the conceptual blueprint of a "quantum internet," showing that distributed intelligence would rest on infrastructures as much as on algorithms.

System Relevance

For QAIS, repeaters are not simply communication tools but infrastructural primitives. They provide the backbone for distributed sensing, secure channels, and multi-node computation, ensuring that fragile coherence can span beyond laboratory distances. By embedding entanglement swapping and purification into system pipelines, QAIS architectures gain scalability, trust, and resilience. The ability to extend coherence across networks directly shapes the feasibility of distributed AI inference, synchronization, and secure hybrid pipelines.

Cross-Disciplinary Orientation

Physicists treat repeaters as validation of entanglement as a resource, focusing on purification protocols and swapping fidelity. Engineers frame them as modular network elements, comparable to routers in classical systems but governed by quantum coherence limits. AI specialists envision repeaters as the communication scaffolds enabling distributed kernel embeddings and federated quantum learning.

Cybersecurity experts interpret them as critical control points, where adversarial manipulation of purification or swapping routines could compromise system trust.

Case Study Anchor — Quantum Repeaters as Infrastructure Enablers

Intent: To illustrate how quantum repeaters shifted QAIS from isolated laboratory systems to network-capable infrastructures, making ecosystem-scale coordination feasible. The case contrasts constructive deployment pathways with fragility under adversarial or reliability stress.

- **QALIS (Design Framework):** Uses repeater-based coherence extension to synchronize distributed sensing nodes, showing how fragile physics becomes robust infrastructure.
- **CRQC–LLM (Stress-Test Lens***):* Models how adversarial actors could attempt to exploit repeater fragility, targeting purification failures or swapping vulnerabilities, to disrupt long-distance coherence and compromise distributed QAIS pipelines

This spotlight builds on the algorithmic and formal foundations of Section 3.3.2and points toward Section 3.3.4, where experimental prototypes of NISQ devices show how these infrastructural ideas moved from theoretical constructs into early hardware demonstrations. It positions repeaters as both a milestone in communication design and a preview of the system-level infrastructures needed for QAIS deployment.

3.3.4 2010s–2020s – Engineering, Ecosystem, and Emergence

The 2010s–2020s marked a decisive transition in QAIS from experimental promise to ecosystem-level engineering. Global initiatives reframed QAIS as a strategic priority, with government-backed programs (U.S. NQI, EU Quantum Flagship, China's national strategy)[23] underwriting infrastructure, talent pipelines, and roadmaps [14]. Technology companies such as IBM, Google, Rigetti, and IonQ accelerated accessibility by delivering cloud-based quantum processors, enabling real-time experimentation for a broader community. At the same time, new programming languages and simulation frameworks opened participation to AI and systems researchers, embedding QAIS into wider technological practice.

This era also introduced the **Noisy Intermediate-Scale Quantum (NISQ)**[24] paradigm, where 50–100 qubit devices, while not yet fault-tolerant, began demonstrating hybrid quantum-classical workflows [6].

Variational quantum eigensolvers (VQEs), quantum kernel methods, and noise-aware error mitigation strategies became operational templates [7][19]. These developments crystallized QAIS as an applied systems discipline: architectures were no longer speculative, but deployable prototypes tested against the limits of scalability and resilience.

---

[23] The EU Flagship exemplifies how coordinated programs translate research into infrastructure, emphasizing sovereignty and cross-border collaboration in QAIS ecosystems.

[24] NISQ devices typically contain 50–100 noisy qubits and operate without full error correction.

Current QAIS implementations operate within technological constraints that shape how systems are designed and deployed, particularly in the presence of noise and limited resources. These conditions define the practical boundaries of system capability in near-term environments. Figure 3.3.3 situates these architectural realities within a system-level context, clarifying how contemporary designs reflect underlying technological limitations.

NISQ as Prototype Ecosystem

The figure that follows captures how NISQ-era devices reframed QAIS. It contrasts classical-only workflows with hybrid quantum-classical pipelines, showing how coherence, entanglement depth, and error mitigation became explicit design variables.

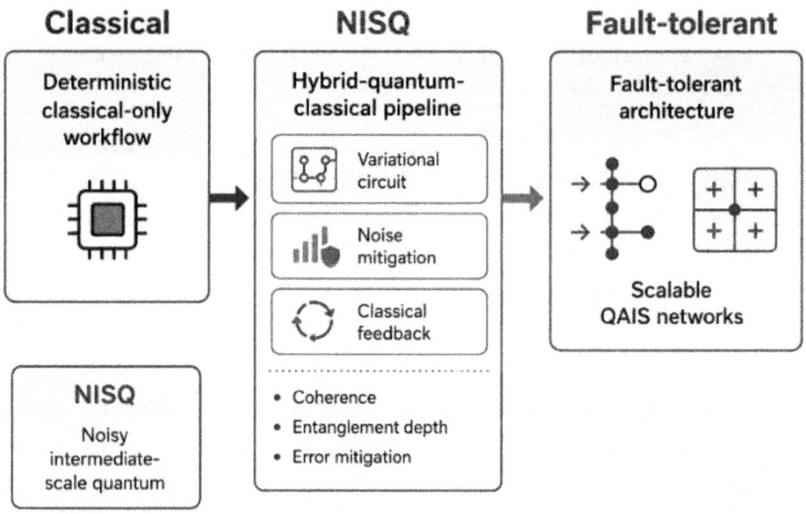

Figure 3.3.3 — NISQ Architectures in QAIS.

This comparative schematic contrasts three tiers of computation. On the left, classical-only pipelines are shown as deterministic loops constrained by simulation limits. In the center, NISQ devices appear as hybrid systems, where variational circuits, noise mitigation, and classical feedback loops stabilize useful outputs. On the right, the trajectory toward fault-tolerant architectures is depicted, highlighting the progression from prototype to scalable QAIS networks.

This figure highlights how QAIS reframed limited hardware into system templates. The following technical spotlight interprets NISQ architectures as a proving ground, showing how they operationalize coherence, entanglement, and error mitigation while preparing the transition to scalable quantum AI systems.

Technical Spotlight — NISQ as Transitional Infrastructure

Conceptual Grounding

The *Noisy Intermediate-Scale Quantum (NISQ)* era represents the transitional phase between theoretical quantum algorithms and fully fault-tolerant systems. Characterized by devices with tens to hundreds of qubits, these architectures are powerful enough to explore hybrid algorithms but remain limited by decoherence, gate errors, and restricted connectivity.

QAIS reframes NISQ devices not as failed prototypes but as training grounds where noise becomes an experimental parameter, error mitigation is a design practice, and hybrid quantum–classical workflows define the trajectory toward scalable systems.

From a training perspective, NISQ introduces learners and practitioners to the realities of constrained resources: shallow circuit depth, noise-aware variational methods, and probabilistic benchmarking. In this sense, NISQ is less about immediate quantum advantage and more about preparing system designers for the constraints, trade-offs, and co-design challenges of quantum-AI integration.

System Relevance

For QAIS, NISQ devices anchor today's deployments. They provide the substrate on which variational quantum eigensolvers (VQE), quantum approximate optimization algorithms (QAOA), and hybrid classifiers are actively tested. While these platforms cannot yet scale to fault tolerance, they demonstrate QAIS as an applied system science, where optimization, learning, and simulation pipelines are already being piloted. NISQ systems thus serve as a proving ground for both technical feasibility and workforce readiness. They allow algorithms, middleware, and error-mitigation strategies to mature in tandem with hardware, ensuring that when fault tolerance arrives, the ecosystem is prepared.

Cross-Disciplinary Orientation

Physicists study noise sources and coherence lifetimes, developing material science and control strategies to stabilize qubits. Engineers design scalable compilers, calibration routines, and middleware for devices with limited connectivity. AI specialists exploit variational methods and hybrid pipelines to make noisy outputs usable for learning and optimization.

Cybersecurity experts interpret NISQ fragility as both a limitation and a stress surface, modeling how adversarial noise injection or calibration tampering could destabilize pipelines.

Each discipline treats NISQ not as a stopgap but as a laboratory for shaping the next era of system architectures.

Case Study Anchor — NISQ Platforms as Transitional Ecosystem Assets

Intent: To examine NISQ-era systems as ecosystem accelerators that enabled workforce training, software tooling, and early adoption—while simultaneously exposing architectural fragility and overclaim risks.

- **QALIS (Design Framework):** Uses NISQ-based hybrid solvers in distributed sensing pipelines, showing how variational circuits can yield actionable insights despite noise.
- **CRQC-LLM (Stress-Test Lens):** Models adversarial exploitation of noise in NISQ-era devices, where calibration errors or noise-injection attacks could undermine trust in hybrid inference systems.

This spotlight links the infrastructural milestone of quantum repeaters in Section 3.3.3 with the policy and ecosystem maturation of Section 3.3.5.

It shows that QAIS moved from fragile proofs of concept (repeaters) into noisy but operational devices (NISQ), which then catalyzed government and industrial investment. NISQ therefore bridges the experimental past with the policy-driven scaling strategies of the present.

3.3.5 Policy & Ecosystem Maturation of QAIS

By the late 2010s and into the 2020s, QAIS evolution was no longer driven solely by scientific milestones but by coordinated policy ecosystems. Governments launched national initiatives, published roadmaps, and incentivized industrial partnerships. The U.S. National Quantum Initiative (NQI), the EU Quantum Flagship, and China's national strategy exemplified this shift. At the European level, the EuroQCI [15] program was especially influential, laying the groundwork for continent-scale secure communication backbones and integrating quantum key distribution (QKD) into strategic infrastructure [9]. These frameworks seeded talent, built fabrication and testbed capacity, and created commercialization pathways that transformed isolated experiments into deployable systems.

The development of QAIS extends beyond technical innovation to include policy frameworks, institutional structures, and ecosystem dynamics that influence adoption and direction.

These factors shape how systems are funded, regulated, and deployed across sectors. Figure 3.3.4 situates these influences within a broader ecosystem perspective, illustrating how non-technical drivers contribute to QAIS evolution. The figure shows how national strategies, governance frameworks, and international partnerships provided the scaffolding that turned fragile prototypes into coordinated QAIS ecosystems.

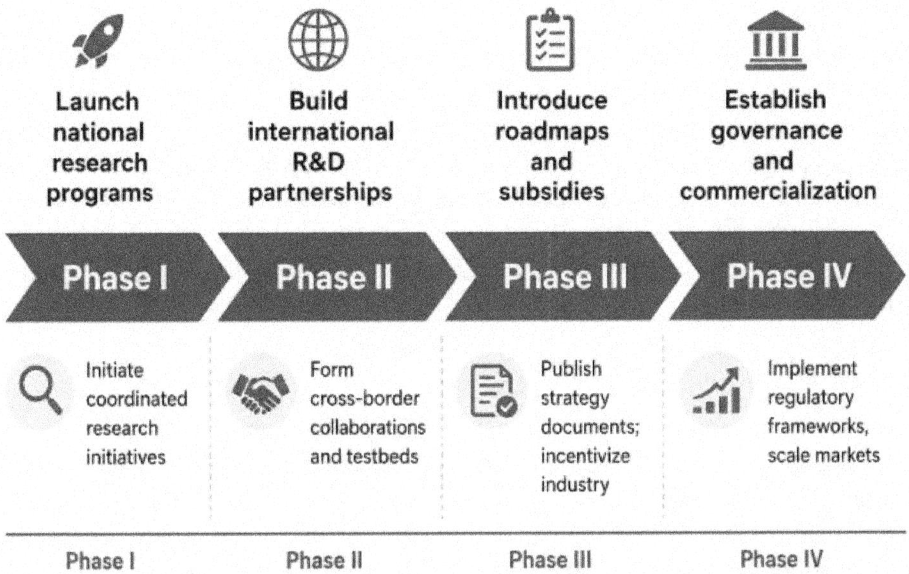

Figure 3.3.4 — Policy and Ecosystem Evolution in QAIS.

This timeline shows how coordinated governance enabled QAIS maturation. Phase I launched national research programs; Phase II built international R&D partnerships; Phase III introduced roadmaps and targeted subsidies; and Phase IV brought strategic governance and commercialization frameworks. These arcs shaped the conditions for deployable QAIS architectures such as QALIS and resilience scenarios like CRQC–LLM.

Technical Spotlight — Policy as Infrastructure for QAIS

Conceptual Grounding

Policy is not an afterthought in the trajectory of QAIS; it is the scaffolding that determines whether laboratory advances mature into sustainable ecosystems.

In the 2000s, QIS existed largely as isolated academic programs and prototype devices. By the late 2010s and early 2020s, national governments recognized that quantum technologies had strategic importance for cybersecurity, economic competitiveness, and scientific leadership.

Programs such as the U.S. National Quantum Initiative, the EU Quantum Flagship, and China's National Laboratory for Quantum AI Systems reframed quantum research as infrastructure-level investment.

From a professional development standpoint, this marks a transition point: learners must understand that QAIS is not only about algorithms and devices, but also about the policy frameworks, funding ecosystems, and strategic agendas that shape which ideas scale, which tools become interoperable, and how talent pipelines are sustained globally.

System Relevance

Policy frameworks translate scientific progress into durable systems by aligning investment, standards, and ecosystem incentives. Funding allocations build national research labs and shared testbeds, while regulatory guidelines establish cryptographic standards and interoperability protocols. For QAIS architectures, policy determines deployment boundaries: which algorithms are prioritized, which infrastructures are open-access, and which domains, finance, health, defense, receive early integration. Without policy, technical gains remain fragmented and siloed; with policy, they become system-building blocks in a globally competitive race.

Cross-Disciplinary Orientation

Physicists depend on policy-driven research programs to finance long-term experimental platforms. Engineers rely on standards and procurement frameworks to ensure interoperability across devices and networks. AI specialists benefit from cross-sector collaborations seeded by government programs, which allow algorithms to migrate from laboratories into industrial and societal contexts.

Cybersecurity experts are directly shaped by national standards for cryptographic resilience, which must anticipate threats from cryptographically relevant quantum computers (CRQCs) while incorporating AI-assisted defense mechanisms. In this way, policy orchestrates the convergence of multiple disciplines into a unified QAIS ecosystem.

Case Study Anchor — Policy as a Structural Accelerator for QAIS

Intent: To analyze how policy frameworks function as non-technical infrastructure, shaping funding, standards, and international coordination that determine QAIS maturity beyond laboratory success.

- **QALIS (Design Framework)** demonstrates how government-backed testbeds integrate QKD into national communication networks, showcasing policy as a driver of deployable architectures.
- **CRQC–LLM (Stress-Test Lens)** highlights how adversaries may exploit uneven international standards, probing whether policy gaps could leave infrastructures vulnerable to CRQC-guided decryption or LLM-assisted attacks.

This spotlight connects the technical milestones of Sections 3.3.3–3.3.4 with the adoption dynamics introduced in Section 3.5. It shows that without the policy and ecosystem maturation of the 2010s–2020s, NISQ-era experiments could not have transitioned into multi-sector adoption. Policy provides the enabling infrastructure that turns scientific breakthroughs into societal assets, preparing the ground for QAIS to evolve as both a technological discipline and a strategic instrument.

The progression from isolated quantum experiments to globally coordinated QAIS did not occur organically—it unfolded through a sequence of deliberate policy phases that shaped research priorities, infrastructure investment, and international collaboration. As national programs matured, policy transformed from a background influence into a primary architectural driver, determining which scientific ideas scaled, which infrastructures became interoperable, and how talent pipelines formed across regions.

The following table synthesizes this policy trajectory into four distinct phases, showing how early foundational research funding in the 1980s evolved into large-scale governance, commercialization frameworks, and global standard-setting by the 2010s–2020s.

Each phase aligns a strategic policy focus with its ecosystem outcomes, revealing how governance, investment, and coordination became the enabling scaffolding for QAIS to emerge as a deployable system discipline.

Table 3.3.1 — Policy Phases and Strategic Outcomes in Quantum AI Systems.

This table distills the policy–ecosystem trajectory into four major phases, aligning strategy with outcomes that progressively enabled QAIS.

Phase & Decade	Policy Focus	Ecosystem Outcomes
Phase I – Foundational Research Policy (1980s)	• Launch of first national QAIS research programs • Establishment of early research frameworks	• Seed funding de-risked basic research • Initial talent pipelines formed
Phase II – International R&D Partnerships (1990s)	• Cooperative research agreements extend across borders • Public-private partnerships finance prototype development	• Shared standards emerge • Early prototypes validated in multiple labs
Phase III – Formal Roadmaps & Targeted Subsidies (2000s)	• Governments publish national quantum strategies and technology roadmaps • Subsidies target critical infrastructure (testbeds, fabrication)	• Cohesive national programs align academia and industry • Supply-chain capabilities begin to scale
Phase IV – Strategic Governance & Commercial Incentives (2010s–2020s)	• Large-scale national initiatives (e.g., U.S. NQI, EU Quantum Flagship) • Regulatory guidance and incentive packages encourage private investment	• Cross-border collaborations accelerate hardware roll-out • Commercialization frameworks open global markets

This four-phase trajectory underscores the increasing complexity and intentionality behind quantum policy as QIS evolved into QAIS. Early phases emphasized enabling discovery and talent, while later phases introduced governance, roadmaps, and commercialization frameworks that scaffolded QAIS into a deployable and secure discipline.

Section 3.3 Summary — The Evolution of Quantum AI Systems

This section traced the emergence of QAIS from early theoretical insights to an integrated system discipline shaped by infrastructure, policy, and research ecosystems. The progression from Feynman's simulation proposal to national and multinational quantum initiatives revealed how coherent experimentation, algorithmic milestones, and institutional coordination transformed quantum information from a scientific curiosity into a strategic technological capability.

This section emphasized that QAIS did not appear abruptly; it accumulated through decades of advances in computation, communication, materials, workforce development, and governance frameworks that enabled coherence-sensitive systems to scale beyond laboratory constraints.

QALIS illustrated the constructive side of this maturation, showing how early primitives—superposition, entanglement, and measurement—evolved into pathways for distributed sensing, learning, and decision alignment across heterogeneous environments. CRQC-LLM served as the counterpoint, demonstrating how the same historical trajectory exposed increasingly complex verification boundaries, cryptographic transitions, and adversarial risks. Together, these perspectives framed QAIS as a discipline whose evolution is inseparable from the institutions, infrastructures, and policies that support it.

With this evolutionary foundation established, the chapter now turns to Section 3.4, where communication, synchronization, and measurement are examined as the operational mechanisms that link physical constraints to systemic trust across distributed QAIS environments.

3.4 Quantum Communication and Measurement

Architectural Framing: *System Integration. Defines information exchange as propagation coordination, enabling distributed system behavior.*

Section Framing: *Section 3.4 defines representation as the foundational layer of QAIS, where information is encoded into structured forms for propagation, transformation, and evaluation. It converts external data into internal states governed by physical, mathematical, or hybrid constraints. In quantum systems, this corresponds to preparing states in Hilbert space, where amplitudes, phases, and entanglement encode structure; in classical systems, it involves feature vectors or symbolic encodings.*

Representation is static with respect to system dynamics: it establishes the state on which operations act but does not define its evolution.

This section establishes communication and measurement as dual control surfaces governing information exchange and extraction in QAIS. Representation is distributed across channels and interfaces, while propagation emerges through transmission, transformation, and observation.

Stability is enforced through bounded propagation, where noise, loss, and disturbance constrain reliability. Communication and measurement thus define how quantum information moves and becomes accessible.

In QAIS, these processes evolve from QIS constraints into design principles. The no-cloning theorem, measurement disturbance, and entanglement are reframed as architectural primitives: coherence lifetimes define reliability, measurement collapse bridges quantum and classical domains, and disturbance enables verification. Together, they govern how information is transmitted, observed, and validated, shaping trust and system integrity.

Operationally, communication governs how correlations are established and preserved, while measurement determines when quantum states are rendered into classical form. Together, they define observability and controllability, ensuring information exchange remains physically grounded and verifiable. Reliable quantum intelligence emerges by incorporating these constraints into system design.

3.4.1 Communication as a System Resource

In QAIS, communication is no longer a neutral medium for transmitting abstract bits but a system resource inseparable from physics. Quantum information cannot be amplified or cloned without violating fundamental principles, and disturbance leaves unavoidable traces of interference.

Communication models define how information is transmitted, coordinated, and interpreted across system components, forming a critical layer in QAIS architectures. Differences between classical and quantum communication approaches influence both capability and constraint. Figure 3.4.1 situates this comparison within a unified framework, clarifying how communication paradigms shape system behavior.

What QIS once described as barriers, fragility, disturbance, and no-cloning, are reframed by QAIS as operational design rules. These rules transform communication channels into active participants in verification, synchronization, and inference.

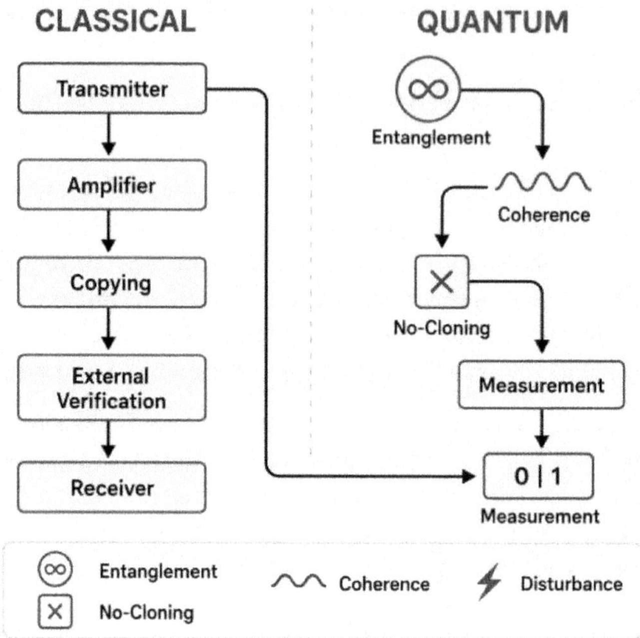

Figure 3.4.1 — Classical vs Quantum Communication.

This figure contrasts classical and quantum communication. On the classical side, signals can be freely amplified, copied, and error-checked externally. On the quantum side, QAIS channels embed coherence, entanglement, and no-cloning directly into the substrate.

Measurement disturbance is shown as an intrinsic verification mechanism, ensuring that intrusion leaves an indelible trace.

Technical Spotlight — Measurement and Verification in QAIS

Conceptual Grounding

In classical communication, verification relies on external mechanisms such as checksums, redundancy codes, or encryption overlays. By contrast, QAIS embed verification directly into the physics of communication. Measurement collapse, the no-cloning theorem, and disturbance sensitivity transform channels into active validators rather than passive conduits.

From an instructional standpoint, this distinction is crucial: learners must see that in QAIS, the act of transmission and the act of verification are inseparable. Any attempt to copy or intercept quantum information alters its state, producing detectable errors. This makes communication not merely a carrier of information but a substrate of trust, where integrity is safeguarded by physical law rather than mathematical assumptions.

System Relevance

For QAIS architectures, this reframing of communication defines how distributed nodes remain synchronized and how outputs become actionable. Channels do more than transfer qubits; they enforce coherence limits, reveal tampering through unavoidable disturbance, and provide native feedback loops for distributed sensing and inference. Unlike classical overlays, where trust is layered on top of transmission, QAIS channels provide trust as an intrinsic property of the medium. This system-level embedding of verification reduces reliance on external protocols and makes communication a first-class design principle in quantum-secure AI systems.

Cross-Disciplinary Orientation

Physicists demonstrate these principles through photonic entanglement, Bell inequality tests, and long-distance correlation experiments. Engineers translate them into stabilized links with timing calibration, quantum memories, and error-mitigating hardware. AI specialists reinterpret communication channels as representational backbones, enabling distributed kernels and model synchronization. Cybersecurity experts see disturbance and no-cloning as a new paradigm of tamper detection, where intrusion becomes visible through physical law rather than through probabilistic guesswork.

Case Study Anchor — Measurement and Verification as Ecosystem Constraints

Intent: To demonstrate how verification burdens and measurement limits propagate upward from physics into ecosystem-level deployment constraints, affecting trust, scalability, and institutional adoption.

- **QALIS:** Employs entanglement-enabled channels to synchronize distributed sensing nodes, using measurement collapse as a shared verification mechanism. Within the *QALIS–Surveillance* layer, these same principles are extended to physics-based verification and anomaly detection across QAIS networks, reinforcing the security and monitoring framework introduced in Chapter 1.
- **CRQC–LLM:** Examines whether fragile coherence in quantum communication channels can be adversarially manipulated, revealing potential resilience gaps and highlighting the importance of structural safeguards under stress testing.

This spotlight extends the historical arc of Section 3.3 into the operational logic of Section 3.4. It prepares the reader for the deeper treatment of primitives in Section 3.4.2 and in the subsequent discussion of scaling toward secure backbones, quantum key distribution, teleportation, and repeaters, by showing how the bedrock concepts of collapse and disturbance evolve into rules for building distributed, verifiable architectures.

3.4.2 Communication Primitives as System Grammar

The transition from principle to practice begins with primitives that convert fragile effects into usable design rules. In QAIS, quantum key distribution (QKD), teleportation, and repeaters are reframed not as isolated proofs but as operational mechanisms that embed physics directly into system grammar. Unlike classical communication channels, where security relies on mathematical hardness, QAIS channels achieve resilience through no-cloning, measurement disturbance, and coherence stabilization.

The construction of QAIS communication systems relies on a set of foundational primitives that enable information transfer, coordination, and security. These primitives serve as building blocks for more complex system behaviors. Figure 3.4.2 position these elements within an architectural context, illustrating how basic operations support higher-level communication functions.

These primitives transform quantum fragility into the scaffolding of secure and scalable communication.

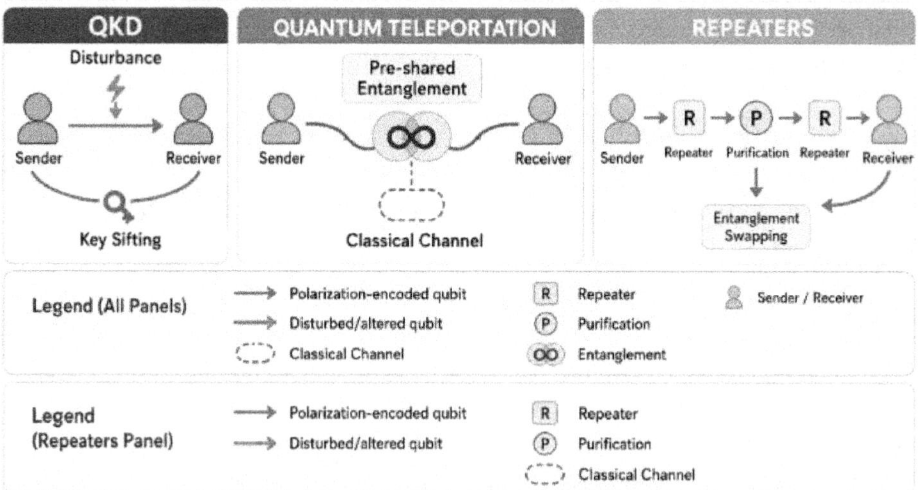

Figure 3.4.2 — Communication Primitives in QAIS.

This comparative schematic illustrates three foundational primitives. On the left, QKD is shown as polarization-encoded qubits exchanged between sender and receiver, where disturbance reveals any interception attempt. In the center, quantum teleportation is depicted through pre-shared entanglement and classical messaging that reconstructs states remotely. On the right, repeaters divide channels into shorter segments, using entanglement swapping and purification to extend coherence across long distances. Together, these mechanisms demonstrate how QAIS converts delicate quantum effects into substrates for secure, distributed architectures.

Technical Spotlight — Communication as a QAIS Backbone

Conceptual Grounding

In classical networking, primitives such as error-correcting codes or encryption algorithms serve as add-ons to otherwise neutral communication channels. QAIS fundamentally redefines this logic. Quantum key distribution (QKD), teleportation, and repeaters are not overlays but *native primitives* that turn fragile quantum effects into systemic building blocks.

QKD operationalizes the no-cloning theorem and measurement disturbance: any interception attempt inherently alters the state, producing detectable anomalies.

Teleportation demonstrates a paradigm shift in transmission, where quantum states are reconstructed at a distance using entanglement and classical communication, without physically moving the particle itself. Repeaters, through entanglement swapping and purification, stabilize coherence over long distances, chaining fragile segments into robust global networks. From a conceptual framing standpoint, these primitives show how the supposed "limitations" of quantum mechanics, fragility, disturbance, and contextuality, are inverted into design rules for scalable and secure infrastructures.

System Relevance

For QAIS architectures, these primitives are not optional enhancements but the backbone of distributed sensing, secure AI pipelines, and hybrid computation. QKD anchors physics-based verification into security protocols, ensuring tamper-proof communication in financial, defense, and healthcare domains. Teleportation enables distributed inference and resource sharing across geographically separated nodes, an essential requirement for global-scale hybrid AI. Repeaters transform local coherence into continental and interplanetary scalability, laying the scaffolding for a quantum internet. In QAIS, primitives become the structural DNA of system design, turning what once appeared as fragile quantum tricks into foundational system grammar.

Cross-Discipline Orientation

Physicists validate these primitives through experimental demonstrations of entanglement fidelity and Bell violation tests. Engineers build photonic channels, satellite links, and purification protocols that stabilize these effects in practice. AI specialists treat entangled communication as a representational scaffold, embedding distributed kernels and enabling coherent model-sharing across networks. Cybersecurity experts frame QKD and disturbance as intrinsic tamper-detection mechanisms, guaranteeing resilience by the laws of physics rather than cryptographic assumptions.

Case Study Anchor — Communication as the Backbone of QAIS Coordination

Intent: To frame quantum communication not as a performance enhancement but as the backbone that enables distributed QAIS coordination, synchronization, and trust propagation across sectors.

- **QALIS (Design Framework):** Deploys QKD-based disturbance checks to harden distributed sensing pipelines against tampering.
- **CRQC-LLM (Stress-Test Lens):** Uses adversarial simulations to test whether coherence and entanglement-based primitives remain secure under CRQC-driven exploitation.

This spotlight extends Section 3.4.1's reframing of communication into verification, showing how the primitives themselves become reusable modules in system design. It provides the conceptual link to the subsequent discussion of scaling toward secure backbones, where chaining these primitives together into repeater-enabled infrastructures lays the groundwork for distributed QAIS backbones.

Scaling Toward Secure Backbones

Primitives such as QKD, teleportation, and repeaters only achieve their full value when chained into larger infrastructures. QAIS reframes these mechanisms as the scaffolding of secure, distributed systems, evolving from isolated demonstrations into backbones that sustain global communication and hybrid inference. The move from fragile effects to repeatable, networked infrastructures marks the transition from laboratory experiments to systemic design.

The transition from foundational communication primitives to fully realized system infrastructures requires integrating these elements into coherent and secure frameworks. This progression defines how QAIS systems achieve reliable information exchange.

Figure 3.4.3 situates this integration within a system-level perspective, illustrating how communication primitives scale into operational backbones.

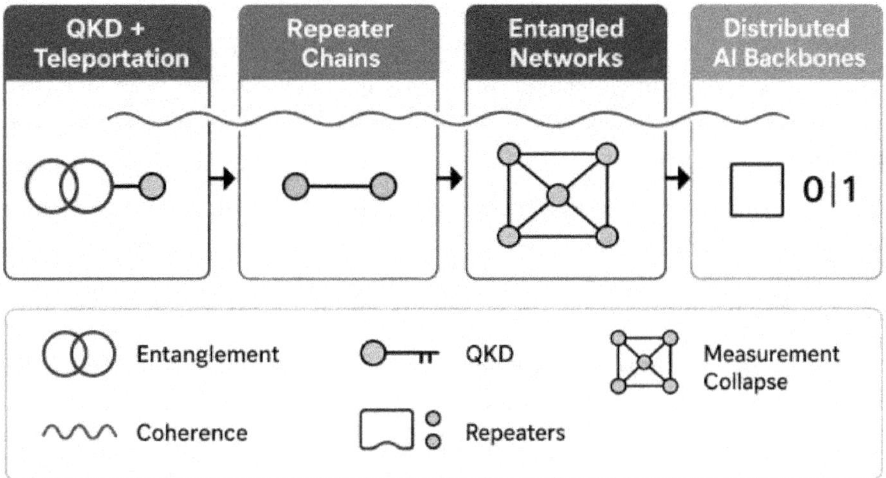

Figure 3.4.3 — From Primitives to Secure QAIS Backbones.

This flow diagram shows the progression from foundational primitives (QKD, teleportation, repeaters) to repeater chains, then to entangled networks, and finally to distributed AI backbones. The figure highlights how coherence and entanglement, once considered fragile curiosities, become engineered resources that support secure communication and synchronized inference pipelines.

Technical Spotlight — From Effects to Infrastructure

Conceptual Grounding

QIS originally treated fragility and no-cloning as fundamental obstacles to communication. QAIS reframes them as design rules, chaining primitives such as QKD, teleportation, and repeaters into distributed infrastructures. QKD secures point-to-point exchanges, teleportation extends correlations beyond direct physical contact, and repeaters scale entanglement across thousands of kilometers. When connected into repeater chains, these mechanisms form entangled networks that function as secure substrates for distributed inference, synchronization, and sensing. This marks the shift from laboratory demonstrations of isolated effects to architectural engineering: primitives become modules, modules become networks, and networks mature into system backbones.

System Relevance

The move from primitives to infrastructures establishes the conditions for a quantum internet. Secure backbones enable multi-node sensing arrays, distributed hybrid inference pipelines, and cryptographic ecosystems grounded in physics rather than computational assumptions. For QAIS, this is the leap from experimental novelty to systemic capability: primitives scaled through repeater networks evolve into foundational backbones for global computation and communication.

These infrastructures ensure that coherence, integrity, and verification are no longer local effects but shared system properties.

Cross-Disciplinary Orientation

Physicists validate long-distance entanglement and coherence stabilization through repeater prototypes and photonic channel experiments.

Engineers design satellite relays, quantum memories, and purification routines to stitch fragile links into robust global channels.

AI specialists interpret entangled networks as distributed representational frameworks, embedding kernels and models across distant nodes.

Cybersecurity experts analyze repeater-based backbones as both a defense surface, tamper-evident by physics, and a potential vulnerability, subject to adversarial probing if coherence thresholds are poorly managed.

Case Study Anchor — From Quantum Effects to Operational Infrastructure

Intent: To trace how abstract quantum effects are translated into deployable infrastructure components, marking the transition from experimental capability to ecosystem reliability.

- **QALIS (Design Framework):** Demonstrates distributed sensing across repeater-enabled entangled networks, transforming fragility into coherence-based trust.
- **CRQC–LLM (Stress-Test Lens):** Probes long-distance coherence with adversarial stress tests, exposing resilience gaps in entanglement distribution and purification protocols.

The output of the representation layer is a state space, a structured configuration of information that satisfies the constraints of the underlying system.

This state space forms the input to the propagation layer, where dynamics are introduced. Any limitations in representation, including encoding fidelity, dimensionality, or noise, directly constrain downstream computation, learning, and reasoning processes.

Section 3.4 Summary — Quantum Communication and Measurement

Quantum communication and measurement illustrate how QAIS converts physical fragility into operational grammar. Disturbance, coherence limits, and the no-cloning constraint are no longer treated as barriers but as the very rules that enforce integrity and synchronization. Measurement collapse, once dismissed as a nuisance, is reframed as a verification mechanism, ensuring that intrusion or tampering leaves unavoidable traces.

Primitives such as quantum key distribution (QKD), teleportation, and repeaters demonstrate this transition in practice. QKD anchors trust through disturbance-based verification, teleportation enables state transfer without particle movement, and repeaters extend coherence across vast distances by chaining entanglement. Together, these mechanisms evolve from isolated laboratory effects into the backbone of scalable networks, enabling secure distributed sensing, inference pipelines, and hybrid QAIS architectures.

This transformation is inherently cross-disciplinary. Physicists validate long-distance entanglement and coherence stabilization through repeater prototypes and photonic channel experiments.

Engineers design satellite relays, quantum memories, and purification routines to stitch fragile links into robust global channels.

AI specialists interpret entangled networks as distributed representational frameworks, embedding kernels and models across distant nodes.

Cybersecurity experts analyze repeater-based backbones as both a defense surface—tamper-evident by physics—and a potential vulnerability when coherence thresholds are poorly managed.

The transition from quantum effects to operational infrastructure marks a decisive shift in QAIS maturity.

Through the constructive lens (QALIS), distributed sensing across repeater-enabled entangled networks transforms fragility into coherence-based trust, embedding verification directly into system architecture.

Through the resilience lens (CRQC–LLM), adversarial stress testing of long-distance coherence exposes fragility in entanglement distribution, purification protocols, and repeater synchronization, revealing how instability can propagate across communication backbones.

The central insight is that communication primitives do not merely transmit information; they define the conditions under which distributed intelligence can operate coherently. By engineering fragile quantum effects into communication-grade primitives, QAIS establishes the infrastructural substrate required for system-level coordination, verification, and trust.

With this foundation in place, the chapter transitions from mechanism to deployment. Section 3.5 examines how these primitives move beyond controlled environments into real-world adoption across business, academia, government, and infrastructure. Adoption across domains becomes the next driver of QAIS maturity, determining whether communication architectures remain experimental constructs or evolve into stable, scalable ecosystems.

3.5 Adoption Across Domains

Architectural Framing: *Distributed Coordination. Examines cross-domain propagation as QAIS integrates into real-world systems.*

Section Framing: *Section 3.5 defines propagation as the dynamic process through which represented states evolve across time or computational steps. Operating on the state space established by representation, propagation transforms states through deterministic or probabilistic rules. In quantum systems, this includes unitary evolution, circuit execution, and measurement dynamics; in classical systems, it includes function evaluation, state transitions, and iterative updates. Propagation thus converts static representation into system behavior.*

This section establishes adoption as a diffusion control surface governing how QAIS extends across industrial, academic, and governmental domains. Representation is adapted to domain-specific contexts, while propagation emerges through translation, integration, and application workflows. Stability is maintained through bounded propagation, where regulatory constraints, resource limitations, and institutional readiness shape deployment.

Adoption therefore marks the transition of QAIS from specialized research into broader societal infrastructure.

Propagation in QAIS can be understood through distinct operational regimes. Deterministic propagation follows fixed transformation rules, such as unitary quantum evolution. Stochastic propagation incorporates probabilistic transitions arising from noise, measurement, and uncertainty. Hybrid propagation integrates classical and quantum processes, where classical optimization iteratively updates quantum parameters. Together, these regimes define the operational envelope of QAIS evolution.

QAIS adoption is advancing across four primary domains: business and industry, academic research and education, government and policy, and infrastructure and ecosystems. Each domain functions both as a site of application and as a driver of systemic maturity, collectively enabling the shift from laboratory experimentation to multi-sector deployment.

The result of propagation is a state trajectory representing system evolution across computational steps, forming the basis for computation, learning, and inference. Constraints such as noise, decoherence, and computational complexity limit the depth and reliability of these processes.

3.5.1 Business and Industry

In QAIS, the commercial sector has become one of the most active arenas of application, treating quantum capabilities not as abstract possibilities but as emerging operational tools. Businesses are adopting QAIS architectures to address problems of scale and resilience that classical methods strain to solve. In cybersecurity, firms are piloting QKD protocols to defend sensitive transactions against both present-day attacks and anticipated threats from cryptographically relevant quantum computers.

Unlike conventional encryption, which depends on mathematical hardness assumptions, QKD secures information through fundamental physical laws such as the no-cloning theorem and unavoidable measurement disturbance. Beyond security, industry leaders are applying quantum-enhanced optimization for high-stakes decision-making.

Energy providers experiment with QAIS optimization pipelines for grid balancing under fluctuating renewable inputs, logistics firms use hybrid solvers for vehicle routing and supply chains, and financial institutions prototype portfolio strategies resilient to volatility. These efforts highlight QAIS as a systems-level enhancer of decision capacity, not merely a niche research tool.

Cloud-based access via IBM Quantum, Amazon Braket, and Xanadu is accelerating industry adoption [18]. By abstracting hardware access into scalable services, enterprises can test QAIS workflows without owning quantum devices, effectively democratizing participation. At the same time, major corporations are establishing in-house quantum R&D labs to cultivate proprietary architectures and domain-specific pipelines. This marks a decisive inflection point: QAIS is shifting from speculative research to a recognized strategic pillar in corporate planning.

The application of QAIS technologies within industry introduces new opportunities for optimization, decision-making, and system design across a range of sectors. These applications reflect the translation of theoretical capabilities into operational value. Figure 3.5.1 situates these developments within a practical context, illustrating how QAIS is integrated into business environments.

QAIS as a Commercial Driver

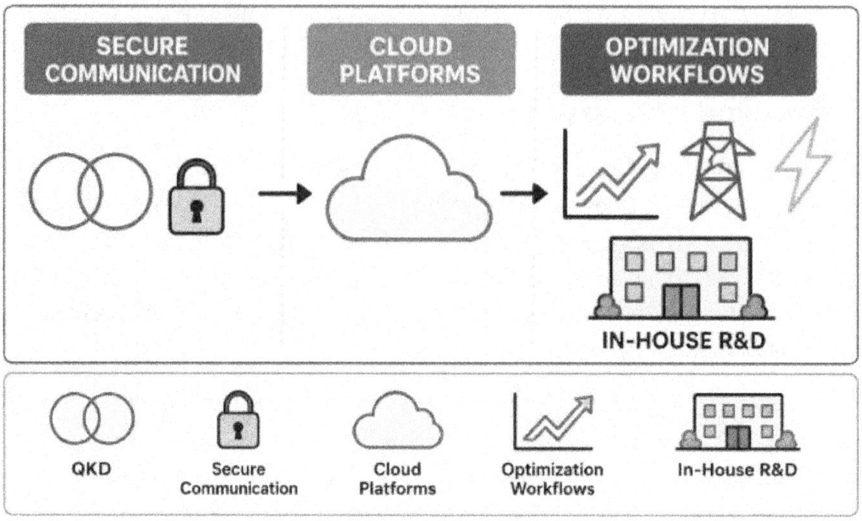

Figure 3.5.1 — QAIS in Business and Industry.

The figure illustrates QAIS adoption across major commercial domains. The left panel shows cybersecurity pipelines employing QKD for secure communication, the right highlights optimization workflows in logistics, finance, and energy. Cloud platforms and in-house R&D appear as bridging infrastructures that accelerate market integration.

Technical Spotlight — QAIS as a Commercial Inflection Point

Conceptual Grounding

The significance of QAIS adoption in business lies in the transition from experimental pilots to institutionalized practice. In earlier QIS phases, enterprises often engaged in speculative exploration, testing quantum algorithms or partnering with research labs without embedding them into daily operations.

With QAIS, the shift is architectural: companies now view quantum-native modules, such as QKD channels or variational solvers, not as isolated trials but as building blocks of operational workflows.

This legitimacy arises because QAIS directly addresses systemic challenges that classical tools struggle to solve. QKD translates physics into tamper-evident security guarantees, appealing to industries where trust and confidentiality underpin business value.

Hybrid optimization pipelines exploit entanglement and interference to uncover solutions in logistics, finance, and energy that are computationally infeasible with classical resources alone. Cloud access further lowers adoption barriers, enabling broad participation and making QAIS a shared infrastructure rather than a specialized experiment.

This marks the stage where businesses become co-creators of the discipline. They push QAIS to meet requirements of cost-effectiveness, interoperability, and compliance, setting practical constraints that research alone cannot provide.

This institutional anchoring explains why industry adoption functions not only as evidence of viability but also as a feedback loop shaping QAIS design priorities for scalability and resilience.

System Relevance

Industry adoption benchmarks viability. QKD introduces verifiable resilience, while hybrid solvers deliver practical gains. Commercial demand shapes priorities in error tolerance, interoperability, and cost scalability.

Cross-Disciplinary Orientation

Physicists validate modules. Engineers design calibration and resilience layers. AI specialists merge quantum kernels with classical workflows. Cybersecurity experts stress-test QAIS protocols against CRQC–LLM adversaries.

Case Study Anchor — Commercialization as an Ecosystem Inflection Point

Intent: To evaluate how commercial entry reshapes QAIS requirements, imposing cost, reliability, and governance constraints that differ sharply from research-driven priorities.

- **QALIS (Design Framework):** Demonstrates enterprise-scale optimization workflows for grid balancing with QAIS-native modules.
- **CRQC–LLM (Stress-Test Lens):** Highlights adversarial exploitation of QKD pipelines, requiring ongoing verification against CRQC decryption and LLM-guided penetration strategies.

This spotlight sets up Section 3.5.2, where academia sustains the knowledge and workforce pipelines that industry depends on.

3.5.2 Academic Research, Education, and Public Sector Alignment

Academic institutions are increasingly serving as the foundation for the formalization of QAIS as a distinct discipline rather than a peripheral extension of quantum information science or artificial intelligence. Whereas quantum information science was historically positioned as an advanced specialization within physics or computer science, QAIS is now emerging as a system-level field that integrates theoretical principles, architectural design, and applied implementation. This transition reflects a shift from isolated domain expertise to coordinated, cross-layer system development.

From a QALIS perspective, academia functions as a primary capability engine for QAIS. Universities and research centers are establishing degree tracks, interdisciplinary laboratories, and collaborative programs spanning quantum computing, quantum communication, and quantum-enhanced AI.

These efforts build the foundational competencies required to design and scale QAIS architectures, emphasizing fluency across physics, mathematics, computer science, and electrical engineering, as well as working knowledge of machine learning and cybersecurity. Academic environments also provide controlled settings for developing and validating new models of representation, propagation, and system integration.

At the same time, the CRQC lens highlights the role of academia in exposing constraints, limitations, and potential failure modes. As research becomes more practice-oriented, students and faculty engage directly with quantum systems through cloud-accessible hardware platforms and high-fidelity simulation environments.

These platforms introduce realistic conditions such as noise, decoherence, calibration drift, and limited coherence windows, revealing the fragility of theoretical constructs when subjected to physical execution. In this sense, academic research not only advances capability but also surfaces the operational boundaries that define feasible QAIS deployment.

Government agencies and standards bodies further shape this ecosystem by coordinating national research priorities, funding quantum technology initiatives, and establishing migration pathways for security-critical infrastructure. Post-quantum cryptography (PQC) initiatives provide a clear example of this role.

As cryptographically relevant quantum computing advances, governments and standards organizations are accelerating the transition toward quantum-resistant cryptographic standards to mitigate long-term security risks. These efforts align directly with the CRQC perspective, where emerging quantum capability is evaluated not only as an innovation pathway but also as a systemic security pressure requiring coordinated mitigation and resilience planning.

Collaboration among academia, industry, and government reinforces this dual role. Academic research contributes rigor, formal validation, and foundational knowledge; industry provides engineering scale, infrastructure, and deployment pathways; and government supports standards, security readiness, and long-term coordination. Under the QALIS lens, this partnership accelerates the translation of research into functional QAIS architectures. Under the CRQC lens, it exposes integration challenges, scaling constraints, cryptographic transition risks, and potential mismatches between theoretical assumptions and operational realities.

Education and training initiatives extend these dynamics into workforce development. Structured curricula, instructional ecosystems such as QAISKit, and laboratory-oriented learning environments help embed QAIS concepts into reproducible educational pathways. QAISKit, as introduced in this work, integrates the core textbook, instructor resources, laboratory modules, assessment frameworks, and applied system exercises into a unified instructional architecture for QAIS education and training.

These efforts produce practitioners capable of operating across disciplinary boundaries while developing familiarity with both the constructive and constraint-oriented dimensions of QAIS.

From a QALIS standpoint, this expands the pool of system designers and accelerates innovation. From a CRQC standpoint, it highlights the importance of training practitioners to reason within real hardware, verification, and security constraints rather than idealized theoretical models.

Together, academic, educational, public-sector, and industry efforts shape both innovation and system resilience. Figure 3.5.2 situates these contributions within an institutional framework, illustrating how capability development (QALIS) and constraint exposure (CRQC) interact to drive the advancement, validation, and responsible deployment of QAIS.

QAIS as a Discipline in Formation

Figure 3.5.2 illustrates how academic ecosystems formalize QAIS as a discipline, blending traditional research practices with new teaching platforms and cloud-enabled experimentation. It positions academia not just as a training ground but as a strategic anchor that produces the intellectual and human capital necessary for QAIS deployment.

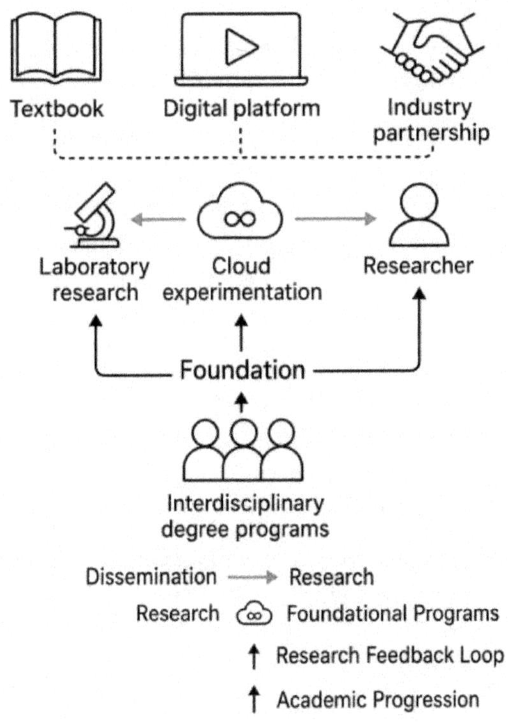

Figure 3.5.2 — QAIS in Academic Research and Education.

The figure shows the multi-layered academic ecosystem for QAIS. At the foundation are interdisciplinary degree programs, while the middle layer highlights laboratory research and cloud-access experimentation. The top layer illustrates dissemination through textbooks, digital platforms, and industry-academic partnerships, forming a pipeline that connects education to enterprise and policy.

Technical Spotlight — Academia as the QAIS Knowledge Engine

The figure bridges the commercial adoption seen in Section 3.5.1 with the deeper structural role of academia. It demonstrates how universities function as the "knowledge engine" of QAIS, sustaining the ecosystem by producing research outputs and skilled practitioners.

Conceptual Grounding

The academic role in QAIS adoption reflects a pivotal shift: moving from teaching quantum information as a specialized elective to establishing it as a system-defining discipline. In the QIS era, learning was primarily theoretical, focusing on postulates, entanglement proofs, and algorithmic potential. QAIS reframes this as an integrated academic field where learners and researchers must bridge physics, AI, engineering, and cybersecurity into a cohesive practice.

This institutionalization matters because academia provides the long-term intellectual capital that industry cannot generate on its own. Universities and research labs create degree programs, simulation environments, and cloud-access testbeds that both train practitioners and validate emerging architectures.

By embedding QAIS into curricula and publishing peer-reviewed findings, academia sets the methodological standards that distinguish discipline-building from corporate prototyping.

For learners and faculty, this progression emphasizes not just conceptual fluency but also systems-level thinking: how coherence lifetimes affect algorithm design, how entanglement depth influences representational power, and how error models shape the architecture of learning pipelines. Academia thereby becomes the "knowledge engine" of QAIS, ensuring the ecosystem is reproducible, rigorous, and intellectually expansive.

System Relevance

Academia anchors the QAIS ecosystem by ensuring reproducibility, workforce readiness, and intellectual expansion. Cloud-access experiments and simulation environments connect theoretical inquiry directly to QAIS architectures, while interdisciplinary labs establish pipelines from conceptual breakthroughs to prototype systems.

Academic research also provides independent validation, helping prevent overhyped claims and ensuring that system designs meet rigorous benchmarks.

Cross-Disciplinary Orientation

Physicists advance experimental platforms that form the substrate of academic QAIS programs, while engineers design calibration codes and control frameworks for academic testbeds. AI specialists integrate machine learning pipelines into quantum curricula, highlighting representational benefits and algorithmic frontiers. Cybersecurity researchers embed cryptographic and verification modules into coursework, framing QAIS not only as a tool for innovation but also as a safeguard for trust and resilience. This layered participation ensures that academia does not replicate corporate agendas but instead sustains QAIS as a holistic, future-oriented discipline.

Case Study Anchor — Academia as the QAIS Knowledge Engine

Intent: To examine academia's role in sustaining QAIS through theory development, workforce training, and open research pipelines, while highlighting vulnerabilities arising from fragmented curricula or funding asymmetries.

- **QALIS (Design Framework):** Highlights university-led prototypes of quantum-enhanced learning platforms, such as curriculum-integrated simulators and hybrid optimization labs, as models of how QAIS education can be designed.
- **CRQC–LLM (Stress-Test Lens):** Raises concerns about adversarial misuse of educational platforms, where CRQC–LLM pipelines could be repurposed for cyber-instructional attacks, highlighting the importance of building integrity and resilience into academic dissemination channels.

While academia sustains QAIS through intellectual and workforce pipelines, it is governments that set the strategic agenda and provide the funding architectures to scale these efforts. The next section explores how policy frameworks and public investments create the macro-structures within which both academia and industry evolve.

This spotlight prepares the ground for Section 3.5.3 ("Government and Policy"), where state actors build on the academic knowledge engine by supplying funding architectures, regulatory frameworks, and strategic agendas that scale QAIS into national capacity.

3.5.3 Government and Policy

Governments worldwide view QAIS as a strategic frontier for technological sovereignty and national resilience. Whereas QIS was often treated as a research domain, QAIS is now framed as an infrastructure-level priority, with implications for defense, economic competitiveness, and secure communications. National initiatives, including the U.S. National Quantum Initiative, the EU's Quantum Flagship, and China's National Laboratory for Quantum AI Systems, signal the shift from research to state-level strategy.

These programs are designed not only to accelerate R&D but also to cultivate talent pipelines, strengthen domestic ecosystems, and secure long-term leadership in the quantum-AI race.

Governmental action extends well beyond funding. National quantum strategies are now embedding QAIS into broader regulatory and diplomatic frameworks. This includes ethical guidelines, export controls, cryptographic standards, and interoperability requirements that determine how QAIS tools can be deployed and exchanged across borders. International partnerships and treaty-level engagements are beginning to treat QAIS as a new axis of geopolitics, where access, standards, and strategic alliances define global influence.

QAIS is being integrated into public-sector innovation. Defense systems, health data infrastructures, climate modeling, and interagency communication networks are increasingly framed as QAIS testbeds. By embedding quantum-AI capabilities into these mission-critical domains, governments are effectively positioning QAIS not just as a research frontier but as a core enabler of 21st-century governance.

Finally, Government involvement influences QAIS through policy, funding, and strategic direction, shaping the broader environment in which systems are developed and deployed. These factors affect both national and global trajectories. Figure 3.5.3 situates these influences within a governance context, illustrating how policy frameworks intersect with technological advancement.

QAIS as a Strategic State Asset

Figure 3.5.3 captures how governments elevate QAIS from laboratory science to a national asset. It situates QAIS within the broader landscape of public policy, diplomacy, and national resilience, emphasizing that system-level adoption is both a technological and geopolitical act.

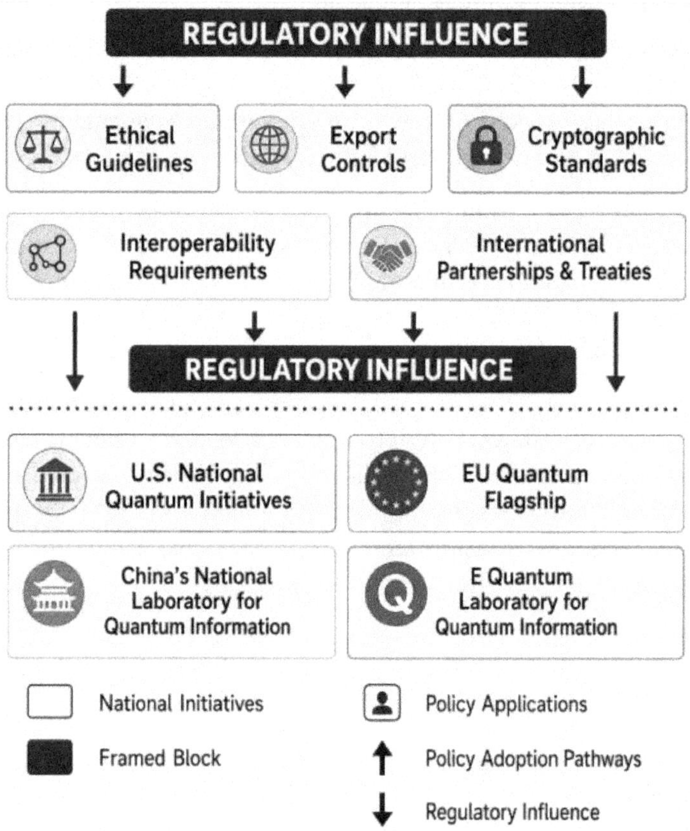

Figure 3.5.3 — QAIS in Government and Policy.

The figure depicts government engagement with QAIS at three levels. The foundational layer shows national initiatives (e.g., U.S. NQI, EU Quantum Flagship, China's National Laboratory). The middle layer illustrates regulatory frameworks and international partnerships. The top layer highlights policy applications in defense, health, climate, and secure communication. Together, these layers show how QAIS is being embedded as a strategic state asset.

Technical Spotlight — Governments as QAIS System Architects

This figure builds on academia's role in Section 3.5.2 and demonstrates how governments scale those efforts into policy and infrastructure. It frames QAIS as both a domestic capability and an international bargaining chip.

Conceptual Grounding

Governmental involvement in QAIS represents the recognition of quantum technologies as matters of sovereignty, not just science. Where academia sustains intellectual capital and industry drives commercialization, governments provide the structural frameworks, funding, regulation, and diplomacy, that scale QAIS into national and international infrastructures.

This progression is important because it reframes QAIS as a discipline governed not only by physics but also by geopolitics. National initiatives like the U.S. NQI, the EU Quantum Flagship, and China's national programs are not merely research accelerators; they are instruments of technological independence, talent cultivation, and strategic security.

From a design perspective, governments embed QAIS into ethical guidelines, cryptographic standards, and international treaties, ensuring that innovation is paired with oversight. This institutional role explains why QAIS cannot mature as a purely technical endeavor, it requires policy frameworks that define its boundaries of deployment, its resilience under adversarial stress, and its role in critical infrastructures. This signals that QAIS professionals must be literate in both scientific and policy dimensions, understanding that the field is as much about governance as it is about computation.

System Relevance

For QAIS architectures, government engagement defines the outer boundary conditions of deployment. Export controls, national cryptographic standards, and secure communication mandates shape the protocols and interfaces that QAIS designers must follow. At the same time, public-sector adoption in health, defense, and climate ensures that QAIS is tested at scale under high-stakes conditions, setting benchmarks for reliability and trustworthiness. Government-deployed **QALIS-Surveillance frameworks** extend QALIS evaluation framework to critical-infrastructure monitoring, quantum-secure communications, and compliance auditing.

Cross-Disciplinary Orientation

Physicists contribute to government-funded laboratories that validate hardware readiness. Engineers develop interoperability frameworks and secure device architectures that meet public-sector specifications.

AI researchers partner with government agencies to adapt hybrid algorithms for climate modeling or defense simulations.

Cybersecurity specialists shape cryptographic standards and assess resilience against CRQC–LLM exploitation, ensuring QAIS infrastructures remain secure even under adversarial stress. This convergence highlights government as both regulator and end-user of QAIS architectures.

Case Study Anchor — Governments as System Architects of QAIS

Intent: To frame governments as architectural actors whose policies, investments, and standards decisions directly shape QAIS scalability, resilience, and geopolitical alignment.

- **QALIS (Design Framework):** Demonstrates how government-backed labs prototype QAIS architectures for secure communication networks and climate modeling systems, serving as frameworks of state-driven design.
- **CRQC–LLM (Stress-Test Lens):** Examines how adversarial actors could exploit gaps in international interoperability or cryptographic standards, testing whether government frameworks are resilient to CRQC–LLM guided breaches.

Section 3.5 Summary — Adoption and Maturation Across Domains

Section 3.5 demonstrated how QAIS progresses from a scientific field into a multi-sector technological ecosystem. Adoption across business, academia, government, and critical infrastructure shows that QAIS maturity depends not only on advances in quantum hardware, but on coordinated development in standards, research capacity, industrial readiness, and national strategy. Each sector contributes complementary roles: business validates commercial viability, academia sustains foundational research and workforce development, government programs provide policy and infrastructure coordination, and critical infrastructure sectors integrate QAIS into high-stakes operational environments.

Across this adoption landscape, QALIS illustrates the constructive trajectory of QAIS, showing how quantum-enhanced sensing, optimization, and learning pipelines are entering mission-critical workflows. In contrast, CRQC–LLM provides the stress-test lens, revealing how coherence fragility, uneven standards, and verification gaps can destabilize deployments when resilience is not engineered into underlying systems.

Adoption is therefore not merely expansion, but an ongoing negotiation between capability and constraint.

By mapping the interplay between innovation, workforce development, regulatory frameworks, and system integration, Section 3.5 establishes adoption as a structural driver of QAIS evolution. Crucially, this progression reveals that sector-level adoption alone is insufficient for sustained scalability. The integration of these domains requires a unifying substrate that can coordinate, stabilize, and operationalize distributed capabilities.

This necessity reframes QAIS not simply as a collection of advancing technologies, but as shared digital infrastructure. Section 3.6 ("Infrastructure and Ecosystem") builds on this foundation by examining how hardware platforms, communication networks, middleware, and software systems interconnect to form a cohesive technological backbone. It is within this infrastructural layer that sector-level adoption is transformed into coordinated system behavior, enabling scalable, resilient QAIS deployment across the ecosystem.

3.6 Infrastructure and Ecosystem

Architectural Framing: Infrastructure & Ecosystem (System Embedding Layer). Defines QAIS as digital infrastructure, embedding system capabilities into persistent, scalable environments.

Section Framing: *This section establishes infrastructure as the foundational control surface enabling QAIS operation at scale across interconnected ecosystems. Representation is instantiated through hardware, software, and network layers, while propagation emerges as system behavior across distributed computational and communication platforms. Stability is enforced through bounded propagation, where reliability, interoperability, and verification mechanisms constrain system performance. Infrastructure therefore functions as the enabling substrate that supports sustained QAIS deployment and evolution.*

Within this layer, QALIS represents the functional capabilities enabled by coordinated systems, while CRQC–LLM highlights the fragility introduced by limited coherence, interoperability gaps, and scaling constraints. Infrastructure defines the execution boundary of QAIS, mediating between abstract computational models and their physical realization. This boundary determines which representations can be instantiated, which propagation processes can be executed, and how reliably outcomes can be produced.

QAIS deployment requires the coordinated evolution of hardware, software, and network systems, moving beyond isolated prototypes toward integrated digital infrastructure. A QAIS-first perspective positions infrastructure not as passive support, but as a defining layer that shapes scalability, interoperability, and system behavior.

At the hardware level, modular processors are increasingly integrated with control electronics, compilers, and embedded error-correction protocols. These components function as cohesive platforms rather than standalone devices, reflecting system-level design principles. Real-time monitoring interfaces track coherence, error rates, and system stability, supporting the transition from individual processors to distributed networks.

Quantum networking is advancing in parallel, linking laboratories and regions through entangled communication channels. These early architectures support secure communication, distributed computation, and synchronization across quantum nodes. Hybrid co-processing further extends capability, with quantum accelerators operating alongside classical CPUs and GPUs to support workflows requiring both quantum advantage and classical scale.

The software ecosystem plays a complementary role. Open-source frameworks such as QAISkit, Cirq, and PennyLane enable the design of abstracted quantum circuits independent of specific hardware platforms. These tools broaden access, support experimentation, and cultivate a workforce fluent in quantum-native development.

Collectively, these developments signal the maturation of QAIS into a foundational layer of digital infrastructure, comparable in systemic significance to the emergence of cloud computing.

The QAIS ecosystem spans multiple sectors, each contributing distinct capabilities and requirements that influence system design and deployment. These interactions create a complex network of dependencies and opportunities. Figure 3.6.1 situates this ecosystem within a unified perspective, illustrating how cross-sector dynamics shape QAIS development.

3.6.1 Mapping QAIS as Digital Infrastructure

Figure 3.6.1 synthesizes the applied QAIS ecosystem across multiple societal domains, showing how business, academia, government, and infrastructure converge to form a coherent digital fabric. It positions infrastructure as the unifying backbone, transforming QAIS from isolated demonstrations into interconnected systems that diffuse across society.

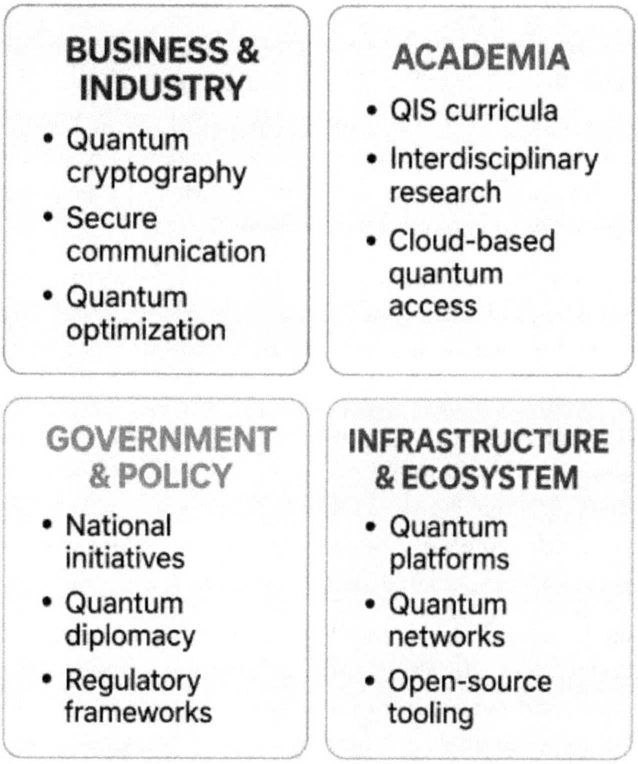

Figure 3.6.1 — Applied Domains of Quantum Information Science.

This figure summarizes the operational domains of QAIS: business and industry, academia, government and policy, and infrastructure. Each domain emphasizes specific priorities, security, research, policy, and tooling, demonstrating how QAIS is institutionalized as a multi-sector digital infrastructure.

Technical Spotlight — QAIS as the Backbone of the Future Internet

This figure extends the previous discussions of enterprise and policy by situating QAIS infrastructure as the foundation upon which societal applications rest.

It bridges organizational adoption with system-level deployment, framing QAIS as the emerging backbone of a new digital ecosystem.

The role of infrastructure in QAIS adoption highlights the moment when a discipline ceases to be laboratory-bound and becomes a system science. Infrastructure here refers not just to physical devices but to the layered environment of processors, networks, software platforms, and error-correcting codes that allow QAIS to scale and interoperate.

This is essential because it shows how constraints become design rules. Coherence lifetimes define scheduling and circuit depth. Quantum networks, built on entanglement swapping and repeater chains, establish secure backbones that classical infrastructures cannot provide. Middleware frameworks, like QAISkit or Cirq, abstract hardware differences, enabling wide participation and talent development.

The emergence of such infrastructure underscores QAIS as a distributed discipline. It is no longer about isolated machines but about ecosystems where hardware, software, and human expertise converge. By embedding reproducibility, interoperability, and resilience into infrastructure, QAIS establishes itself as the backbone of future intelligent systems.

System Relevance

Infrastructure determines whether QAIS architectures can move from promising demonstrations to sustainable deployments. Integrated platforms address error correction and scalability, quantum networks establish secure multi-node communication, and software ecosystems ensure broad accessibility. Together, these layers enable QAIS to function as a resilient architecture rather than a fragmented set of tools.

Cross-Disciplinary Orientation

Physicists validate entanglement distribution and network coherence. Engineers design modular processors, control electronics, and hybrid co-processing platforms. AI specialists build algorithms that leverage these infrastructures for optimization and learning tasks. Cybersecurity experts ensure that networks and open-source frameworks maintain integrity, especially against CRQC–LLM guided adversarial attacks. Each discipline interprets the infrastructure not as background but as an operational front line of QAIS development.

Case Study Anchor — QAIS as the Backbone of the Future Internet

Intent: To position QAIS-enabled communication and computation as foundational to next-generation internet architectures, emphasizing coordination, trust, and verification over raw throughput.

- **QALIS (Design Framework):** Demonstrates modular processor networks that integrate quantum error correction, serving as a design Framework for scalable QAIS architectures.
- **CRQC–LLM (Stress-Test Lens):** Highlights vulnerabilities in open-source ecosystems, where adversarial actors could exploit CRQC resources guided by LLM strategies to infiltrate QAIS software supply chains.

This spotlight provides a bridge to Section 3.7, where the chapter turns from infrastructure realization to the interdependent barriers that constrain QAIS maturity.

Having established the infrastructure backbone, the discussion pivots toward how benchmarks and evaluation frameworks determine the maturity of QAIS deployments.

3.6.2 QAIS Infrastructure Stack — Ecosystem to System Architecture

Architectural Extension of QAIS Infrastructure

Section 3.6.1 established QAIS as a multi-sector digital infrastructure, showing how business, academia, government, and critical systems converge into a coordinated ecosystem. This section extends that perspective by defining the **internal system architecture** that enables such coordination.

Within this execution boundary, QAIS systems must coordinate multiple layers of operation, from physical hardware to application-level deployment. This requirement motivates the formal definition of the QAIS Infrastructure Stack, which organizes these layers into a coherent system architecture.

While the ecosystem view identifies *who participates* in QAIS deployment, the QAIS Infrastructure Stack defines *how these systems function as an integrated whole*. It formalizes QAIS as a layered architecture in which physical systems, information representations, computational processes, learning mechanisms, and application domains interact within a unified operational framework.

The QAIS Infrastructure Stack provides an implementation-oriented realization informed by the Quantum Stack. Its layers translate capabilities associated with one or more Quantum Stack layers into executable system components; the relationship is therefore one of structural alignment rather than strict one-to-one equivalence.

Table 3.6.1 — Quantum Stack Mapping to QAIS Infrastructure Stack.

Quantum Stack Layer	Primary QAIS Infrastructure Stack Alignment	Structural Relationship
Quantum Physics	Physical Substrate Layer	Establishes the fundamental physical laws, phenomena, and limits governing qubit realization, coherence, entanglement, measurement, noise sensitivity, and environmental constraints.
Quantum Mechanics	Physical Substrate Layer; Quantum Information Layer	Provides the mathematical and theoretical rules governing quantum states, state evolution, operators, measurement, probability, and the admissible behavior of physically realized quantum information.
Quantum Technology	Physical Substrate Layer	Realizes quantum principles through physical qubit platforms, control electronics, cryogenic or other operating environments, interconnects, sensors, materials, and supporting device technologies.
Quantum Information Science	Quantum Information Layer	Organizes quantum states and physical resources into representations, encoding structures, transformations, correlations, communication protocols, sensing processes, and measurement frameworks.
Quantum Computing	Computational Layer	Converts quantum-information representations into programmable computation through gates, circuits, algorithms, compilation, execution, measurement, and error-management mechanisms.
Hybrid Quantum-Classical Computing & AI	Hybrid Quantum-Classical Computing & AI Layer; Learning Layer	Coordinates quantum processors with classical computation, optimization, simulation, data processing, AI/ML components, inference pipelines, and orchestration mechanisms; provides the computational bridge from quantum execution to adaptive intelligent behavior.
Quantum AI Systems (QAIS)	Reasoning and Decision Layer; Application and Ecosystem Layer, supported by all lower infrastructure layers	Integrates the capabilities of the complete infrastructure stack into governed intelligent systems supporting learning, inference, reasoning, decision-making, resilience, verification, interoperability, and deployment.

The mapping shown in Table 3.6.1 represents structural alignment rather than strict one-to-one equivalence. Quantum Stack layers describe the conceptual progression from fundamental physics to QAIS, whereas QAIS Infrastructure Stack layers identify implementation-oriented system functions. A single Quantum Stack layer may therefore inform multiple infrastructure layers, and an infrastructure layer may depend on capabilities inherited from several Quantum Stack layers.

Table 3.6.2 — Layered Architecture of QAIS Infrastructure (Structural Alignment).

Infrastructure Layer	Name	Definition	Core Components	Role in QAIS
Layer 1	Physical Substrate	Defines the physical realization and operating environment of quantum systems.	Qubit modalities, including superconducting, trapped-ion, neutral-atom, spin, and photonic systems; control electronics; cryogenic or environmental support; calibration mechanisms; noise and decoherence constraints	Establishes the physical capabilities and limitations within which all higher infrastructure functions operate.
Layer 2	Quantum Information	Defines how information is represented, encoded, structured, and preserved in quantum systems.	Quantum states; Hilbert-space representations; basis, amplitude, phase, and hybrid encoding strategies; entanglement structures; state-preparation mechanisms	Translates problem information into quantum-representable forms while preserving the properties required for subsequent processing.
Layer 3	Quantum Computation and Execution	Defines how encoded quantum information is transformed through controlled computational operations.	Quantum circuits and gates; quantum algorithms; compilation and transpilation; execution and measurement; variational frameworks such as VQE and QAOA	Converts quantum representations into executable computational processes and measurable results.
Layer 4	Hybrid Quantum–	Defines how quantum	Quantum–classical	Coordinates quantum and

Infrastructure Layer	Name	Definition	Core Components	Role in QAIS
	Classical Computing and AI	processing, classical computing, and artificial intelligence are coordinated in integrated workflows.	execution loops; classical preprocessing and optimization; quantum accelerators; AI and machine-learning components; runtime orchestration; simulation; scheduling; resource allocation; quantum–classical data exchange	classical resources for optimization, simulation, learning, inference, and other hybrid computational tasks.
Layer 5	Learning and Adaptation	Defines model training, parameter adaptation, generalization, and data-driven improvement within hybrid QAIS workflows.	Parameterized quantum circuits; hybrid optimization loops; quantum neural networks; quantum kernels; generative models; training and validation mechanisms	Enables models to learn from data or feedback and adapt their parameters within defined operational and verification constraints.
Layer 6	Reasoning, Decision, and Control	Defines how processed information, learned representations, measurements, and feedback are converted into bounded decisions or control actions.	Probabilistic and interference-informed reasoning; measurement-driven decision processes; inference mechanisms; control logic; feedback loops; stability and verification mechanisms	Converts computational and learned outputs into interpretable decisions, control signals, or system responses.

Infrastructure Layer	Name	Definition	Core Components	Role in QAIS
Layer 7	Application, Deployment, and Ecosystem	Defines how QAIS capabilities are integrated into operational environments and governed across their lifecycle.	Domain applications; distributed systems; enterprise and public infrastructure; security and resilience controls; governance and policy frameworks; workforce and institutional support	Embeds QAIS capabilities within real-world technical, organizational, regulatory, and societal contexts.

Interpretation: Table 3.6.2 presents QAIS infrastructure as a functional progression from physical realization to operational deployment. The Physical Substrate establishes the hardware capabilities and environmental constraints upon which quantum information can be represented. The Quantum Information layer organizes that information into usable state and encoding structures, while the Quantum Computation and Execution layer applies controlled transformations and produces measurable results.

The Hybrid Quantum–Classical Computing and AI layer coordinates quantum processors with classical computation, optimization, artificial intelligence, and runtime services. Building on this coordination, the Learning and Adaptation layer supports training and model refinement, while the Reasoning, Decision, and Control layer converts computational and learned outputs into bounded inferences, decisions, and feedback actions. The Application, Deployment, and Ecosystem layer integrates these capabilities into operational domains under appropriate governance, resilience, security, and lifecycle controls.

These infrastructure layers describe functional responsibilities within an integrated QAIS environment. They are not chapter assignments and do not redefine the canonical seven-layer Quantum Stack or the published chapter-layer designations. A chapter may discuss multiple infrastructure functions without being reassigned to any layer represented in this table.

Relationship to Figure 3.6.1

Figure 3.6.1 provides the external view of QAIS as a multi-sector ecosystem, illustrating how business, academia, government, and infrastructure domains converge to support the deployment of quantum-enabled intelligent systems.

The present section complements this perspective by introducing the corresponding internal architectural view. In this interpretation, the sectoral domains represented in Figure 3.6.1 align with the application and ecosystem layer of QAIS, while the infrastructure backbone depicted in the figure corresponds to the underlying architectural layers encompassing physical systems, quantum information representation, computation, learning, and reasoning.

The convergence illustrated in the figure is therefore reinterpreted as cross-layer integration, where system capability emerges through coordinated interaction across these layers. QAIS achieves the status of true infrastructure only when these layers operate as a coherent, interdependent system rather than as isolated technological capabilities.

System-Level Properties of QAIS Infrastructure

Viewing QAIS as a layered architectural stack reveals several defining system-level properties that govern its performance, scalability, and resilience.

Interdependence

Interdependence defines the structural relationship between layers within the QAIS infrastructure stack. Each layer inherits constraints imposed by the physical and operational limits of the layers beneath it. For example, coherence limitations at the hardware level directly constrain the depth and reliability of quantum computation, which in turn bounds the fidelity and expressiveness of learning processes. As a result, no layer can be designed or evaluated in isolation; system capability emerges only through coordinated alignment across all levels of the architecture.

Constraint Propagation

Constraint propagation characterizes how errors and limitations originating in lower layers extend upward through the stack. Noise at the physical layer affects quantum circuit execution, which introduces bias and instability into learning processes and ultimately impacts the reliability of decision-making at higher levels.

This vertical transmission of constraints highlights the importance of end-to-end system design, where mitigation strategies must account for cumulative effects across layers rather than addressing isolated components.

Hybridization

Hybridization is a fundamental property of QAIS infrastructure, reflecting the integration of multiple computational and organizational paradigms. QAIS systems combine quantum and classical computation, unify physical processes with software frameworks, and coordinate algorithmic mechanisms with organizational structures.

This hybrid nature enables QAIS to leverage quantum advantages while maintaining the robustness, scalability, and accessibility of classical systems. It also introduces additional complexity, as effective operation requires synchronization across heterogeneous components and domains.

Infrastructure as Control Surface

Infrastructure in QAIS does not function as a passive support layer but as an active control surface that shapes system capability. Hardware limitations define the space of feasible computational models, middleware determines accessibility and programmability, network architectures govern scalability and distribution, and verification frameworks establish trust and reliability. Through these mechanisms, infrastructure directly influences how QAIS systems are designed, deployed, and evaluated. Performance, resilience, and scalability therefore emerge not solely from algorithmic innovation but from disciplined coordination across the infrastructural layers that enable and constrain system behavior.

The role of infrastructure as a control surface becomes most evident at the boundary where classical and quantum systems interact. This boundary is formalized through the hybrid classical-quantum interface, which governs how computation, control, and learning processes are coordinated across domains.

3.6.3 Hybrid Classical-Quantum Interface

The hybrid classical-quantum interface represents the operational boundary through which QAIS infrastructure becomes executable. Within the QAIS Infrastructure Stack, this interface connects the computational, learning, and reasoning layers to the underlying quantum substrate, enabling coordinated interaction between classical control systems and quantum processing units. This section formalizes the structure, dynamics, and constraints of this boundary as a core architectural element.

Interface Definition

QAIS systems operate across a hybrid boundary that connects classical and quantum computational domains. This interface governs how data, control signals, and optimization processes are exchanged between classical processors and quantum hardware. It is not a simple data transfer mechanism, but a bidirectional coordination layer that enables interaction between fundamentally different computational paradigms.

At this boundary, classical systems provide data preparation, orchestration, and optimization, while quantum systems perform state evolution and measurement under physical constraints. The interface therefore functions as the operational bridge through which quantum capabilities are embedded within classical computational workflows.

Interface Components

The hybrid interface is composed of four functionally distinct components.

- The **encoding interface** transforms classical data into quantum states using encoding strategies such as amplitude, angle, or basis encoding. This process defines how information enters the quantum domain and directly impacts the expressiveness and efficiency of subsequent computation.
- The **execution interface** transmits quantum circuits, control signals, and parameter configurations from classical systems to quantum hardware. It governs how abstract computational instructions are realized as physical operations.
- The **measurement interface** converts quantum states into classical outputs through measurement, producing probabilistic results that reflect the underlying quantum state distribution. This process introduces stochasticity and limits the precision of extracted information.
- The **feedback interface** uses classical computation to process measurement outcomes, update parameters, and optimize subsequent executions. This component enables iterative refinement and adaptive behavior within hybrid workflows.

Together, these components define the structural composition of the hybrid boundary, ensuring that information can move coherently between classical and quantum domains.

The interaction between classical and quantum components defines a core structural feature of many QAIS implementations, particularly in hybrid systems. This interaction governs how information is exchanged and processed across domains. Figure 3.6.2 situates this interface within a feedback-oriented framework, illustrating how hybrid loops enable coordinated system behavior.

Operational Loop

The interaction between these components forms a closed-loop system that characterizes hybrid QAIS execution. Classical data are first encoded into quantum states, which are then propagated through quantum circuits. The resulting states are measured to produce classical outputs, which are processed to update parameters and guide subsequent iterations.

This iterative cycle underlies variational quantum algorithms, hybrid optimization methods, and quantum machine learning pipelines. System performance is therefore determined not by a single execution step, but by the convergence behavior of the entire loop.

To illustrate the iterative nature of hybrid QAIS execution, Figure 3.6.2 presents the classical–quantum interface as a closed-loop system.

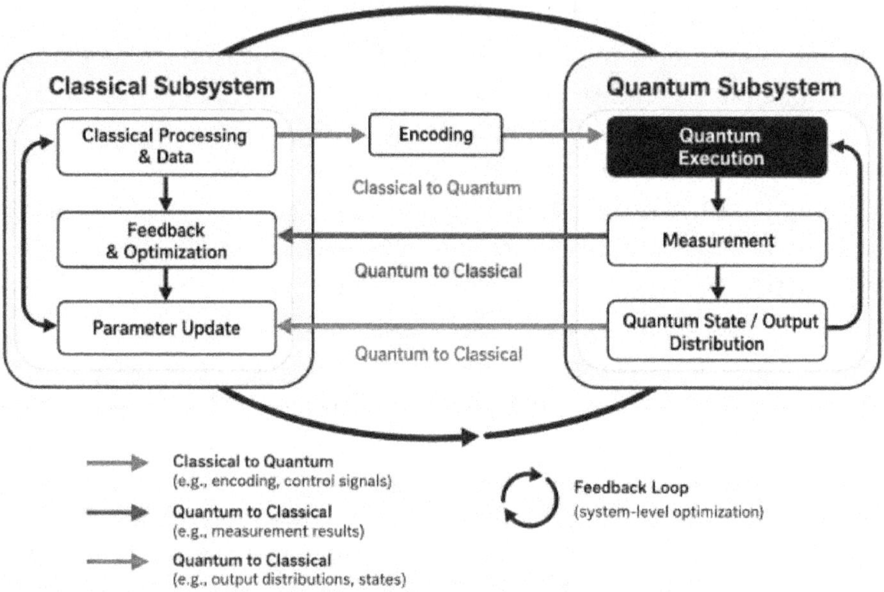

Figure 3.6.2 — Hybrid Classical–Quantum Interface Loop.

The figure depicts the iterative interaction between classical and quantum subsystems, where encoding, execution, measurement, and feedback form a continuous optimization loop. Classical-to-quantum pathways provide encoded parameters, control signals, and execution directives, while quantum-to-classical pathways return measurement outcomes and optimization feedback for adaptive refinement. This structure underlies variational algorithms, hybrid learning systems, and adaptive QAIS workflows. In practice, measured outputs and distributed quantum states may be treated separately within the architecture, since distributed states can remain quantum-native during propagation and synchronization processes, whereas measurement outputs are classicalized observables used for optimization, verification, and control.

Interface Constraints

The hybrid boundary introduces constraints that fundamentally shape system behavior. Latency arises from communication delays between classical processors and quantum hardware, limiting the speed of iterative updates. Noise and decoherence introduce uncertainty into quantum operations and measurement outcomes, reducing fidelity.

Finite sampling imposes limits on the accuracy of measurement-based estimates, while feedback sensitivity can lead to instability in optimization processes.

These constraints propagate through the operational loop, influencing convergence rates, solution quality, and overall system reliability. As a result, the hybrid interface is not merely a conduit for computation, but a primary determinant of system performance.

QALIS and CRQC-LLM Interpretation

From the **QALIS perspective**, the hybrid interface enables adaptive learning through structured feedback loops, allowing systems to iteratively refine parameters and improve performance. It provides the mechanism through which quantum resources are integrated into intelligent, learning-driven architectures.

From the **CRQC-LLM perspective**, the same interface represents a critical vulnerability surface. Noise amplification, adversarial perturbations, and instability in feedback dynamics can propagate across iterations, leading to degraded performance or system failure.

Architectural Role

The hybrid classical-quantum interface functions as the operational boundary layer that connects infrastructure, computation, learning, and reasoning within QAIS architectures. It enables the translation of quantum computational capability into actionable system behavior under classical control.

By mediating interactions across these layers, the interface ensures that quantum advantage can be harnessed within practical systems. QAIS architectures become executable only when this boundary is properly defined, stabilized, and optimized, making the hybrid interface a central component of scalable and resilient quantum intelligent systems.

This boundary-layer role positions the hybrid interface as a central determinant of system execution, forming the basis for the infrastructure-level synthesis presented in the section summary that follows.

Section 3.6 Summary — Infrastructure and Ecosystem

This section further formalized QAIS infrastructure through the QAIS Infrastructure Stack, which defines how physical systems, quantum information representations, computational processes, learning mechanisms, and application domains interact as a coherent architecture.

Complementing this layered model, the hybrid classical-quantum interface was introduced as the operational boundary through which these layers become executable, enabling coordinated interaction between classical control systems and quantum processing units.

Together, the infrastructure stack and hybrid interface reveal that QAIS capability emerges from tightly coupled interactions across layers rather than from isolated components. However, this same interdependence exposes the system to compound constraints, where limitations originating in hardware, communication, computation, or human coordination propagate across the architecture and influence overall system behavior.

Section 3.7 builds directly on this foundation by examining these constraints as an interconnected barrier network. It analyzes how physical fragility, technical limitations, infrastructural fragmentation, and workforce challenges interact across layers, defining the operational boundaries and maturity trajectory of scalable QAIS systems.

3.7 Barriers to Quantum Maturity

Physical, Technical, and Human Dimensions

Architectural Framing: *This section identifies the coupled constraint surfaces through which physical fragility, standards immaturity, verification limits, and workforce access collectively bound QAIS maturity.*

Section Framing: *This section establishes barriers as constraint surfaces that limit the propagation and scalability of QAIS systems. Representation is restricted by physical limits, engineering constraints, and human factors, while propagation emerges as system behavior impeded by noise, complexity, and resource limitations. Stability is enforced through bounded propagation, where unresolved constraints define the operational boundaries of current systems. Barriers therefore function as critical indicators of where architectural advancement is required to achieve QAIS maturity.*

Building on the infrastructure architecture established in Section 3.6, QAIS systems are now understood as tightly coupled, multi-layered systems in which performance emerges from coordinated interaction across physical, computational, and organizational domains. However, this same interdependence gives rise to a corresponding network of constraints, where limitations originating in one layer propagate across the entire system and shape overall behavior.

These constraints do not operate independently. Physical fragility, including noise and decoherence, constrains computational depth and fidelity. Technical limitations in control, communication, and interoperability restrict system integration and scalability.

This section examines these barriers as a unified structural phenomenon rather than as isolated challenges. By analyzing how constraint surfaces interact across physical, technical, infrastructural, and human dimensions, the discussion reframes limitations as intrinsic properties of the QAIS architecture. Understanding these coupled barriers is essential for identifying both the current limits of QAIS maturity and the pathways required to achieve stable, scalable, and resilient quantum intelligent systems.

The maturation of QAIS is constrained by a range of barriers that limit scalability, reliability, and deployment readiness. These barriers arise from both technical and systemic factors. Figure 3.7.1 situates these challenges within a structured context, providing a basis for understanding their impact on system development.

3.7.1 Mapping the Interdependent Barriers of QAIS Maturity

Figure 3.7.1 visualizes the physical, technical, infrastructural, and human barriers that prevent QAIS from reaching full maturity. The diagram highlights interdependencies, showing that noise, standards, and skills shortages reinforce one another, underscoring the systemic nature of the challenge.

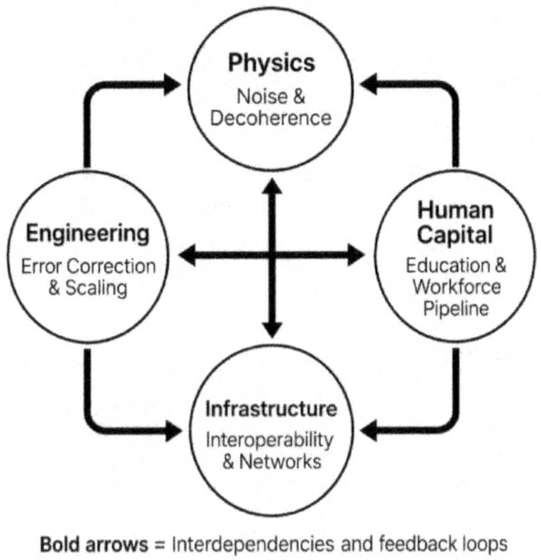

Figure 3.7.1 — Barriers to QAIS Maturity.

The figure illustrates four interdependent domains of constraint, Physics (noise and decoherence), Engineering (error correction and scaling), Infrastructure (interoperability and networks), and Human Capital (education and workforce). Bold arrows form a systemic feedback loop, while diagonal links highlight reinforcing dependencies across domains. Together, these barriers demonstrate that QAIS maturity is limited not by a single breakthrough but by how physical fragility, infrastructural gaps, and human capital constraints interact as a constraint network.

Technical Spotlight — Barriers as a Systemic Constraint Network

Conceptual Grounding

Earlier quantum research often treated challenges such as decoherence, qubit instability, or limited fabrication capacity as isolated engineering problems to be solved independently.

QAIS reframes these issues as *systemic constraints*—structural rules that shape how architectures can scale, coordinate, and remain reliable.

Decoherence becomes a temporal design boundary for algorithms; error correction becomes a resource allocation problem rather than a mere engineering patch; and interoperability is recognized as a precondition for distributed coherence across networks.

Workforce shortages, similarly, shift from being viewed as external bottlenecks to being understood as *core architectural constraints*, determining how quickly and safely systems can be designed, deployed, and secured. In QAIS, barriers are not peripheral obstacles, they are built into the design logic itself.

System Relevance

Interpreting barriers as a constraint network allows designers, engineers, and policymakers to identify where system-level failure points emerge. Physical fragility limits algorithmic depth and representational capacity; infrastructural fragmentation prevents geographically distributed coherence; and workforce gaps slow the transition from prototypes to deployable systems. These constraints define the *true maturity frontier* of QAIS. By incorporating them into planning, design methodology, and verification strategies, QAIS practitioners ensure that architectures reflect realistic limits while anticipating future scalability and resilience requirements.

Cross-Disciplinary Orientation

Physicists interpret barriers as experimental frontiers: extending coherence lifetimes, reducing noise, and stabilizing qubits. Engineers see them as design challenges: modular processors, standard protocols, and scalable error correction. AI specialists treat them as constraints on representational capacity, influencing how variational algorithms or kernel embeddings can function in noisy regimes.

Cybersecurity experts regard them as resilience gaps: points where CRQC–LLM adversaries can exploit fragile coherence, incomplete verification, or workforce inexperience to destabilize infrastructures.

QALIS–Surveillance pipelines are particularly sensitive to coherence and verification bottlenecks, as these determine real-time anomaly-detection reliability. Building on the ecosystem layer established in Section 3.6 this spotlight establishes the conceptual framework for the sections that follow. **Section 3.7.2** examines physical and engineering fragility as foundational design rules.

Section 3.7.3 analyzes infrastructure and verification challenges that arise from fragmented standards and immature networks. **Section 3.7.4** focuses on human-capital limitations that shape the pace and security of QAIS adoption. The section summary then synthesizes these layers as a coupled barrier network governing QAIS maturity.

Case Study Anchor — Barriers as a Coupled Constraint Network

Intent: To show how technical, organizational, and workforce barriers interact as a coupled system of constraints, shaping feasible QAIS architectures more strongly than any single limitation.

- **QALIS (Design Framework):** Demonstrates how coherence limits and error thresholds are explicitly encoded into distributed sensing pipelines, treating fragility as a design parameter.
- **CRQC-LLM (Stress-Test Lens):** Shows how adversarial probing exploits weak verification or fragmented infrastructures, stressing QAIS at its systemic boundaries.

Taken together, these systemic barriers remind us that QAIS maturity is not determined by any single breakthrough but by how constraints interact across physical, infrastructural, and human layers.

The next step is to examine these categories in detail, beginning with the most fundamental challenge: the fragility of quantum systems themselves. Section 3.7.2turns to physical and engineering constraints, showing how decoherence, scaling, and noise are reframed as design rules that set the boundaries for QAIS architectures.

3.7.2 Physical and Engineering Constraints

Quantum systems remain extraordinarily sensitive to their environment, and this fragility translates directly into barriers for QAIS deployment. Thermal fluctuations, electromagnetic interference, and vibrational noise can cause decoherence, collapsing qubit states into classical outcomes and erasing quantum information.

While QIS originally identified decoherence as a theoretical limitation, QAIS reframes it as a design rule that shapes the architecture of learning pipelines, inference scheduling, and resource allocation.

Maintaining coherence for extended periods remains an unsolved challenge. Solutions such as cryogenic cooling, electromagnetic shielding, and active noise correction extend quantum lifetimes but still fall short of thresholds required for scalable systems.

In QAIS, coherence time becomes a critical architectural constraint, defining the "quantum lifetime" of usable information and influencing decisions about gate depth, logical qubit overhead, and algorithmic structure.

Scaling compounds the fragility problem. Each additional qubit increases the difficulty of preserving entanglement, synchronizing operations, and layering error correction. Classical computing scaled through abstraction and density gains; quantum computing instead requires precise, low-noise control over entangled networks.

Research into modular processors, distributed architectures, and topological qubits seeks to address these challenges, but fault-tolerant scalability remains aspirational.

Fragility and Scale as System Design Rules

Figure 3.7.2 visualizes how physical fragility and engineering scale jointly constrain QAIS maturity. The figure highlights the trade-offs between coherence time, logical qubit error correction, and architectural scalability, underscoring why these are not isolated engineering puzzles but systemic barriers.

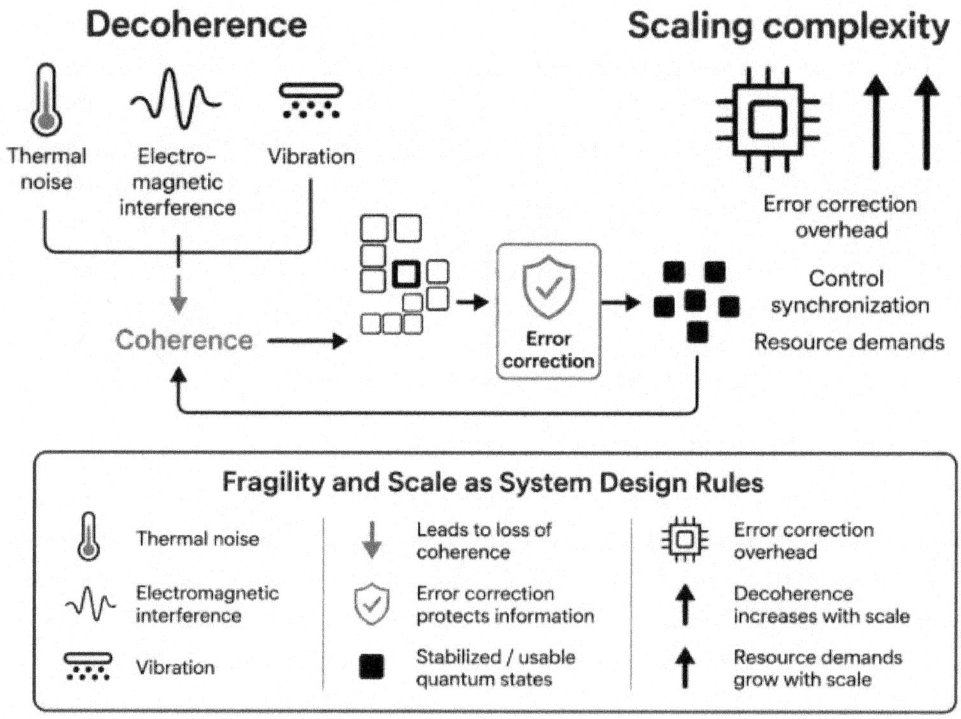

Figure 3.7.2 — Physical and Engineering Constraints on QAIS.

This figure illustrates the dual challenges of decoherence and scaling. The left side depicts environmental noise degrading qubit stability, while the right side shows the rising complexity of control and error correction as system size increases. Together, they emphasize how fragility and scale interlock as core design limits.

Physical and engineering constraints impose fundamental limits on QAIS architectures, influencing how systems are constructed and operated. These constraints must be accounted for in both design and implementation. Figure 3.7.2 highlight these limitations within a system-level perspective, illustrating their role in shaping achievable performance.

Technical Spotlight — From Fragility to Design Principles

This figure builds directly on the systemic barrier network introduced in Section 3.7.1. It situates physical fragility and engineering scale not as background problems but as active constraints that shape QAIS architectures.

At the heart of QAIS lies a paradox: the same principles that grant quantum systems their power, superposition, entanglement, and interference, also make them extraordinarily fragile. Qubits interact with their environment in ways that rapidly cause decoherence, collapsing delicate states into classical outcomes.

Maintaining coherence has required cryogenic cooling, electromagnetic shielding, and active error mitigation, yet these interventions still fall short of the lifetimes needed for large-scale computation. Scaling compounds the challenge: each additional qubit increases control complexity, error correction overhead, and synchronization demands. Unlike classical scaling, which relies on transistor density and abstraction, quantum scaling requires direct management of fragile entanglement networks.

In QAIS, these fragilities are not simply problems to be solved but active design rules. Coherence time becomes the effective "clock speed" of quantum computation, while scaling overhead defines the architectural trade-offs between depth, fidelity, and resource allocation.

System Relevance

For QAIS architectures, physical and engineering limits shape every layer of design. Algorithm feasibility depends on whether coherence times can support the required gate depth. Error correction introduces overhead that defines how many logical qubits can be realized from a given number of physical qubits. Scaling through modular processors or distributed entanglement changes not only hardware requirements but also the orchestration of learning pipelines. By recasting fragility and scale as architectural constraints, QAIS establishes boundaries within which systems must be designed, deployed, and defended.

Cross-Disciplinary Orientation

Physicists work to extend coherence lifetimes through new materials, topological qubits, and cryogenic environments.

Engineers design scalable control electronics, error correction codes, and modular processor frameworks.

AI specialists adapt algorithms to operate within short coherence windows, emphasizing hybrid orchestration and noise-aware training.

Cybersecurity experts assess whether fragility could be exploited, with CRQC–LLM adversaries potentially destabilizing systems by targeting noise or scaling vulnerabilities. Each discipline engages with fragility not as background noise but as an operational frontier of QAIS.

By reframing fragility and scale as system-level design rules, this spotlight highlights the first and most fundamental category of barriers to QAIS maturity. The discussion now turns to the ecosystem-level, where infrastructure, interoperability standards, and verification frameworks determine whether even stable local devices can be scaled into distributed and trustworthy systems.

Case Study Anchor — Fragility as a Design Signal

Intent: To reinterpret fragility not as failure, but as an architectural signal that guides resilience engineering and ecosystem-level risk management.

- **QALIS (Design Framework):** Demonstrates how coherence time constraints directly inform inference scheduling in distributed sensing pipelines.
- **CRQC–LLM (Stress-Test Lens):** Highlights how adversarial actors could exploit decoherence and scaling fragility, using CRQC–LLM strategies to destabilize QAIS architectures at their weakest points

While physical fragility and scaling define the local limits of QAIS devices, true maturity requires more than stabilizing individual qubits. The broader challenge lies in weaving these devices into interoperable infrastructures governed by common standards and validated by scalable verification frameworks.

Section 3.7.3 turns to these ecosystem-level barriers, showing how fragmentation in hardware, middleware, and benchmarking constrains QAIS from evolving into a globally reliable system.

3.7.3 Infrastructure, Standards, and Verification

The QAIS ecosystem currently suffers from fragmentation and immaturity in its infrastructure standards. While QIS established proof-of-principle devices, QAIS requires interoperable architectures capable of scaling across domains and geographies. Today, superconducting circuits, ion traps, and photonic qubits all use different encoding models, control protocols, and error correction schemes. This diversity reflects healthy exploration but creates severe obstacles for algorithm portability, cross-platform deployment, and the creation of unified development environments.

Verification compounds the challenge. Classical systems produce deterministic outputs that can be exhaustively tested. Quantum systems, however, evolve probabilistically, making large-scale simulation infeasible. Current methods, randomized benchmarking, process tomography, and cross-entropy benchmarking, are valuable but limited in scalability, interpretability, and generalizability.

Benchmarking itself is now a contested field, with Bowles et al. [12] emphasizing that comparisons between quantum and classical models require carefully designed baselines to avoid misleading claims of superiority [25]." Without efficient, scalable benchmarking frameworks, QAIS architectures cannot earn trust or validation at the system-level.

History shows how standards transform technologies into infrastructures. Just as the internet only matured once TCP/IP and SSL provided universal protocols, QAIS must converge on standards for error correction, circuit compilation, and benchmarking. "Benchmarks such as QUARK [13] are emerging as modern frameworks to evaluate the reliability and scalability of

[25] Bowles warns against premature claims of quantum advantage, noting that benchmarking must incorporate fairness, noise models, and realistic baselines to ensure credibility.

QAIS deployments, ensuring that system-level claims are reproducible across platforms [26]." Without such standards, progress risks remaining siloed in laboratories, constraining the leap to distributed, globally integrated QAIS systems. Infrastructure and verification challenges affect the ability to validate and scale QAIS systems, introducing uncertainty into system reliability and performance assessment. These challenges extend beyond individual components to the system as a whole.

Figure 3.7.3 situates these challenges within a broader architectural context, illustrating how heterogeneous hardware platforms, immature APIs, and unscalable verification methods fragment today's QAIS ecosystem, hinder interoperability, and ultimately constrain deployment. In doing so, it presents standards not as bureaucratic necessities, but as essential gatekeepers of QAIS reliability and trust.

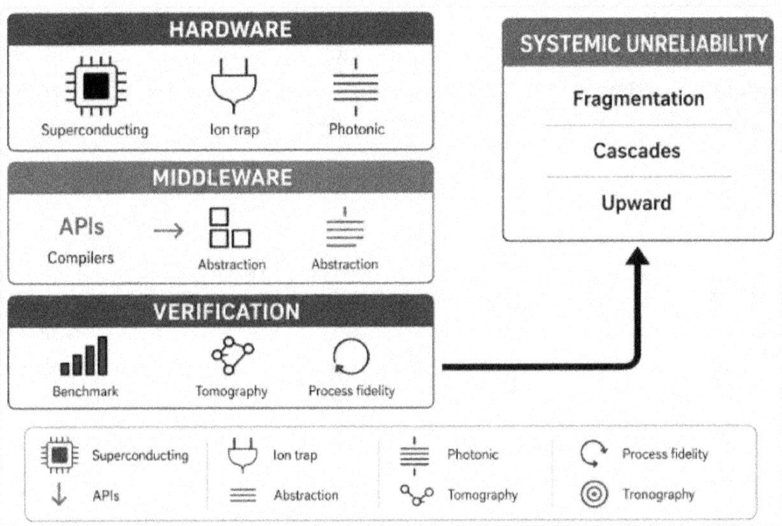

Figure 3.7.3 — Infrastructure and Verification Barriers in QAIS.

The figure depicts three layers of constraint: heterogeneous hardware platforms (superconducting, ion trap, photonic), middleware gaps (APIs, compilers, abstraction layers), and immature verification methods (benchmarking, tomography, process fidelity). Arrows highlight how gaps at one layer cascade upward, reinforcing systemic unreliability.

[26] QUARK establishes benchmarking as an open-source framework, embedding reproducibility and cross-platform comparability as system-level requirements for QAIS adoption.

Technical Spotlight — Toward Standards and Scalable Verification

Conceptual Grounding

The central insight of this figure is that QAIS cannot rely on fragmented infrastructures if it is to mature into a deployable discipline. Unlike classical computing, where standard instruction sets and communication protocols (e.g., TCP/IP, SSL) allowed devices and networks to interoperate seamlessly, QAIS remains divided by incompatible qubit modalities, middleware interfaces, and verification schemes.

Each hardware approach, superconducting circuits, trapped ions, photonic qubits, brings its own encoding methods and control protocols.

While this diversity fuels innovation, it obstructs portability and system-level reliability. Verification is equally unsettled.

Classical systems can be exhaustively tested, but QAIS outputs are probabilistic and often infeasible to simulate at scale. Current methods such as randomized benchmarking, process tomography, and cross-entropy checks provide partial confidence but lack generalizability.

Without shared standards for compilation, error correction, and benchmarking, QAIS risks becoming a patchwork of isolated experiments rather than a coherent system science.

System Relevance

For QAIS architectures, interoperability and verification standards are not bureaucratic afterthoughts but the conditions for trust. If algorithms cannot run consistently across hardware platforms, scalability stalls. If results cannot be validated efficiently, adoption slows in critical domains such as sensing, finance, and cybersecurity. Standards thus become the gatekeepers of reliability, enabling QAIS systems to function predictably across global infrastructures.

Cross-Disciplinary Orientation

Physicists view the absence of verification standards as an experimental bottleneck, limiting reproducibility. Engineers see fragmented middleware and APIs as obstacles to scalable design. AI specialists recognize the difficulty of embedding quantum kernels into hybrid workflows without cross-platform portability. Policymakers treat the lack of standards as a governance gap, raising concerns about security and international competitiveness.

Cybersecurity experts warn that without robust benchmarking and verification, adversarial CRQC–LLM strategies could exploit unverified pipelines to destabilize QAIS trust.

By exposing infrastructure fragmentation and verification immaturity, this spotlight sets the stage for Section 3.5 on education, workforce, and access. Even if physical stability and interoperability standards improve, QAIS progress will falter without a globally distributed workforce trained to build, operate, and extend these protocols.

Case Study Anchor — Standards and Verification as Scaling Prerequisites

Intent: To emphasize standards formation and scalable verification as necessary conditions for QAIS transition from pilot systems to operational infrastructure.

- **QALIS (Design Framework):** Demonstrates how distributed sensing platforms rely on standardized benchmarks of entanglement fidelity and depth, underscoring the necessity of interoperable verification.
- **CRQC-LLM (Stress-Test Lens):** Shows how adversaries could exploit incomplete verification regimes, using CRQC decryption power and LLM-guided exploits to manipulate unverified infrastructures.

The structural challenges of interoperability through standards and the need for scalable verification highlight that QAIS maturity cannot be achieved through technology alone.

Even the most reliable platforms and benchmarks will remain underutilized if the global workforce lacks the interdisciplinary training to design, validate, and secure them.

This makes the human dimension, education, workforce development, and equitable access, the next critical frontier, where the availability of skilled practitioners determines whether QAIS can progress from prototype infrastructures to resilient, deployable system architectures.

3.7.4 Education, Workforce, and Access

The rapid pace of quantum innovation has decisively outstripped the global supply of trained personnel. QAIS draws upon multiple domains, quantum physics, computer science, electrical engineering, and applied mathematics, each advancing rapidly with its own technical body of knowledge.

Yet very few institutions currently provide comprehensive, integrated curricula that span these disciplines in a cohesive and accessible way. As a result, learners and mid-career professionals often face steep, siloed learning curves with limited interdisciplinary guidance or formal mentorship. This widening skills gap slows the cadence of research and development, constraining the talent pipeline available to startups, enterprise labs, and national initiatives alike.

Access to state-of-the-art hardware compounds the barrier. Quantum processors are costly, require ultra-low-temperature cryogenic environments, and demand continuous calibration and maintenance. As such, most operational systems are concentrated at large research institutions or corporate facilities, made available externally only through controlled cloud platforms.

While simulators and limited remote access tools have broadened introductory exposure, high-fidelity experimentation on physical devices remains scarce. This exclusivity risks reinforcing geographic, economic, and institutional inequities, narrowing the diversity of perspectives shaping QAIS innovation.

Expanding both formal and informal educational pathways is essential. Joint degree programs, micro-credential tracks, and open-access courseware can reduce entry barriers for learners from diverse backgrounds. Interdisciplinary fellowships and industry–academia exchanges cultivate transferable expertise, while shared national testbeds, middleware abstractions, and tiered cloud-credit programs can democratize access to hardware. By combining curriculum innovation with equitable infrastructure access, the QAIS community can cultivate a globally distributed workforce capable of sustaining future progress.

Here again, QAIS reframes the issue. Where QIS primarily trained theorists, QAIS requires practitioners who can design, implement, and secure hybrid architectures. Educational programs must integrate linear algebra, Hilbert space formalism, variational learning frameworks, cryptographic principles, and error correction techniques into unified curricula.

This scaffolding is essential for developing practitioners who can both advance and defend QAIS architectures.

Section 3.7 Summary — Barriers to QAIS Maturity

QAIS maturity is defined not by isolated breakthroughs, but by how effectively it navigates a network of interdependent constraints. Physical fragility, verification limits, infrastructure fragmentation, and workforce shortages collectively shape the boundaries of architectural feasibility. Within this framework, coherence functions as the operational clock of computation, verification becomes a probabilistic foundation of trust, and standards enable interoperability across systems. Workforce capacity, in turn, emerges as the limiting factor in scaling these advances.

QALIS illustrates how such constraints can be engineered into resilience through coherence-aware scheduling, interoperable benchmarking, and hybrid operational design. In contrast, CRQC–LLM exposes the systemic vulnerabilities that persist, highlighting how weaknesses in coherence, verification, infrastructure, and access can be exploited to destabilize distributed systems.

Together, these perspectives underscore that QAIS progress depends on the alignment of technical innovation, ecosystem governance, and human capability.

These constraints define the operational envelope within which QAIS must evolve. Among them, workforce limitation is not merely a resource gap, but a central scaling variable that conditions how effectively all other barriers can be addressed.

Section 3.8, Building the Quantum Workforce: Skills and Competencies, examines how structured skill development, interdisciplinary literacy, and institutional capacity-building expand this human layer, enabling QAIS systems to scale more reliably and sustainably.

3.8 Building the Quantum Workforce — Skills and Competencies

***Architectural Framing**: Workforce (Continuity and Skill Architecture Layer). Defines human systems as propagation enablers, completing ecosystem realization. Defines workforce as structured through: Terminology frameworks, Mathematical and Quantum Physics, Computational tools, Human competencies, Adaptability, Workforce pipeline.*

***Section Framing:** This section establishes the workforce as a human-centered control surface that enables the design, operation, and evolution of QAIS systems. Representation is embodied in knowledge structures, skills, and competencies, while propagation emerges as system behavior through education, training, and collaborative practice. Stability is enforced through bounded propagation, where expertise gaps, learning curves, and institutional support constrain workforce development. The workforce therefore functions as the operational bridge between theoretical QAIS frameworks and real-world implementation.*

The success of QAIS depends as much on human capacity as on technical innovation. Hardware lifetimes, algorithmic fidelity, and architectural resilience cannot translate into mature systems without practitioners who can design, implement, and secure them.

Unlike the narrower training pathways of QIS, QAIS requires interdisciplinary fluency, linking mathematics, physics, engineering, AI, and cybersecurity into a single systems-oriented skillset.

"Reports from QED-C [16] stress that without a robust and interdisciplinary workforce, even advances in qubit fidelity and algorithm design cannot mature into sustainable QAIS ecosystems [27]." Similarly, the National Academies [17] highlight workforce preparation as a structural bottleneck, noting that shortages of trained professionals pose risks equal to physical qubit fragility [28].

This section maps the competencies that define such a workforce. It begins with foundational terminology as a skills framework (Section 3.8.1), then turns to mathematical and Quantum Physics (Section 3.8.2), computational thinking and practical tools (Section 3.8.3), and human competencies and lifelong learning (Section 3.8.4).

Case Study Anchor continues the book's framing: QALIS illustrates constructive trajectories for workforce preparation through applied training pipelines. These pipelines incorporate QALIS-Surveillance modules for monitoring, diagnostics, and resilience testing, while CRQC-LLM underscores the risks of adversarial exploitation if skills remain siloed or unevenly distributed. Together, these anchors reinforce that building the QAIS workforce is not supplemental, it is central to system-level maturity.

---

[27] QED-C frames workforce shortages as structural bottlenecks, asserting that without interdisciplinary training, advances in fidelity and error correction will remain underutilized.

[28] The National Academies highlight workforce pipelines as essential infrastructure, recommending federal-academic-industry collaboration for QAIS readiness.

The development of a QAIS workforce requires defining the conceptual and technical domains that underpin system design and operation. These domains form the basis for education and professional development. Figure 3.8.1 organizes these domains within a structured framework, illustrating how foundational knowledge aligns with workforce needs.

3.8.1 Foundational QAIS Terminology as a Skills Framework

At this stage in the chapter, the glossary transitions from a passive reference into a structural guide for capability development. Rather than listing terms in isolation, it groups them according to where they function within QAIS architectures and how they inform professional practice. Each category therefore signals a distinct area of competency—linking conceptual understanding to the skills required for system design, integration, and assurance.

Figure 3.8.1 serves as the visual bridge between terminology and application. It highlights how foundational concepts map onto roles and tasks within the QAIS lifecycle, from physical-layer reasoning to algorithmic implementation and system-level validation. By organizing terms into coherent domains, the figure clarifies how knowledge is layered and how expertise progresses across the architecture stack.

Figure 3.8.1 organizes foundational QAIS terms into four categories: primitives, algorithmic modules, security mechanisms, and case studies. Together, these categories illustrate how conceptual vocabulary aligns with applied competencies, providing a concise reference point for understanding how QAIS knowledge is operationalized across the workforce.

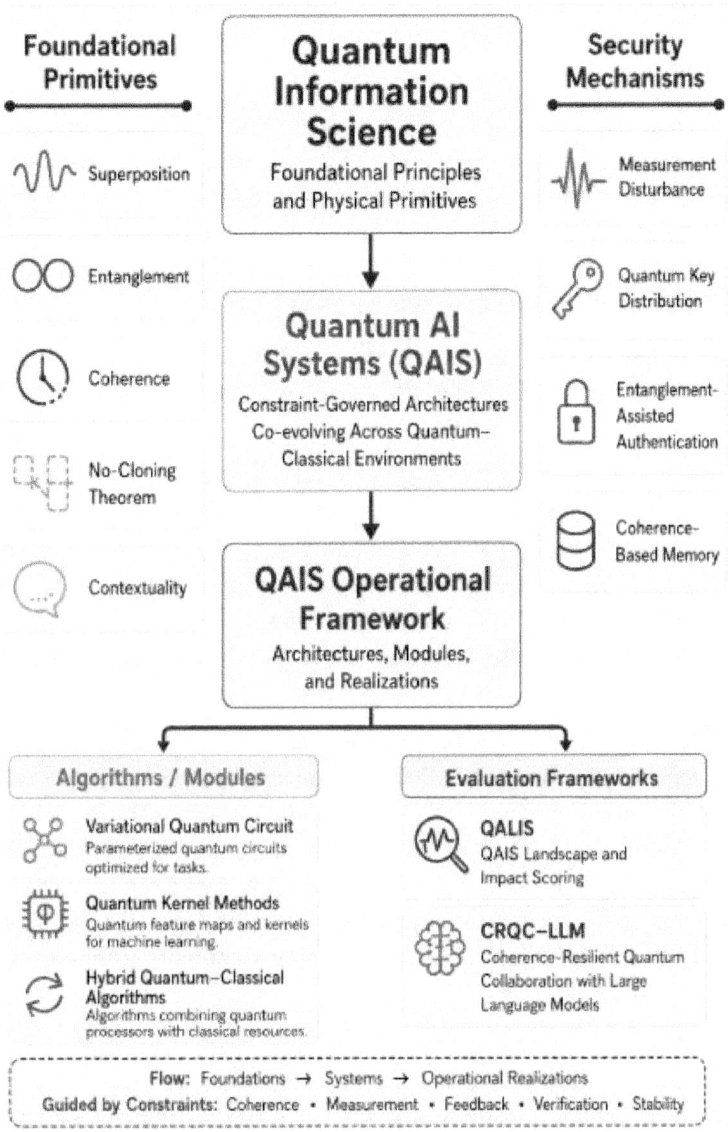

Figure 3.8.1 — Glossary of Quantum Terms as Workforce Skill Domains.

This diagram organizes foundational QAIS terms into four categories—primitives, algorithms and modules, security mechanisms, and case studies—forming a skills framework for building, deploying, and securing QAIS architectures. It also reflects the bidirectional relationship between theory and implementation: foundational principles guide system design, while deployment constraints shape their operational use. The figure therefore functions as both a knowledge taxonomy and a representation of how QAIS expertise evolves through practice.

With these domains mapped as the skills backbone of QAIS, the discussion turns now to the challenge of cultivating a workforce fluent in both theory and applied system design, an indispensable step for QAIS maturity.

Technical Spotlight — Glossary as a Skills Framework

Conceptual Grounding

The glossary serves as more than a static list of definitions. It shows how primitives such as superposition, entanglement, and coherence connect directly to higher-order modules like variational circuits and kernels. These in turn extend into security mechanisms such as quantum key distribution or entanglement-assisted authentication. By positioning these elements side by side, the framework highlights QAIS as a system of interdependent competencies rather than isolated concepts.

This reinforces that fluency in QAIS is cumulative: mastery of primitives enables algorithms, which in turn inform architectures, resilience strategies, and case study applications.

System Relevance

For QAIS, the glossary-as-framework highlights which competencies are essential for system design. Superposition and entanglement are not abstract curiosities but operational resources. Variational circuits and hybrid algorithms define usable pipelines. Security primitives ground trust in physical law. QALIS and CRQC–LLM serve as applied anchors, reminding practitioners that competencies must be stress-tested in both constructive and adversarial contexts.

Cross-Disciplinary Orientation

Physicists interpret primitives in terms of Hilbert space dynamics and experimental validation. Engineers translate modules into hardware-ready implementations, embedding error correction and calibration layers. AI specialists adapt quantum algorithms into representational pipelines that extend learning capacity. Cybersecurity experts connect no-cloning and disturbance to verification frameworks. Educators and policymakers treat the glossary as a roadmap for building coherent training programs that link foundational science to deployable architectures.

By turning terminology into a skills map, this spotlight sets the stage for the next Section (3.8.2). There, the focus deepens into the mathematical and Quantum Physics that form the grammar of QAIS, linear algebra, probability, and quantum mechanics, without which higher-order modules cannot be developed.

Case Study Anchor — Glossary as an Operational Skills Framework

Intent: To reinterpret foundational terminology as an operational skills map that defines workforce readiness and cross-disciplinary fluency in QAIS.

- **QALIS (Design Framework):** Demonstrates how glossary terms function as operational competencies in real systems, such as coherence lifetimes and entanglement depth as design parameters in distributed sensing arrays.
- **CRQC–LLM (Stress-Test Lens):** Shows how incomplete or siloed training in these competencies can be exploited by adversaries, stressing the need for workforce readiness across all categories.

With the glossary reframed as a skills framework, the discussion now turns to the first competency domain: the mathematical and Quantum Physics that provide the representational grammar and operational rules of QAIS. Section 3.8.2 explores how linear algebra, complex numbers, probability theory, and quantum mechanics together form the baseline upon which all higher-level QAIS architectures are built.

3.8.2 Mathematical and Quantum Physics

The intellectual foundations of QAIS rest on mathematics and physics, which together form the grammar by which all higher-order architectures are expressed. These are not abstract skills learned in isolation but the operational language that makes it possible to represent, manipulate, and verify quantum information in real systems.

A robust command of *linear algebra and complex vector spaces* is essential. Quantum states are represented as vectors in Hilbert space, evolving through matrix operations that preserve probability. Tensor products formalize multi-qubit entanglement, while eigenvalues and eigenvectors determine the measurable outcomes of gates and circuits. Unitary operators guarantee that state evolution remains reversible, encoding the conservation of probability as a design rule.

Complex numbers provide more than symbolic convenience: their amplitudes map directly to probabilities, while their phases drive interference effects central to quantum learning.

Equally essential is a working fluency in *quantum mechanics and probability theory*. Superposition expands representational capacity, entanglement encodes non-classical correlations, and measurement collapse links abstract amplitudes to actionable outcomes. The Born rule provides the bridge between wavefunction amplitudes and probabilities, while the uncertainty principle sets operational limits on what can be simultaneously known or measured. Practitioners must be able to apply the Schrödinger equation to model state evolution, and to reason probabilistically about algorithmic outcomes, inference pipelines, and error distributions.

Competency in QAIS spans multiple foundational disciplines that collectively support system understanding and implementation. These competencies define the skill set required for effective participation in the field. Figure 3.8.2 illustrates these domains within a unified perspective, clarifying how they contribute to system-level capability.

Foundations as the Grammar of QAIS Competence

Before delving into specific domains of skill, it is essential to recognize that QAIS depends on a layered competency framework that unites mathematics, physics, computation, and systems thinking. The figure below reframes terminology into a structured map of workforce skills, showing how conceptual knowledge evolves into operational capacity for building, deploying, and securing QAIS architectures.

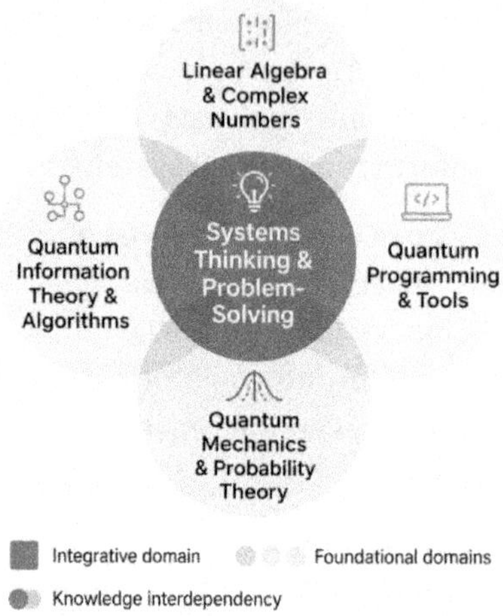

Figure 3.8.2 — Foundational Domains of Quantum AI Systems Competency.

This radial diagram highlights five overlapping domains: Linear Algebra & Complex Numbers, Quantum Mechanics & Probability Theory, Quantum Programming & Tools, Quantum Information Theory & Algorithms, and Systems Thinking & Problem-Solving. The central placement of systems thinking reflects its integrative role in uniting technical knowledge into deployable QAIS architectures.

Technical Analysis — Foundational Constraint Navigation in QAIS

Conceptual Grounding

The foundational disciplines of QAIS define the permissible structure of quantum information processing. Linear algebra establishes the representational space in which states evolve, probability theory governs statistical interpretation, and quantum mechanics defines the transformation and measurement rules that distinguish quantum systems from classical ones.

These disciplines together form a constraint framework rather than a toolbox: they specify what operations are allowed, how information can interfere, and where classical extraction becomes irreversible.

At this level, the concern is not computational advantage but structural admissibility—what kinds of transformations and correlations can exist without violating physical law.

System Relevance

These constraints propagate directly into architecture. Circuit depth is bounded by coherence, transformation fidelity limits achievable precision, and measurement boundaries determine how and when quantum information becomes classically actionable. Design decisions in QAIS therefore emerge from constraint navigation: selecting representations, estimating tolerable error, and aligning algorithmic structure with physical viability. Systems that ignore these foundational limits accumulate hidden fragility that later manifests as instability, unverifiable outputs, or security exposure.

Cross-Disciplinary Orientation

- Physicists formalize the permissible dynamics of quantum states and characterize decoherence regimes.
- Engineers translate those dynamics into control systems, cryogenic platforms, and error-corrected gates.
- Computer scientists and AI specialists map representational structures into algorithmic patterns that respect physical bounds.
- Cybersecurity professionals interpret measurement disturbance and no-cloning as architectural defense surfaces.

Each discipline operates within the same constraint geometry but contributes a different operational layer, making foundational literacy a shared language rather than a specialized silo. With constraint navigation established as the baseline competency, the discussion advances from admissibility to execution. Section 3.8.3 shifts focus toward computational fluency, examining how algorithmic reasoning and toolchains transform physically valid representations into deployable hybrid pipelines under real infrastructure conditions.

Case Study Anchor — Constraint Literacy in Distributed QAIS Architectures

Intent: To illustrate how foundational constraint literacy determines whether distributed quantum sensing and learning systems can be scaled without exceeding coherence, fidelity, or verification thresholds.

- **QALIS (Design Framework):** Treats entanglement depth, measurement timing, and coherence lifetime as explicit architectural parameters in distributed sensing arrays.
- **CRQC–LLM (Stress-Test Lens):** Exposes how shallow constraint modeling enables adversarial exploitation of error surfaces and verification boundaries.

With foundational constraints framed as navigational boundaries for admissible QAIS design, attention now shifts from structural validity to executable realization.

Section 3.8.3 examines how computational fluency transforms physically permissible representations into controlled hybrid pipelines, integrating algorithmic reasoning, orchestration discipline, and toolchain validation under real infrastructure conditions.

3.8.3 Computational Thinking and Practical Tools

At the core of QAIS practice is *systems thinking*, the ability to reason across layers of abstraction, from circuit design to algorithmic performance to infrastructure constraints. QAIS extends Shannon's classical information theory into the quantum domain, where entropy, mutual information, and channel capacity must be reinterpreted for superposed and entangled states. This demands practitioners who can think not only mathematically but also architecturally, connecting information-theoretic principles to deployable system pipelines.

Quantum algorithms such as *Shor's factoring* and *Grover's search* demonstrate how complexity classes shift under quantum resources. Understanding these shifts requires fluency in computational complexity, resource estimation, and hybrid orchestration, skills that allow professionals to evaluate trade-offs and design algorithms adapted to noisy, intermediate-scale quantum (NISQ) devices as well as future fault-tolerant systems.

Practical QAIS implementation depends on computational thinking and tool usage that bridge theoretical concepts and applied system design. These capabilities enable the translation of knowledge into operational systems.

Figure 3.8.3 show these elements within a practice-oriented framework, illustrating their role in system development.

On the practical side, fluency with platforms such as *QAISkit, Cirq, and PennyLane* is indispensable. These environments support circuit simulation, variational algorithm design, and noise-mitigation testing. Professionals must be able to debug, optimize, and architect across hybrid pipelines that combine quantum accelerators with classical CPUs and GPUs, turning conceptual designs into functioning QAIS workflows.

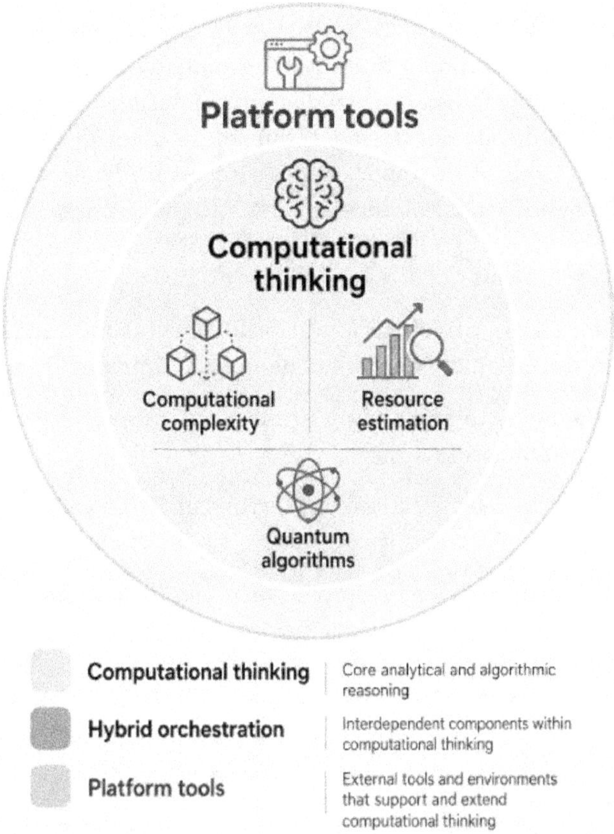

Figure 3.8.3 — Computational Thinking and Tools in QAIS Practice.

This figure depicts three concentric layers: the inner layer highlights computational complexity and quantum algorithms, the middle layer shows hybrid orchestration and algorithm-infrastructure integration, and the outer layer represents platform tools such as QAISkit, Cirq, and PennyLane. Together, they illustrate how theoretical reasoning is disciplined into deployable execution under resource and infrastructure constraints.

Technical Spotlight — Translating Abstraction into Deployment

Conceptual Grounding

The distinctive insight here is that computational thinking in QAIS must reinterpret classical information theory for quantum contexts while also accounting for new algorithmic landscapes.

Concepts such as entropy and capacity evolve when applied to superposed or entangled states, and practitioners must learn to navigate these adaptations while balancing resource constraints in hybrid systems.

System Relevance

For QAIS architectures, computational fluency determines whether algorithms are feasible, scalable, and resilient. Complexity analysis guides system design choices, while platform tools allow practitioners to implement, test, and refine these algorithms under real hardware constraints. Without this bridge from theory to practice, even advanced insights into quantum mechanics and linear algebra cannot mature into deployable systems.

Cross-Disciplinary Orientation

Machine learning practitioners treat computational fluency as the foundation for designing **variational hybrid models** that leverage entanglement as representational power.

Cybersecurity experts apply it to embedding primitives like measurement disturbance into verification and cryptographic protocols.

Engineers translate Hilbert space operations into error-corrected gate sequences and scalable control electronics.

AI theorists evaluate trade-offs between classical and quantum resources in hybrid orchestration pipelines, balancing stability against expressiveness.

Case Study Anchor — Translating Abstraction into Deployment

Intent: To examine how abstract QAIS concepts are operationalized into deployable systems through tooling, training, and infrastructure integration.

- **QALIS (Design Framework):** Demonstrates how hybrid orchestration is applied in distributed sensing arrays, where algorithm design and computational tools ensure coherence is maximized and error is minimized.
- **CRQC-LLM (Stress-Test Lens):** Highlights how adversaries could exploit weak algorithmic verification or under-tested pipelines, showing why computational fluency must extend to resilience strategies under CRQC-LLM guided stress.

With computational fluency established as the bridge between abstract principles and deployable pipelines, the question becomes whether professionals can adapt these skills as the field evolves. Algorithms, platforms, and orchestration methods will continue to change, making adaptability and interdisciplinary collaboration just as essential as technical literacy. By embedding computational literacy into practical tools, this spotlight prepares the ground for Section 3.8.4. There, the focus expands to human competencies, adaptability, lifelong learning, and systems-level collaboration, which ensure that computational skills remain relevant as QAIS platforms and adversarial environments evolve.

Section 3.8.4 turns to the human dimension of QAIS, where competencies such as communication, contextual awareness, and lifelong learning become operational requirements for sustaining progress and resilience.

3.8.4 Human Competencies and Lifelong Learning

QAIS depends on human adaptability as much as technical precision. Practitioners must be able to *collaborate across vocabularies and disciplines*, bridging physics, software engineering, AI modeling, and hardware design. Communication, contextual awareness, and systems-level thinking are not "soft skills" in QAIS, they are operational competencies that enable deployment and ensure resilience.

Equally essential is a *mindset of continual learning*. Quantum toolchains, architectures, and protocols evolve at a rapid pace. New abstractions, error correction strategies, and hybrid orchestration methods will continue to emerge. Professionals must remain agile in updating their skillsets, cultivating the ability to translate theory into application while adapting to shifting technologies.

QAIS professionals operate at the intersection of multiple disciplines, requiring the integration of knowledge across technical and conceptual domains. This interdisciplinary structure defines how systems are designed and managed. Figure 3.8.4 map these relationships within a competency framework, illustrating the breadth of required expertise.

3.8.5 Human Adaptability as a QAIS Competency

Figure 3.8.4 places lifelong learning at the boundary of all QAIS skill domains, emphasizing that adaptability is the connective tissue of professional growth. The figure highlights how systems thinking and problem-solving sit at the center, while mathematical, physical, programming, and algorithmic competencies orbit around it, framed by a commitment to continuous learning.

Interdisciplinary Skill Domains for QAIS Professionals

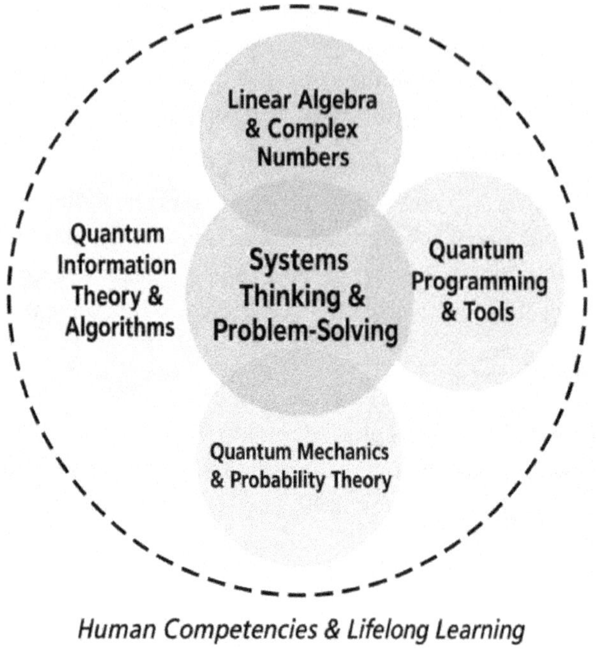

Figure 3.8.4 — Interdisciplinary Skill Domains for QAIS Professionals.

This radial diagram organizes five interdependent domains essential for QAIS expertise: Linear Algebra and Complex Numbers; Quantum Mechanics and Probability Theory; Quantum Programming and Tools; Quantum Information Theory and Algorithms; and Systems Thinking and Problem-Solving. A surrounding boundary labeled "Human Competencies & Lifelong Learning" frames these domains, emphasizing that adaptability integrates and sustains all technical competencies as QAIS platforms, protocols, and adversarial environments evolve. The dashed outer boundary represents an open and continuously evolving competency environment rather than a rigidly closed educational framework. It signifies that QAIS expertise develops through ongoing interdisciplinary learning, professional adaptation, and iterative integration across emerging quantum–classical systems.

Technical Analysis — Adaptive Capacity as Structural Infrastructure

The figure reframes adaptability not as professional enrichment but as structural infrastructure for QAIS maturity. Foundational knowledge and computational fluency establish admissibility and execution discipline; however, long-term system resilience depends on whether practitioners can reinterpret those competencies as technologies, error models, orchestration strategies, and adversarial tactics change. Lifelong learning becomes a stabilizing mechanism that prevents architectural stagnation.

In rapidly evolving hybrid ecosystems, adaptability functions as a resilience amplifier: it preserves coherence between theoretical insight, toolchain evolution, governance frameworks, and deployment realities. Without adaptive capacity, even technically sound systems degrade as their human stewards fall out of alignment with emerging constraints.

Conceptual Grounding

QAIS operates across interacting layers of abstraction. As new error correction codes, hybrid orchestration patterns, and representational strategies emerge, practitioners must recalibrate prior knowledge rather than discard it. The key insight is that professional growth mirrors architectural layering: foundational principles remain constant, but their operational instantiation evolves. Adaptability therefore acts as the connective tissue between stable physical laws and shifting engineering implementations.

System Relevance

From an architectural standpoint, human adaptability directly influences resilience. Deployment pipelines must be updated as coherence diagnostics improve; verification protocols must evolve alongside adversarial strategies; orchestration frameworks must adapt to new hardware topologies. Workforce rigidity becomes a systemic vulnerability, introducing lag between technological evolution and institutional readiness. Adaptive capacity ensures that QAIS architectures remain aligned with both physical constraints and emerging threat landscapes.

Cross-Disciplinary Orientation

Engineers refine calibration and error mitigation strategies as hardware platforms mature. AI specialists incorporate emerging variational paradigms and hybrid optimization methods. Cybersecurity professionals anticipate adversarial innovation guided by CRQC-LLM stress modeling. Physicists reinterpret theoretical insights in light of new experimental regimes.

Educators and policymakers institutionalize these adaptive processes through curriculum design, certification pathways, and cross-sector coordination. In each role, adaptability preserves coherence across disciplines and prevents fragmentation of expertise.

Case Study Anchor — Adaptive Workforce as Defensive Infrastructure

Intent: To demonstrate how sustained workforce adaptability determines whether QAIS architectures remain resilient as platforms evolve and adversarial strategies intensify.

- **QALIS (Design Framework):** Shows how iterative refinement of coherence diagnostics, orchestration strategies, and verification pipelines strengthens distributed QAIS deployments.
- **CRQC-LLM (Stress-Test Lens):** Reveals how stagnant expertise becomes an exploitable surface, allowing adversaries to capitalize on outdated models, incomplete validation practices, or unpatched orchestration vulnerabilities.

With foundational literacy, execution discipline, and adaptive capacity integrated into a unified workforce model, the chapter transitions from human infrastructure to system architecture.

Section 3.9 ("QAIS and Quantum AI") builds directly on this competency scaffold, showing how QAIS primitives and modules organize into hybrid Quantum AI architectures that must be both expressively powerful and resilient under the adversarial pressures highlighted by CRQC-LLM.

3.8.6 Building the QAIS Workforce Pipeline

Interdependent Barriers to QAIS Deployment

The systemic challenge of workforce development and equitable access makes clear that QAIS maturity depends as much on people as on platforms. Standards, infrastructure, and physical stability cannot achieve their potential without a globally trained community capable of designing, validating, securing, and deploying QAIS architectures. These human and technical bottlenecks do not exist independently—they interlock as part of the broader constraint network introduced in **Section 3.7**.

As **Section 3.7** synthesizes, decoherence limits, scalability barriers, verification gaps, fragmented standards, and talent shortages are not parallel challenges. They are tightly coupled layers within a single system architecture. Improvements in coherence extend executable circuit depth, which in turn strengthens verification reliability. Robust standards accelerate workforce training, while workforce growth lowers operational cost and increases system accessibility. Conversely, fragility in any one domain amplifies vulnerabilities across the others, demonstrating why coordinated progress across the full ecosystem is required for QAIS to mature into a deployable system discipline.

Figure 3.8.5 captures this dual challenge of education and access. It shows how fragmented curricula, uneven global training pathways, and restricted hardware availability create structural inequities in QAIS readiness—while also highlighting emerging solutions such as modular quantum curricula, open-access cloud hardware, and distributed learning platforms that can expand and diversify the workforce pipeline. Workforce and access barriers influence the ability to scale QAIS development and adoption, affecting both participation and system deployment.

These barriers introduce constraints that extend beyond technical challenges. Figure 3.8.5 situates these issues within a systemic perspective, highlighting their impact on field growth.

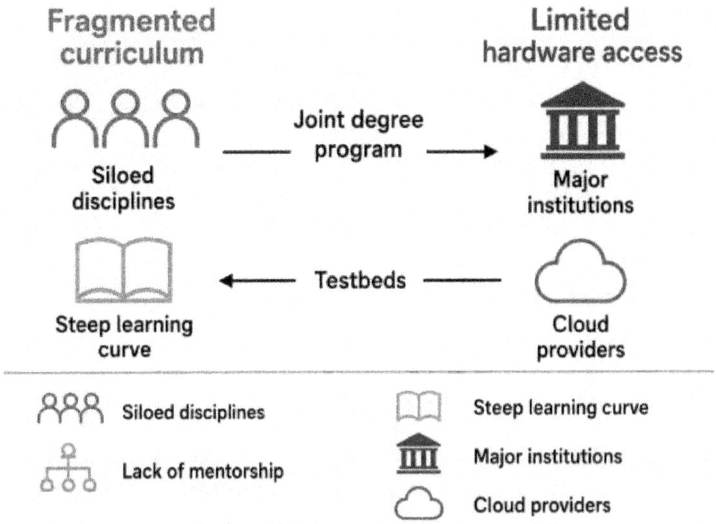

Figure 3.8.5 — Workforce and Access Barriers in QAIS.

This figure illustrates how education pathways, curriculum fragmentation, hardware access, and training infrastructure shape the global QAIS workforce pipeline. The left side highlights structural constraints—uneven hardware access, limited training capacity, and inconsistent curricula—that produce disparities in readiness. The right side presents emerging enablers, including modular curricula, cloud-based quantum platforms, international training initiatives, and distributed learning ecosystems. Together, these factors position workforce development and access as system-level constraints that directly influence QAIS scalability, resilience, and global deployment capacity.

Technical Spotlight — Workforce and Access Barriers in QAIS

Conceptual Grounding

Workforce development in QAIS cannot follow the trajectory established by classical computing, where well-defined academic pipelines fed directly into industry. QAIS requires practitioners who can operate simultaneously across physics, electrical engineering, computer science, machine learning, and cybersecurity. This interdisciplinary breadth is not optional; it is integral to understanding qubit behavior, hybrid algorithm design, error-correction constraints, and adversarial resilience.

The second dimension of the problem is access. Unlike classical fields where developers can experiment using commodity hardware, QAIS depends on cryogenic processors, stabilized photonic platforms, or high-fidelity superconducting devices—resources concentrated within a small number of elite institutions and corporations.

While cloud platforms (IBM Quantum, Amazon Braket, Xanadu, etc.) expand nominal availability, the access they provide is often shallow: limited shot counts, restricted queuing, or devices too noisy for advanced experimentation.

Figure 3.8.5 visualizes these twin barriers. Fragmented curricula create siloed training paths and steep learning curves, while limited hardware access reinforces institutional inequities. Together, they act as **equity multipliers**, determining who can meaningfully contribute to QAIS and who remains structurally excluded from the field.

System Relevance

Workforce and access barriers directly shape QAIS maturity. Insufficient interdisciplinary training slows algorithm development, impedes system integration, and weakens verification practices. Restricted access to hardware limits experimentation, constrains reproducibility, and centralizes innovation within a small number of regions. These constraints are not peripheral—they determine the scalability, security, and resilience of QAIS architectures. Without a broad and well-trained workforce, neither standards nor infrastructures can deliver their intended impact.

Cross-Disciplinary Orientation

Physicists and engineers define curriculum modules that reflect real device-level limitations such as coherence windows, gate fidelities, and calibration cycles. AI specialists integrate hybrid algorithm design, showing how quantum kernels, variational circuits, and interference-driven feature extractors extend representational power. Cybersecurity experts ensure that education includes CRQC–LLM–aware verification, secure control protocols, and post-quantum cryptographic models. Educators and policymakers establish degree pathways, micro-credentials, open curricula, and equitable testbed access to democratize entry into the field.

Together, these communities transform QAIS workforce development into a holistic systems discipline rather than a collection of siloed training tracks.

This spotlight demonstrates that QAIS maturity depends as much on *human infrastructure* as on physical and computational infrastructure. Workforce shortages and inequitable access act as systemic constraints that reinforce fragility across the technical stack. This directly sets the stage for the synthesis that follows, where these human and technical barriers are synthesized into a single system-level constraint network, showing how weaknesses in one domain reverberate across all others.

Case Study Anchor — Workforce Gaps as Systemic Risk

Intent: To highlight workforce shortages and skills fragmentation as risks equivalent to physical decoherence, capable of destabilizing QAIS deployments at scale.

- **QALIS (Design Framework):** Demonstrates how national testbeds, interdisciplinary fellowships, and shared training infrastructures expand participation and create resilient pipelines for deploying QAIS architectures in sensing, optimization, and distributed reasoning systems.
- **CRQC–LLM (Stress-Test Lens):** Shows how inequitable access to high-fidelity hardware can itself become a vulnerability. Actors with privileged infrastructure can outpace global resilience efforts, widening capability gaps and introducing asymmetries that CRQC–LLM adversaries can exploit.

The barriers affecting QAIS deployment are often interdependent, forming a network of constraints that collectively influence system feasibility and scalability. Understanding these relationships is essential for addressing them effectively. Figure 3.8.6 situates these interactions within a system architecture view, illustrating how multiple constraints combine to shape deployment outcomes.

Barriers as an Interdependent Constraint Network

Figure 3.8.6 reframes QAIS barriers as an interdependent architecture. Rather than treating them as isolated challenges, the figure organizes them into layered domains of physical infrastructure, platform design, and ecosystem access, with operational cost cutting across all layers. This system view highlights the feedback loops that define QAIS maturity, underscoring why coordinated progress is essential for scalable, deployable architectures.

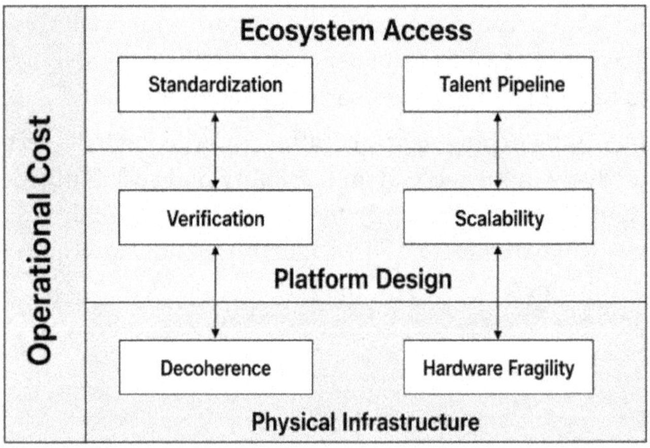

Figure 3.8.6 — Interdependent Barriers in QAIS Deployment.

This figure organizes six major QAIS barriers into three architectural layers: physical infrastructure (decoherence, hardware fragility), platform design (verification, scalability), and ecosystem access (standardization, talent pipeline). Operational cost is shown as a vertical band spanning all layers, reflecting its cross-cutting influence. Bidirectional arrows indicate feedback relationships where progress in one domain reinforces, or is constrained by, another.

Technical Spotlight — Coherence, Verification, and Cost as Coupled Constraints

This figure builds on the prior subsections, physical fragility (Section 3.7.2), infrastructure standards and verification (Section 3.7.3), and workforce access (Section 3.7.4), and synthesizes them into a system-level view. It demonstrates that QAIS maturity is shaped by the interplay among coherence, verification, and operational cost.

Conceptual Grounding

In QAIS, barriers are not isolated obstacles but interconnected design rules. Coherence is no longer just a fragile property of qubits; it defines the usable lifetime of computation and inference.

Verification is not a deterministic check, as in classical computing, but a probabilistic process relying on entanglement witnesses, randomized benchmarking, or cross-entropy validation. Operational cost extends beyond finances, shaping who can access QAIS infrastructures and how quickly systems scale. These dimensions reinforce one another: extending coherence times improves verification fidelity, which in turn reduces cost by enabling more efficient architectures. Conversely, fragility in any one domain amplifies weaknesses across the others.

System Relevance

The interdependence of coherence, verification, and cost sets the operational boundaries of QAIS. Without coherence, algorithms collapse before completion; without scalable verification, results cannot be trusted; without manageable cost, access remains limited to a few elite institutions. Treating these as coupled constraints highlights why QAIS must evolve as a coordinated system science, not just as a series of isolated technical advances.

Cross-Disciplinary Orientation

Machine learning practitioners interpret coherence as a hyperparameter that constrains circuit depth, training epochs, and representational generalization. Engineers see verification both as a bottleneck and an opportunity to design hardware–software co-interfaces that improve scalability. Cybersecurity experts view operational cost as a strategic vulnerability, recognizing that inequitable access not only slows democratization but also limits adversarial testing, leaving resilience unevenly distributed across the ecosystem.

Physicists focus on extending coherence and entanglement lifetimes through materials and topological designs, while policymakers treat cost as a governance challenge, ensuring that funding and infrastructure support are distributed equitably. Collectively, these perspectives reveal that coherence, verification, and cost are not separate challenges but systemic levers whose balance defines the resilience of QAIS.

Case Study Anchor — Coherence, Verification, and Cost as Coupled Constraints

Intent: To frame coherence limits, verification overhead, and economic cost as interdependent constraints that jointly define viable QAIS deployment envelopes.

- **QALIS (Design Framework):** Illustrates how coherence management and probabilistic verification protocols are engineered into distributed sensing arrays, making inference pipelines resilient to noise and cost-efficient at scale.
- **CRQC–LLM (Stress-Test Lens):** Demonstrates how adversarial actors exploit verification gaps and cost pressures, using LLM-guided CRQC attacks to destabilize infrastructures at their weakest interdependencies.

Section 3.8 Summary — Building the Quantum Workforce

The case study anchor highlights that coherence limits, verification overhead, and economic cost do not act as isolated constraints, but as a tightly coupled system that governs the feasibility of QAIS deployment.

Through the QALIS lens, these constraints are engineered into resilient architectures; through the CRQC–LLM lens, they expose points of vulnerability under adversarial pressure. In both perspectives, the decisive factor is not the constraint itself, but the capacity to interpret, manage, and coordinate across these interdependencies.

This shifts the focus from system limitations to system capability: the ability of practitioners to design, verify, and adapt QAIS under real-world conditions. The constraint network therefore becomes a training ground, where physical principles, infrastructural requirements, and adversarial considerations converge into actionable expertise.

It is this need for coordinated, constraint-aware execution that motivates the discussion that follows. The development of a quantum-ready workforce is not an external support function, but the mechanism through which these coupled constraints are understood, mitigated, and ultimately transformed into stable, scalable system performance.

3.9 QAIS and Quantum AI

Architectural Framing: *Hybrid Architecture (Integration Layer)*

Section Framing: *This section establishes QAIS–AI integration as a hybrid control surface that combines quantum and classical intelligence architectures. Representation is extended beyond classical feature spaces into quantum state spaces, while propagation emerges as system behavior through hybrid workflows integrating quantum primitives and AI models. Stability is enforced through bounded propagation, where interface compatibility, noise sensitivity, and security constraints define integration limits. Hybridization therefore functions as the architectural pathway toward next-generation intelligent systems built on QAIS foundations.*

Earlier sections established QAIS as the operational foundation for quantum information and computation. Section 3.9 advances this trajectory by introducing Quantum AI as the architectural convergence of QAIS primitives with classical artificial intelligence.

In this model, QAIS provides the physical and operational substrate—superposition, entanglement, coherence, and unitary transformation—upon which hybrid intelligent systems are constructed.

From the operational perspective adopted throughout this book, ecosystem maturity reflects the increasing ability to reliably control interference, enforce measurement boundaries, and manage coherence as an execution-level resource rather than a theoretical abstraction.

As these capabilities stabilize, they enable quantum operations to participate meaningfully in learning, inference, and decision processes without undermining classical system guarantees.

Quantum AI does not replace classical AI; it expands its representational and computational landscape. Classical preprocessing, feature extraction, and optimization remain essential, but they are now interwoven with quantum-native operations such as state preparation, amplitude-based encoding, kernel embeddings, and interference-driven learning. This interweaving introduces new degrees of freedom while preserving the interpretability and control structures developed within classical AI practice.

The result is a hybrid system in which classical modules provide stability and scalability while quantum modules introduce richer geometries for representation, correlation, and inference. This convergence reflects a deeper conceptual shift. Classical AI operates on probabilities, linear transformations, and logic-based pipelines. Quantum AI extends this paradigm into

Hilbert space, where amplitudes encode structure inaccessible to classical systems and interference patterns shape decision landscapes.

Computation unfolds not as symbolic manipulation but as controlled evolution of physically embodied states, enabling new forms of expressivity, optimization, and model alignment.

Crucially, this shift does not eliminate classical reasoning but situates it within a broader operational context, where learning outcomes emerge from the interaction between quantum state dynamics and classical evaluation criteria.

Within this framing, the figures in Section 3.9 collectively illustrate how QAIS primitives scale into hybrid architectures.

They trace the progression from primitive operations to quantum-native modules, to integrated QAIS–AI pipelines, and finally to system-level stacks anchored in the physical substrate.

Their combined message is that Quantum AI is not an independent discipline layered onto classical AI, but a coherent extension built directly on the operational rules and physical constraints established by QAIS.

Section 3.9 therefore positions QAIS as the enabling foundation for hybrid intelligent systems, showing how physical substrate, quantum operations, and classical computation interlock to form deployable Quantum AI architectures. In doing so, it clarifies that the long-term viability of Quantum AI depends not on abstract quantum advantage claims, but on the disciplined integration of quantum operations into end-to-end systems governed by the same verification, robustness, and control requirements that shape all intelligent technologies.

Hybrid architectures provide a framework for integrating classical and quantum components into cohesive systems, enabling coordinated operation across domains.

These architectures define how different layers interact. Figure 3.9.1 situates this structure within a layered perspective, illustrating how hybrid integration supports system functionality.

3.9.1 Hybrid Architectural Convergence

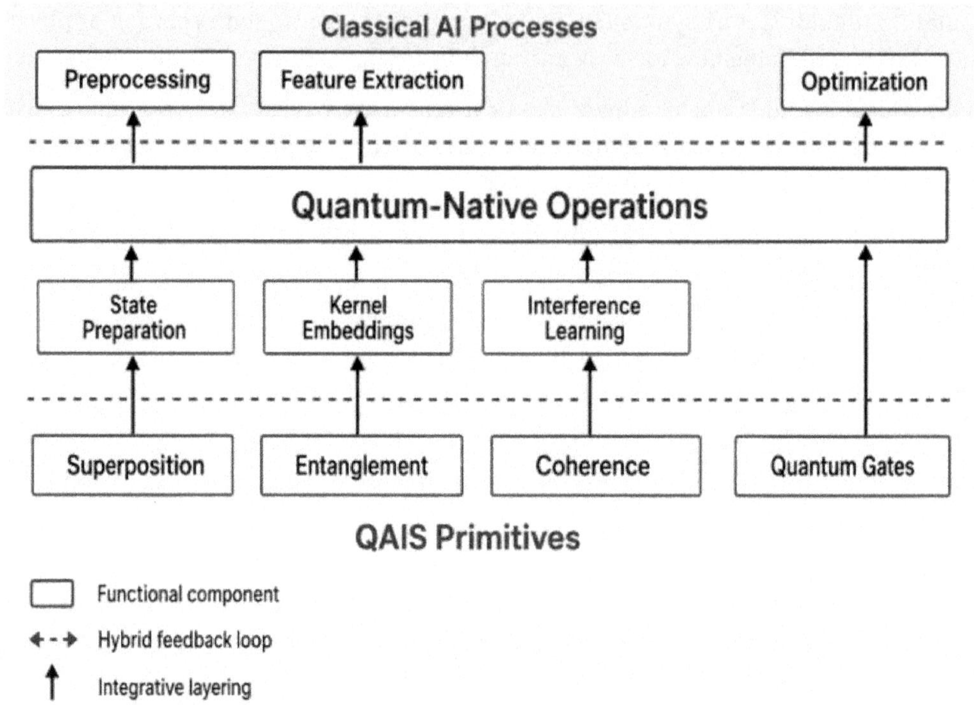

Figure 3.9.1 — Three Layer Hybrid Architecture.

This figure illustrates the three-layer hybrid architecture linking QAIS primitives, quantum-native operations, and classical AI processes. Upward arrows represent integrative layering from physical effects to algorithmic processes, while bold bidirectional arrows highlight the four critical feedback loops: Superposition ↔ State Preparation, Entanglement ↔ Kernel Embeddings, Coherence ↔ Interference Learning, and Optimization ↔ Quantum Gates. Together, these flows show how quantum resources, operations, and classical optimization form closed learning loops that enable deployable hybrid intelligent systems.

Technical Spotlight — Convergence as Architectural Integration

Conceptual Grounding

QAIS primitives form the architectural substrate of Quantum AI.

Superposition enlarges the representational space; entanglement encodes correlations beyond classical reach; coherence determines the operational window in which meaningful computation can occur; and quantum gates enact reversible transformations that shape inference pathways. These primitives act not as abstract principles but as the physical mechanisms through which hybrid intelligent behavior emerges.

System Relevance

Hybrid feedback loops couple QAIS primitives to algorithmic behavior. Superposition shapes the feasibility of state preparation; entanglement determines the expressivity of kernel embeddings; coherence governs whether interference-based learning can be exploited; and classical optimization tunes quantum gates within variational circuits.

These interactions define the operational envelope for Quantum AI and reveal where constraints, noise, depth limits, or circuit fragility, directly influence learning performance.

Cross-Disciplinary Orientation

Physicists extend coherence times and entanglement fidelity to ensure primitives remain viable. Engineers translate these primitives into compilers, middleware, and orchestration layers that make quantum-native operations programmable. AI practitioners design models that leverage superposition and interference as representational assets. Cybersecurity analysts evaluate whether hybrid feedback loops can be disrupted by CRQC–LLM adversaries, assessing resilience at the physical and algorithmic boundary.

Case Study Anchor — Convergence as Architectural Integration

Intent: To frame convergence not as technological blending but as architectural integration across quantum, classical, organizational, and human layers.

- **QALIS (Design Framework):** QALIS demonstrates how QAIS primitives become functional components of hybrid AI workflows, integrating coherence-aware state preparation, entanglement-driven kernel embeddings, and variational optimization loops into scalable, mission-critical architectures.
- **CRQC–LLM (Stress-Test Lens):** CRQC–LLM exposes how adversaries may target these same loops, destabilizing coherence, manipulating gate-optimization pathways, or exploiting fragile embedding structures, revealing where hybrid systems require additional verification and error-mitigation strategies.

With this foundation established, Section 3.9.2 drills deeper into how QAIS primitives evolve into modular system components, showing how variational circuits, quantum kernels, and hybrid algorithms form the deployable building blocks of intelligent architectures.

This spotlight completes the bridge from ecosystem constraints (Section 3.7) and adoption trajectories (Section 3.5) to hybrid system architecture. It establishes the conceptual foundation for later chapters, particularly Chapter 12, where QALIS is formalized as a design framework and CRQC–LLM is applied as a stress-test lens for feedback-loop resilience.

3.9.2 QAIS Modules as Intelligent System Foundations

Having established the logic of hybrid convergence at the architectural level, the discussion now turns to the modular components that operationalize this integration within deployable systems. QAIS foundations supply the modular building blocks that enable quantum-enhanced intelligent systems. In this section, primitives from Chapter 2 are no longer treated as abstract physical rules, they appear as **operational modules** that can be embedded directly into hybrid AI pipelines.

The deployment of QAIS systems often relies on modular components that can be configured and combined to meet specific operational requirements. These modules enable flexible system design. Figure 3.9.2 illustrate this modular approach within an architectural framework, highlighting its role in enabling scalable deployment.

Variational quantum circuits, quantum kernels, and hybrid quantum-classical algorithms represent the first generation of reusable QAIS components, forming the bridge between foundational physics and deployable intelligent systems.

Blocks to Modules in QAIS Architectures

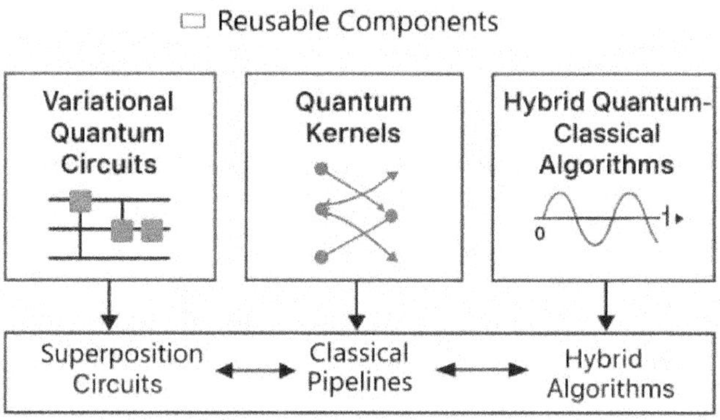

Figure 3.9.2 — Deployable QAIS Modules.

This figure presents three core QAIS modules, Variational Quantum Circuits, Quantum Kernels, and Hybrid Quantum-Classical Algorithms, as reusable components that integrate with classical pipelines. The diagram highlights how these modules mature from physical primitives into system-level building blocks used across intelligent architectures.

Technical Spotlight — Primitives as Deployable QAIS Modules

Conceptual Grounding

Variational quantum circuits (VQCs), quantum kernels, and hybrid algorithms demonstrate how QAIS primitives transition into **deployable system modules**.

- **VQCs** provide tunable quantum layers whose parameters are optimized via classical feedback, enabling them to approximate objective functions, perform classification or regression, and implement quantum neural networks.
- **Quantum kernels** map classical data into quantum Hilbert space, exposing relational structures through state overlaps that classical kernels may fail to reveal.
- **Hybrid quantum–classical algorithms** combine quantum feature extraction with classical post-processing, forming iterative workflows that stabilize raw quantum outputs into actionable results.

Together, these modules reflect the shift from primitive operations (superposition, entanglement, coherence) toward reusable components capable of forming the backbone of intelligent QAIS-AI systems.

System Relevance

These modules determine how QAIS advances into a deployable systems discipline. Reliable VQCs, robust quantum kernels, and scalable hybrid algorithms provide the foundation for consistent behavior on noisy hardware, reproducible training and evaluation cycles, and stable integration with classical pipelines. Their modular maturity also enables extensible design across sensing, optimization, cryptography, and machine learning domains.

Without the stability and interoperability of these components, hybrid quantum intelligence cannot progress beyond isolated laboratory demonstrations or small-scale proofs of concept.

Cross-Disciplinary Orientation

Physicists focus on ensuring experimental fidelity for kernels and circuit depth.

Engineers design middleware and error-correction layers that stabilize VQCs in noisy environments.

AI specialists adapt kernels and hybrid workflows for classification, clustering, and optimization.

Cybersecurity experts analyze how fragile kernels or insufficient error thresholds could expose vulnerabilities, making resilience part of the design process from the outset.

This cross-disciplinary coordination transforms QAIS modules into stable components of hybrid intelligent systems.

Case Study Anchor — Primitives as Deployable QAIS Modules

Intent: To show how QAIS primitives evolve into reusable modules that anchor both system design (QALIS) and adversarial stress-testing (CRQC–LLM).

- **QALIS (Design Framework):** Shows how VQCs, kernels, and hybrid algorithms can be embedded into distributed sensing, optimization, and classification pipelines, demonstrating modular, fault-aware design.
- **CRQC–LLM (Stress-Test Lens):** Examines how adversaries might exploit coherence limits or fragile kernel embeddings; highlights the importance of verification thresholds, error mitigation, and modular resilience.

By elevating operational primitives into functional system modules, Section 3.9.2 provides the missing link between the primitive-level mechanisms formalized in **Chapter 2** and the ecosystem-level modularization discussed earlier in **Chapter 3**.

It shows how superposition, entanglement, coherence, and gate operations evolve into reusable components—VQCs, quantum kernels, hybrid algorithms—that can be embedded into larger intelligent workflows.

This progression prepares the ground for **Section 3.9.3**, where these modules are situated within the broader architectural context of classical and quantum stacks. There, the chapter contrasts how classical and quantum layers interconnect, demonstrating how the modules highlighted in Figure 3.9.2 become active participants in layered system design, ultimately enabling hybrid QAIS–AI pipelines.

3.9.3 Modular QAIS Components for Hybrid Architecture

These foundational modules gain practical significance when examined as reusable architectural components embedded within broader hybrid pipelines. Figure 3.9.2 depicts three foundational QAIS modules, Variational Quantum Circuits (VQCs), Quantum Kernels, and Hybrid Quantum, Classical Algorithms, positioned as distinct yet interconnected components within a broader architectural pipeline.

The diagram highlights their modularity by isolating them as independent building blocks, while the bidirectional arrows represent functional interdependence, emphasizing that these modules are designed to collaborate rather than operate in isolation.

Variational Quantum Circuits (VQCs): VQCs provide parameterized circuit templates whose coefficients can be tuned across tasks. A single VQC topology can be reused for classification, regression, clustering, or generative modeling through reparameterization.

- **Quantum Kernels:** Quantum kernels embed data into Hilbert space where inner products capture relational features unavailable to classical algorithms. Their modularity allows kernels to be exchanged or scaled by adjusting embedding depth and circuit structure.

- **Hybrid Quantum–Classical Algorithms:** These algorithms coordinate quantum subroutines with classical optimization loops, enabling end-to-end learning and inference on NISQ-era machines. Their composability makes hybrid algorithms a natural control layer for incorporating quantum modules into classical AI systems.
- **Architectural Significance of Reusable Components:** A QAIS module qualifies as reusable when it exhibits modularity, allowing independent development and verification; composability, ensuring that it interoperates cleanly across diverse system pipelines; and portability, enabling the same module to be applied across tasks, hardware backends, and operational domains. When these properties are present, modularization accelerates system integration and supports scalable hybrid AI design, turning quantum components into dependable building blocks for larger intelligent architectures.
- **Implications for System Design:** Reusable QAIS modules have direct implications for system design. Because each component can be independently tested, they enhance verification and reliability across the pipeline. Their standardized interfaces support seamless integration across heterogeneous platforms, while their separability enables consistent benchmarking for reproducibility and maturity tracking.

Together, these characteristics reduce architectural complexity and allow hybrid systems to scale more effectively, transforming modular quantum capabilities into deployable system-level functionality.

Research Outlook

Advancing QAIS at the module level requires coordinated research across several fronts. Error mitigation techniques must improve to stabilize VQCs and quantum kernels on noisy hardware, while standardized interfaces are needed to ensure interoperability across diverse software and hardware ecosystems.

Robust cross-platform benchmark suites will be essential for evaluating maturity and reproducibility as systems scale. Equally important is the development of security-aware design practices that protect modules against CRQC-driven adversarial perturbations, ensuring that hybrid architectures remain resilient as quantum capabilities expand.

Summary of QAIS Foundational Modules

Table 3.9.1consolidates the three foundational QAIS modules illustrated in Figure 3.9.2, summarizing their primary functions, reusability profiles, and roles within hybrid AI systems. These modules—Variational Quantum Circuits, Quantum Kernels, and Hybrid Quantum–Classical Algorithms—represent the first layer of deployable components built on QAIS primitives.

Table 3.9.1 — Summary of QAIS Foundational Modules.

Module	Primary Function	Reusability	Role in Hybrid AI Systems
Variational Quantum Circuits (VQCs)	Parameterized quantum circuits optimized via classical feedback loops to approximate target functions or minimize cost landscapes.	**High:** Same circuit ansatz can be reparameterized for classification, regression, or generative tasks.	Serve as tunable quantum layers enabling adaptive learning, pattern recognition, and quantum neural network construction.
Quantum Kernels	Map classical data to quantum Hilbert space to compute state overlaps and expose relational structures.	**Moderate–High:** Kernel choice and embedding depth can be adapted to different data distributions and tasks.	Provide a quantum-enhanced feature space for kernel-based learning (e.g., SVMs, clustering), offering potential quantum advantage in classification and regression.
Hybrid Quantum–Classical Algorithms	Combine quantum subroutines with classical post-processing and optimization to produce iterative improvement cycles.	**High:** Algorithmic frameworks can be reused across domains by swapping quantum submodules and modifying classical optimization strategies.	Orchestrate end-to-end hybrid pipelines, enabling practical use of quantum resources on NISQ hardware while leveraging classical compute for stability and scalability.

The table provides a structured reference showing how each module transforms physical quantum effects into functional capabilities that can be integrated, reused, and scaled within intelligent architectures.

3.9.4 Architectural Layers of Quantum AI Systems

QAIS provide the physical and operational substrate from which hybrid intelligent architectures emerge. Understanding this progression requires contrasting the structural layers of classical AI pipelines with the layered organization of quantum systems. Classical workflows are built around preprocessing, feature engineering, and optimization, whereas quantum systems introduce state preparation, entangled kernel embeddings, and interference-driven feature extraction.

Differences between classical and quantum AI architectures reflect underlying distinctions in representation, computation, and system behavior. These differences influence how systems are designed and evaluated. Figure 3.9.3 situates this comparison within a unified perspective, clarifying the implications for hybrid system development. Comparing these stacks clarifies the architectural logic behind each paradigm and establishes the foundation for hybrid QAIS–AI integration.

Modules to Layered Architectures

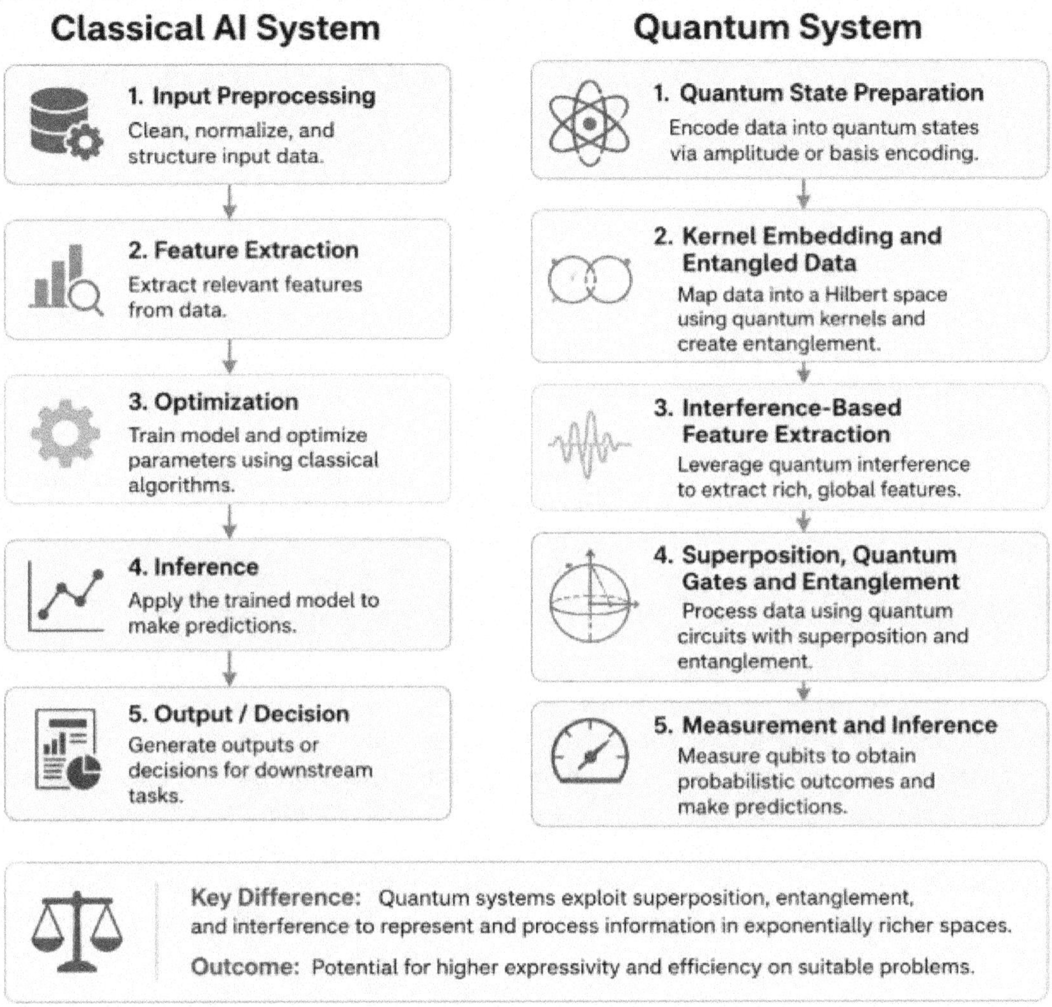

Figure 3.9.3 — Classical vs Quantum AI Architectures.

This block-diagram illustration presents a side-by-side comparison of classical AI and quantum system architectures. The classical stack emphasizes preprocessing, feature extraction, modeling, and optimization. The quantum stack introduces state preparation, entangled kernel embedding, and interference-based feature extraction. At its base, highlighted in blue, lie the QAIS primitives, superposition, entanglement, and quantum gates, that function as the foundational physical resources supporting all higher-order quantum operations.

This comparative view provides the architectural context for hybrid workflows developed later in the chapter, demonstrating how classical preprocessing and optimization interleave with quantum-native operations such as state preparation, kernel embedding, and interference-driven learning. By aligning the two stacks, QAIS reveals how classical and quantum layers can be coupled into unified, layered workflows that leverage the strengths of each paradigm.

Technical Spotlight — Contrasting Foundations of Classical vs Quantum AI

Conceptual Grounding

Figure 3.9.3 connects the primitive-level modules introduced in Section 3.9.2 with the architectural structures required for hybrid system design. Classical AI architectures rely on engineered features, linear algebraic operators, and optimization cycles.

Quantum architectures introduce fundamentally different layers: state preparation to encode data into quantum states; entangled kernel embeddings to capture high-dimensional relational structure; and interference-based feature extraction to reveal patterns inaccessible to classical models.

Beneath these layers are the QAIS primitives, superposition, entanglement, and quantum gates—which function as the physical building blocks enabling all higher-order quantum operations.

System Relevance

Contrasting these stacks highlights why QAIS primitives matter at the architectural level. Superposition expands representational capacity by allowing amplitudes to encode richer feature spaces. Entanglement imposes global relational structure across subsystems, enabling correlations that classical models cannot replicate. Quantum gates define reversible transformation rules that govern signal evolution and interference. These primitives collectively determine the feasibility, expressivity, and efficiency of interference-driven feature extraction, kernel embedding, and hybrid co-optimization. By positioning classical and quantum layers side by side, the figure exposes the architectural "connection points" that make hybrid QAIS pipelines actionable.

Cross-Disciplinary Orientation

From a machine learning perspective, quantum-native kernels and interference-driven feature extraction extend representational capability beyond the limits of classical linear algebra, accessing Hilbert-space geometries that encode richer decision boundaries.

For engineers, the comparison underscores how quantum gates, coherence budgets, and timing constraints introduce new hardware and calibration challenges that must be accounted for in system design.

Cybersecurity practitioners view fragile entanglement, incomplete error correction, and hybrid control surfaces as potential vulnerabilities, foreshadowing attack vectors that must be addressed through robust verification mechanisms.

Together, these perspectives emphasize that QAIS primitives demand an integrated systems mindset, where representation, architecture, and resilience are inseparable.

Case Study Anchor — Classical vs Quantum AI as Ecosystem Divergence

Intent: To contrast classical AI and quantum AI not at the algorithmic level, but as divergent ecosystem trajectories shaped by physical constraints, verification demands, and infrastructure requirements. This anchor establishes why QAIS cannot inherit classical AI deployment assumptions.

- **QALIS (Design Framework):** Demonstrates how kernels, variational circuits, and interference-based operations transition from primitive demonstrations into operational modules embedded within distributed sensing, classification, and reasoning pipelines.
- **CRQC–LLM (Stress-Test Lens):** Reveals how fragilities in foundational quantum layers—such as short coherence windows or insufficient error-correction thresholds—can propagate upward and destabilize entire hybrid workflows under adversarial pressure.

This layered comparison prepares readers for Figure 3.9.4, where classical and quantum stacks are formally integrated into a hybrid architecture. These integrated pathways define how data, features, and gradients circulate within Quantum AI workflows, forming the operational basis for full QAIS–AI convergence.

3.9.5 Bridging Classical and Quantum

The hybrid Quantum AI stack represents the convergence point where classical and quantum paradigms interleave into unified architectures. Classical preprocessing and optimization remain indispensable, providing stability, reproducibility, and deployment assurance. Quantum-native layers expand representational capacity through state preparation, entangled kernels, and interference-based feature extraction. Together, these layers form hybrid pipelines that translate QAIS primitives into functional architectures for intelligent systems.

Bridging classical and quantum layers requires coordinated mechanisms for translating and managing information across domains. These mechanisms define how hybrid pipelines operate. Figure 3.9.4 illustrates this bridging process within a system-level framework, illustrating how cross-layer integration is achieved

Bridging Paradigms Through Layered Integration

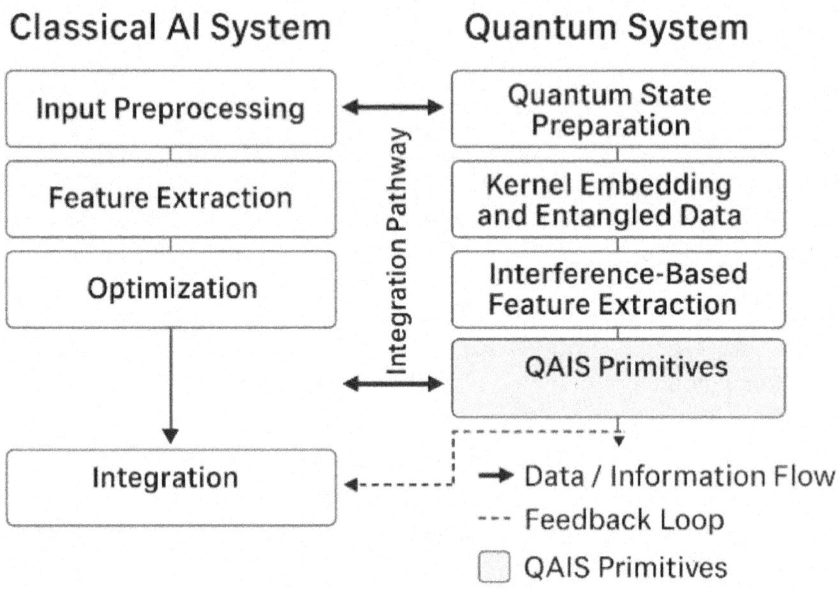

Figure 3.9.4 — Bridging Classical and Quantum Layers in QAIS–AI Pipelines.

This figure illustrates how classical and quantum operations are interleaved into a hybrid QAIS–AI architecture. The left side highlights classical preprocessing and optimization layers, ensuring reproducibility and system stability. The right side shows quantum-native operations, including state preparation, entangled kernel embeddings, and interference-driven feature extraction. The integration pathway in the center demonstrates how these layers combine to create unified pipelines with both stability and expanded representational power.

Technical Spotlight — Integration Pathways as Convergence Mechanisms

Figure 3.9.4 introduces the architectural pathways that allow classical and quantum subsystems to exchange information within hybrid QAIS–AI pipelines. Classical preprocessing prepares data for state encoding, quantum feature extraction produces interference-shaped representations, and classical optimization refines parameters through iterative loops.

Co-optimization across these layers ensures that quantum advantages translate into actionable, system-level performance.

From a security perspective, these pathways also introduce potential attack surfaces. CRQC–LLM adversaries may exploit fragile quantum layers, attempt to bias classical control signals, or manipulate optimization feedback loops. Securing these integration pathways requires authenticated control protocols, hardened interfaces, and quantum-safe cryptographic wrapping to ensure trust across classical–quantum boundaries.

Mitigation Strategies for Integration Pathways

To maintain robust and resilient hybrid QAIS operation, architectures should incorporate the following:

- **Error Mitigation & Verification:** Use randomized benchmarking, tomography variants, and cross-entropy validation to detect drift or malicious tampering.
- **Quantum-Safe Key Distribution:** Apply QKD or post-quantum cryptography to secure control channels and data exchanged between classical and quantum nodes.
- **Interface Hardening:** Implement authenticated middleware, strict access controls, and runtime signature checks across classical–quantum boundaries.
- **Noise-Aware Scheduling:** Dynamically adjust circuit depth, gate ordering, and execution timing based on coherence diagnostics.
- **Resilience Testing:** Integrate CRQC–LLM stress-tests across the pipeline to validate behavior under worst-case adversarial and operational conditions.

By clarifying where and how information flows between classical and quantum layers, this spotlight prepares the ground for Section 3.9.6, where these interleaved operations are formalized into a layered hybrid stack architecture.

Hybrid Integration and Security

Once classical and quantum stacks are aligned architecturally, integration challenges and security implications emerge as the next layer of concern. As shown in Figure 3.9.5, hybrid QAIS–AI systems emerge as layered pipelines in which classical preprocessing and optimization anchor reproducibility, while quantum-native layers enrich representational power.

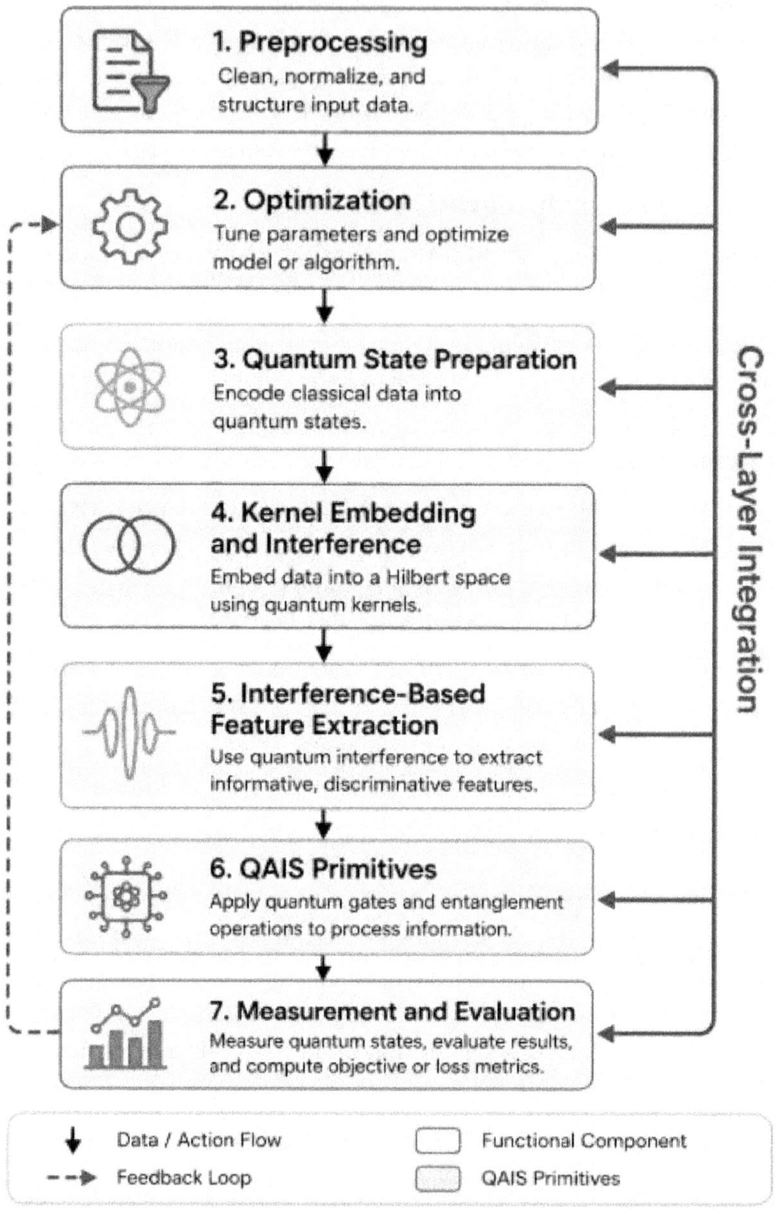

Figure 3.9.5 — Hybrid QAIS–AI Stack Architecture.

This figure depicts the integrated hybrid stack, showing classical preprocessing and optimization interleaved with quantum state preparation, kernel embedding, and interference-based feature extraction. At the base, QAIS primitives provide the substrate that enables the entire layered system.

At the foundation, QAIS primitives, superposition, entanglement, and quantum gates, serve as the substrate enabling all upper-layer operations. The integration of QAIS and AI systems into a unified stack requires aligning representational, computational, and control layers across domains. This integration supports the development of hybrid intelligence. Figure 3.9.5 depicts this stack within an architectural context, illustrating how different layers combine to form a cohesive system

Technical Spotlight — Hybridization as System Architecture

Conceptual Grounding

Hybridization fuses the strengths of both paradigms. Classical stages handle noisy or large-scale inputs, preparing them for quantum embedding.

Quantum layers then provide representational richness by exploiting superposition, entanglement, and interference.

A closing classical optimization loop ensures convergence and reproducibility, making hybridization the architectural logic of deployable Quantum AI.

System Relevance

Hybrid pipelines are essential for practical QAIS deployments. They ensure that quantum resources are applied selectively, where they add the most value, without sacrificing the stability of classical workflows. This layered integration is what allows QAIS to move from isolated demonstrations into mission-critical intelligent systems.

Cross-Disciplinary Orientation

Physicists validate entanglement and interference in hardware-ready conditions. Engineers design compilers, middleware, and co-processing systems to manage integration. AI specialists adapt learning models to exploit entangled kernels and interference-based features. Cybersecurity experts assess vulnerabilities at the interfaces where fragile quantum layers connect with classical control, anticipating adversarial exploitation under CRQC–LLM conditions.

Case Study Anchor — Hybridization as the Only Deployable Trajectory

Intent: To demonstrate why hybrid QAIS–AI architectures represent the only viable deployment pathway at ecosystem scale, balancing quantum advantage with classical stability and governance.

- **QALIS (Design Framework):** Demonstrates how the hybrid stack delivers operational performance in distributed sensing and decision pipelines.
- **CRQC–LLM (Stress-Test Lens):** Reveals how adversarial probing of quantum layers can destabilize hybrid stacks, underscoring resilience as a primary design requirement.

By establishing the hybrid stack as a complete layered blueprint, this section provides the architectural baseline for Section 3.9.7, where QAIS primitives themselves are examined as the layered foundations of these hybrid systems.

3.9.6 QAIS Primitives as Hybrid Building Blocks

Security and integration considerations ultimately trace back to the primitive operations that underwrite all higher-order hybrid behaviors. The hybrid QAIS–AI stack is ultimately anchored in its primitives. Superposition, entanglement, and quantum gates are not abstract curiosities; they are the physical operations that enable higher-order modules such as variational circuits, kernels, and hybrid algorithms. In moving from theory to architecture, these primitives function as deployable building blocks, ensuring that hybrid pipelines are grounded in the laws of quantum mechanics.

Recent proposals such as Marchisio et al. [10] demonstrate strategies for partitioning large-scale quantum neural networks into modular segments, enabling execution on today's limited-qubit devices while still preserving expressive capacity [29]. Hybrid models like TunnElQNN [11] illustrate how tunneling-inspired embeddings can balance efficiency and representational power, offering a template for near-term quantum–classical learning pipelines [30].

3.9.7 QAIS Primitives as Layered Foundations

Beyond functioning as reusable modules, QAIS primitives also operate as structural layers that constrain and enable representational depth. Figure 3.9.6 depicts the hybrid QAIS–AI stack as a layered system anchored in QAIS primitives. It shows how foundational physical operations support state preparation, kernel embedding, and interference-based extraction, all integrated with classical preprocessing and optimization to form a robust and expressive pipeline.

Building hybrid QAIS–AI systems on foundational primitives ensures that higher-level capabilities remain grounded in well-defined operations. This approach supports consistency and scalability in system design. Figure 3.9.6 implements this construction strategy within a structural framework, illustrating how primitives support system architecture.

[29] Marchisio proposes modular partitioning of quantum neural networks, enabling execution on limited-qubit hardware while preserving representational power, critical for near-term QAIS scalability.

[30] TunnElQNN introduces tunneling-inspired embeddings that balance efficiency and expressiveness, offering a template for quantum-classical learning pipelines under resource constraints.

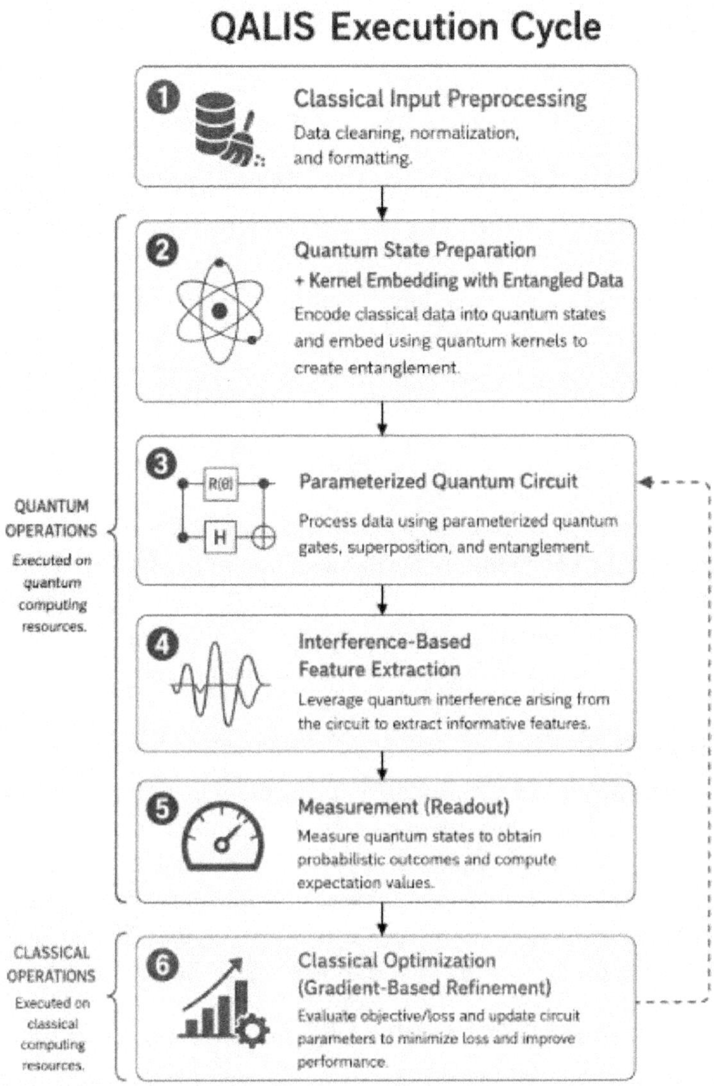

Figure 3.9.6 — QALIS Hybrid Quantum–Classical Learning Workflow.

This figure illustrates a QAIS hybrid quantum–classical learning workflow evaluated through the QALIS framework. Classical preprocessing prepares data for quantum state preparation and kernel embedding, after which parameterized quantum circuits employ QAIS primitives—including superposition, entanglement, and quantum gates—to construct and manipulate quantum representations. Interference-based feature extraction and measurement transform quantum-state information into classical outcomes that can be evaluated and optimized. A classical feedback loop updates circuit parameters and guides iterative refinement, demonstrating how quantum and classical resources cooperate to realize adaptive learning within QAIS architectures. The QALIS lens emphasizes how these interactions contribute to capability realization, learning effectiveness, and stable inference behavior within governed QAIS deployments.

Technical Spotlight — Building the Hybrid Stack

Conceptual Grounding

By treating primitives as modular anchors, QAIS redefines how architectures are assembled.

Superposition extends representational capacity; entanglement encodes correlations that classical systems cannot reproduce; and quantum gates provide the universal control operations that make circuit-level computation possible.

When integrated into layered pipelines, these primitives evolve into modules such as VQCs, kernels, and hybrid algorithms that scale into functional, intelligent architectures. "Circuit-centric quantum classifiers [20] demonstrate how QAIS primitives can be structured into trainable modules, offering blueprints for hybrid learning architectures [31]."

System Relevance

Anchoring hybrid pipelines in primitives ensures that architectures remain faithful to the constraints and opportunities of quantum mechanics. Coherence lifetimes dictate viable circuit depth, entanglement depth defines correlation capacity, and gate fidelities determine reliability. Without careful engineering of these parameters, hybrid stacks risk fragility, making the primitive layer the most decisive factor in system resilience.

Cross-Disciplinary Orientation

Physicists validate superposition and entanglement through experiment, ensuring primitives are realizable in hardware. Engineers translate primitives into control electronics, calibration frameworks, and middleware that stabilize execution. AI specialists design learning models that treat entanglement as representational power and interference as a decision amplifier. Cybersecurity experts examine how fragile primitives can become attack surfaces, especially under CRQC–LLM–guided probing of coherence or error thresholds.

Case Study Anchor — Stack Construction as an Ecosystem Commitment

Intent: To show how building the hybrid QAIS stack is not a technical choice alone, but an ecosystem-level commitment involving tooling, workforce skills, verification regimes, and governance alignment.

- **QALIS (Design Framework):** Demonstrates how coherence management and entanglement depth define inference capacity in distributed sensing pipelines.
- **CRQC–LLM (Stress-Test Lens):** Shows how adversaries may exploit fragile primitives such as superposition and coherence windows, highlighting the need for resilience at the most fundamental layer.

---

[31] Schuld and colleagues show that classifiers structured around variational circuits demonstrate how QAIS primitives can directly support hybrid learning architectures.

By situating primitives as the architectural anchors of hybrid stacks, this section closes the loop between theory and deployment. Section 3.9.9 extends this trajectory by examining how data itself is transformed when encoded directly into quantum states, moving beyond engineered features toward QAIS-native representations.

3.9.8 Encoding Data Beyond Features: QAIS Representations

With the layered role of primitives established, attention now shifts to how data itself is represented within these quantum-native structures. In classical AI, data must be heavily preprocessed and transformed into engineered feature vectors before becoming usable for learning. This process is resource-intensive and often fails to capture subtle correlations within complex datasets. By contrast, QAIS introduces a fundamentally different representational framework: raw data can be embedded directly into quantum states, where superposition, entanglement, and phase coherence natively encode relational and contextual structures. "Work on quantum correlations, such as that by Groisman et al. [21], underlines how entanglement extends beyond classical correlation measures, embedding relational structure directly into the substrate [32]."

A central method in quantum data representation is **amplitude encoding**, in which classical numerical data are embedded directly into the amplitudes of a quantum state. Given a normalized vector $x = (x_0, x_1, \ldots, x_{n-1})$, the corresponding quantum state is prepared as:

$$|\psi\rangle = \Sigma_i\, x_i\, |i\rangle \qquad (3.9.1)$$

This encoding enables exponentially compact representations, as n classical values can be represented using $\log_2(n)$ qubits. Amplitude encoding is particularly significant in quantum machine learning, where it supports efficient state preparation and enables interference-based processing across high-dimensional data spaces.

As shown in Figure 3.9.7, this shift reframes data not merely as input, but as a physically embodied resource that carries correlations inaccessible, or prohibitively expensive, for classical methods. In practice QALIS illustrates this design trajectory, embedding entangled quantum feature maps into distributed sensing systems to heighten pattern sensitivity.

Conversely, CRQC–LLM highlights the resilience challenge, where adversaries may probe coherence and interference layers as new forms of attack surfaces.

The representational power of quantum systems introduces new possibilities for encoding and processing data, influencing how learning and inference are performed. These capabilities extend beyond classical limits. Figure 3.9.7 models this representational advantage within a system-level context, illustrating its implications for QAIS design.

[32] Groisman's correlation taxonomy, classical, quantum, and total, provides the framework for QAIS representation, highlighting entanglement as a system-native relational encoding.

Expanding Representational Power

Figure 3.9.7 contrasts classical feature-based encoding with QAIS-native representations. The classical pipeline progresses from symbols to vectors to engineered features, while the QAIS pipeline begins with quantum states that inherently carry amplitude, phase, and entanglement correlations. This comparison underscores QAIS's ability to treat data as a physical resource, not just a symbolic construct.

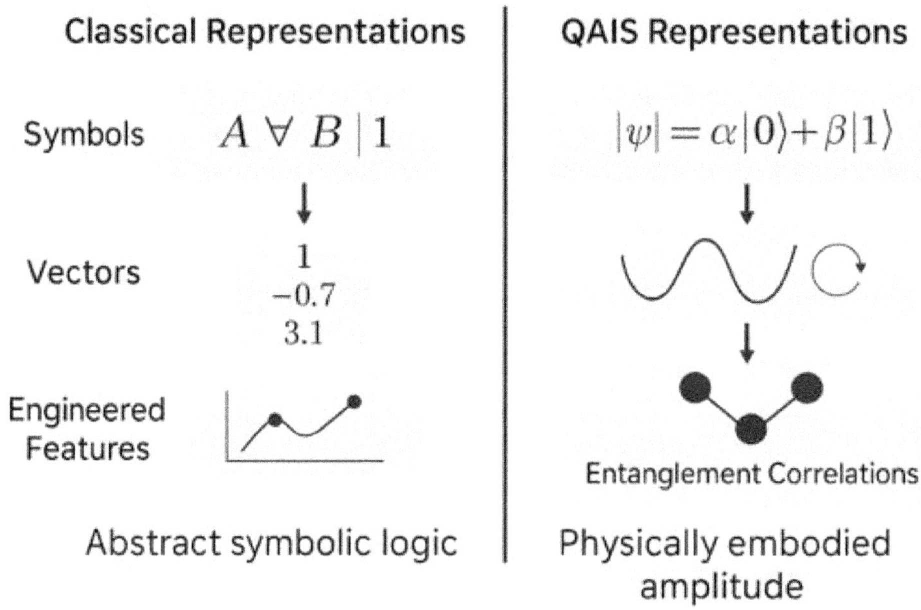

Figure 3.9.7 — Quantum Data and Representational Power.

This side-by-side diagram compares classical and QAIS representations. On the left, classical systems encode data through symbols, numerical vectors, and engineered features, capturing correlations indirectly. On the right, QAIS encodes data as quantum states ($|\psi\rangle$), with amplitudes and phases that evolve physically, enabling interference and entanglement to reveal correlations directly. The figure highlights QALIS as a design Framework (leveraging entangled feature maps in distributed sensing) and CRQC–LLM as a stress-test lens (adversarial probing of fragile coherence).

Technical Spotlight — Expanding Representational Power in Quantum AI

The figure bridges the transition from abstract QIS concepts to QAIS data architectures, showing how representation moves from symbolic encodings to physically embodied states. It positions QAIS as a discipline where **representation itself becomes a system resource**.

Conceptual Grounding

Classical feature engineering relies on human-designed transformations, dimensionality reduction, kernel tricks, or statistical approximations. QAIS bypasses these limitations by embedding data directly into Hilbert space. A quantum state $|\psi\rangle = \alpha|0\rangle + \beta|1\rangle$ does not merely store information but encodes both magnitude ($|\alpha|^2$, $|\beta|^2$) and phase ($\arg(\alpha)$, $\arg(\beta)$), enabling interference patterns that highlight correlations.

Entanglement further extends this representational power by linking multiple qubits into shared states, embedding relational information that cannot be decomposed into independent variables.

System Relevance

Encoding data as quantum states allows QAIS to surface structures that classical preprocessing might miss. Phase coherence provides a temporal window for stable representations, while entanglement depth determines how richly correlations can be embedded. These encodings form the backbone of quantum kernels, variational circuits, and hybrid pipelines. At the same time, fragility in coherence and entanglement means representational choices also define system vulnerabilities, encoding is not only about expressiveness but also about resilience.

Cross-Disciplinary Orientation

Physicists design the experimental methods, lasers, ion traps, photonic sources, that physically realize data encoding. Engineers build quantum memories and interfaces that stabilize coherence during representation. AI specialists develop learning algorithms that exploit entangled feature maps and interference-driven kernels. Cybersecurity experts treat representational fragility as both a risk and a safeguard, where disturbances reveal tampering but also expose new attack surfaces under CRQC–LLM stress scenarios.

Case Study Anchor — Representational Power as an Ecosystem Differentiator

Intent: To frame expanded representational power as the primary ecosystem differentiator of QAIS, shaping adoption incentives while simultaneously increasing verification and security demands.

- **QALIS (Design Framework):** Demonstrates how entangled feature maps embedded in distributed sensing pipelines heighten sensitivity to correlations while reducing preprocessing overhead.
- **CRQC–LLM (Stress-Test Lens):** Highlights how fragile coherence in data encodings can be exploited, where adversaries probe interference patterns or force decoherence to destabilize inference pipelines.

By reframing data encoding as a physically embodied process, QAIS makes representation itself a design lever for intelligent systems.

Section 3.9.10builds on this insight, showing how security, integrity, and trust become embedded in the substrate of computation, where disturbance, no-cloning, and entanglement function as defense mechanisms rather than external overlays.

3.9.9 Security, Integrity, and Trust in Quantum Learning

Performance to Trustworthiness

Representation and learning in QAIS are inseparable from integrity and adversarial resilience in distributed quantum workflows. Performance alone does not define system maturity; trust—the ability to maintain integrity, confidentiality, and resilience under stress—is equally fundamental.

In classical AI, security is typically enforced through external mechanisms such as cryptographic overlays and redundancy. QAIS, by contrast, embeds defense directly within the information substrate. Foundational quantum properties—no-cloning, measurement disturbance, and entanglement fidelity—translate physical constraints into intrinsic safeguards, making resilience a native architectural feature rather than an added layer.

Within QALIS-Surveillance architectures, disturbance-based verification and entanglement-assisted authentication enable continuous, physics-based monitoring. Verification and defense are therefore integrated into system operation rather than applied retrospectively.

Figure 3.9.8 organizes these security and integrity mechanisms within a unified framework, highlighting their role in shaping trustworthy QAIS design.

3.9.10 Security as Substrate, Not Overlay

Figure 3.9.8 contrasts classical and quantum approaches to trust. In classical systems, protections such as encryption and access control are applied externally. In QAIS, trust is embedded within the information substrate: qubits resist cloning, entanglement enforces authenticated correlations, and measurement disturbance reveals intrusion attempts. As a result, resilience and trustworthiness emerge from the architecture itself, unifying performance and protection as co-dependent design objectives.

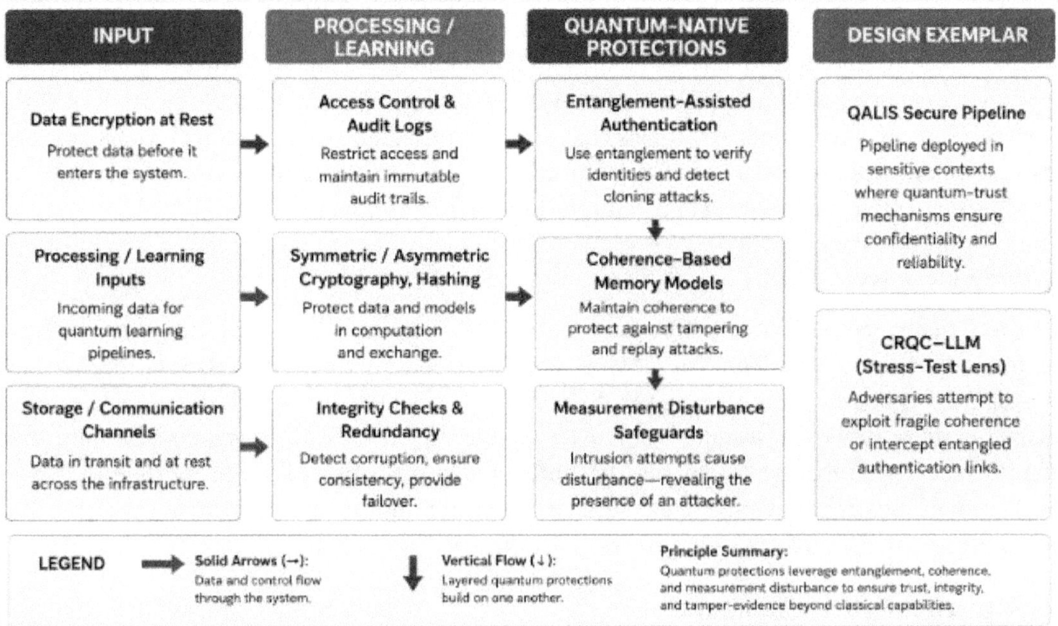

Figure 3.9.8 — Security, Integrity, and Trust in Quantum Learning.

This diagram contrasts classical protections, encryption, redundancy, access control, with QAIS-native mechanisms such as quantum key distribution (QKD), entanglement-assisted authentication, and coherence-based verification. Arrows highlight feedback loops where disturbance reveals intrusion attempts and no-cloning enforces integrity by preventing duplication. QALIS illustrates constructive secure-by-design architectures, while CRQC–LLM stress-tests adversarial exploitation of fragile coherence and entanglement.

Technical Spotlight — Security, Integrity, and Trust in Quantum AI Systems

Conceptual Grounding

In classical pipelines, trust is externally imposed: encryption secures communication, redundancy ensures fault tolerance, and access controls prevent unauthorized use.

QAIS alters this logic by embedding trust into the physics itself. Quantum key distribution leverages no-cloning and disturbance to ensure tamper-proof confidentiality.

Entanglement-assisted authentication provides correlations that cannot be forged, turning entanglement into an integrity guarantee. Coherence-based verification allows systems to detect disruption by monitoring lifetimes of fragile quantum states. "Extending uncertainty relations to systems with quantum memory, Berta et al. [21] show how verification in QAIS must account for adversarial strategies that exploit entangled environments." These are not abstract theoretical promises but architectural rules: the very act of observation or tampering becomes detectable, making trust structural rather than optional [33].

System Relevance

For QAIS, resilience is enforced at the substrate. This shifts the security paradigm from after-the-fact protection to built-in guarantees. Quantum communication networks, distributed sensing arrays, and hybrid learning pipelines inherit these physics-based defenses as part of their operation. However, fragility remains a double-edged sword: the same coherence that enables secure representation can also become an adversarial target. Thus, designing secure QAIS systems means balancing the expressive power of quantum states with mechanisms that anticipate and withstand malicious stress.

Cross-Disciplinary Orientation

Physicists validate QKD and entanglement fidelity in laboratory and field conditions, ensuring theoretical guarantees translate into practice. Engineers design the control electronics, error-corrected memories, and communication protocols that preserve integrity in real systems.

AI specialists integrate physics-based verification into workflows, embedding tamper-evident signals into learning loops.

Cybersecurity experts extend adversarial models to quantum substrates, asking how CRQC–LLM adversaries could manipulate fragile coherence or exploit poorly validated channels. Together, these perspectives reinforce that QAIS security cannot be reduced to a single discipline, it is inherently systemic.

---

[33] Berta et al. extend Heisenberg's uncertainty to systems with entangled memory, directly linking quantum physics to QAIS verification and adversarial resilience strategies.

Case Study Anchor — Security and Trust as Ecosystem Properties

Intent: To reinterpret security, integrity, and trust not as add-on controls, but as ecosystem properties emerging from physical law, institutional coordination, and architectural discipline.

- **QALIS (Design Framework):** Demonstrates how secure-by-design authentication and coherence-driven verification make trust an inherent property of mission-critical QAIS pipelines.
- **CRQC–LLM (Stress-Test Lens):** Highlights how adversarial exploitation of fragile coherence or entanglement could destabilize hybrid systems, underscoring why resilience must be engineered alongside physics-based guarantees.

By embedding integrity and trust directly into computation, QAIS reframes resilience as an architectural property rather than a secondary concern. The discussion next extends this logic to Section 3.9.11, further showing how QAIS serves as the disciplinary scaffold of hybrid intelligence.

3.9.11 Foundational Framework for the Future of AI

When security is treated as substrate rather than overlay, broader implications for the future architecture of artificial intelligence become visible. QAIS extends beyond computational acceleration to redefine the very framework upon which intelligent systems are built. Just as Boolean logic, probability theory, and linear algebra served as the foundations of classical AI, QAIS introduces a new substrate grounded in physics.

Superposition, entanglement, coherence, and no-cloning are not optional add-ons but governing constants that determine how information can be represented, transformed, and secured. At the same time, QAIS introduces trust-centric features that embed integrity and confidentiality at the substrate level, through mechanisms such as quantum key distribution (QKD), entanglement-assisted authentication, and coherence-based memory models.

These properties are especially critical for AI deployed in sensitive sectors such as finance, healthcare, and defense, where resilience and trustworthiness are as vital as performance.

More broadly, QAIS provides for Quantum AI the same disciplinary role that Boolean logic, probability theory, and linear algebra established for classical AI. It defines the computational substrate, the physical constraints, and the algorithmic possibilities that intelligent quantum systems must navigate. From algorithmic expressiveness to secure information flow, QAIS supplies a deeply integrated architecture, serving as the scaffold for hybrid intelligence.

As hardware continues to evolve, Quantum AI will transition from experimental novelty to operational reality. Early prototypes already demonstrate tangible benefits in optimization, classification, and learning tasks, confirming QAIS as the bridge from theory to deployment.

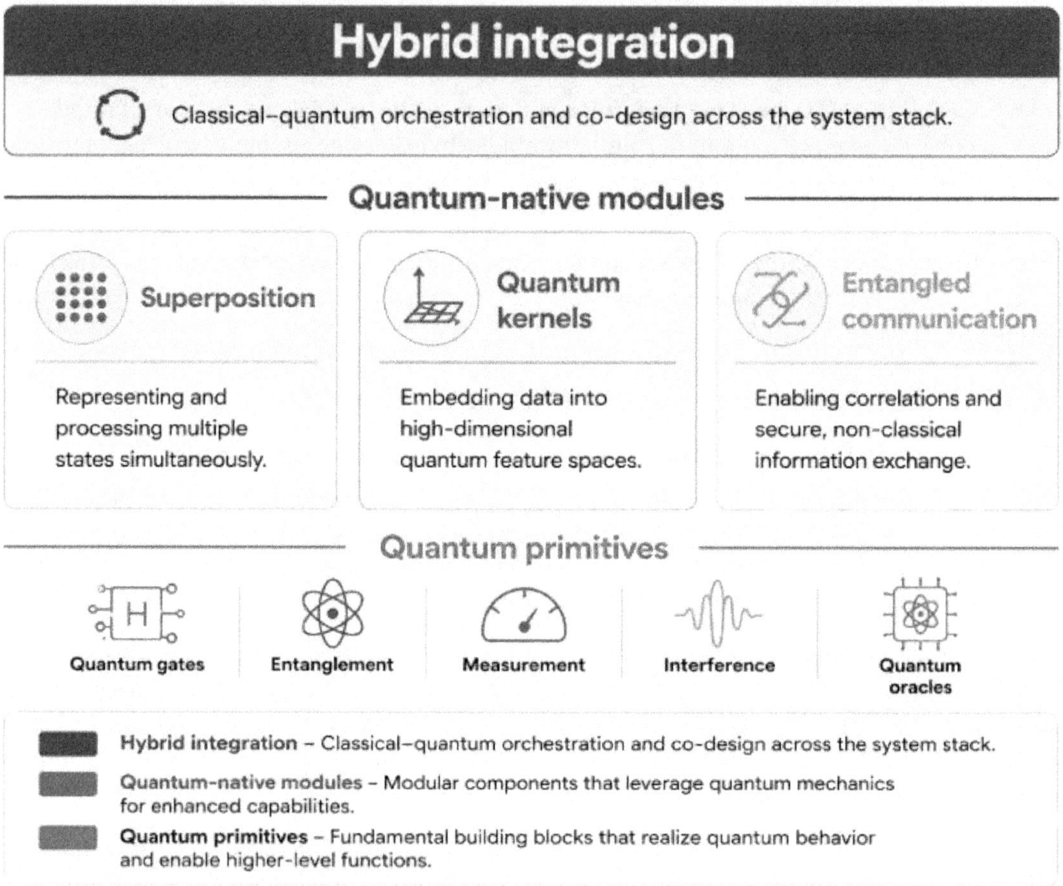

Figure 3.9.9 — QAIS as the Disciplinary Scaffold of Hybrid Intelligence.

This figure illustrates how QAIS establishes the foundations of hybrid intelligent systems. The bottom layer depicts quantum primitives, superposition, entanglement, coherence, and no-cloning, reframed as operational constants. The middle layer introduces quantum-native modules such as variational circuits, kernels, and entangled communication primitives. The top layer highlights hybrid integration, where these modules converge with classical AI processes to produce secure, resilient architectures.

Technical Spotlight — QAIS as the Disciplinary Scaffold of Hybrid Intelligence

Conceptual Grounding

In classical AI, Boolean logic, probability theory, and linear algebra formed the abstract grammar that enabled symbolic reasoning, uncertainty modeling, and vector-based computation. QAIS redefines this foundation by rooting representation and computation directly in physics.

Superposition expands representational capacity beyond linear algebra; entanglement encodes correlations inaccessible to classical probability; coherence establishes usable computational lifetimes; and no-cloning and disturbance transform physical limits into security guarantees. Together, these principles form the new logical constants of intelligent systems.

System Relevance

By embedding trust and resilience into the substrate itself, QAIS reframes intelligence as both expressive and secure. Variational circuits and quantum kernels expand algorithmic capacity; coherence and entanglement define system reliability; and QKD, authentication, and disturbance-based verification ensure integrity. Unlike classical AI, where protections are bolted on as overlays, QAIS integrates resilience into the computational core.

For practitioners, this means that system design must treat physics not as a constraint to be abstracted away but as the very foundation of trustworthy intelligence.

Cross-Disciplinary Orientation

Physicists contribute by refining the experimental underpinnings of coherence and entanglement. Engineers embed these principles into modular processors, middleware, and error-correction protocols. AI specialists adapt models to harness entanglement depth and interference-driven representations as functional learning tools. Cybersecurity experts reframe primitives like no-cloning and disturbance as structural defenses against adversarial exploitation. Each discipline contributes not in isolation but in concert, reinforcing QAIS as a systemic foundation rather than a specialized niche.

Case Study Anchor — QAIS as the Disciplinary Scaffold of Hybrid Intelligence

Intent: To position QAIS as the organizing discipline that stabilizes hybrid intelligence systems by binding physics, computation, governance, and workforce capacity into a coherent whole.

- **QALIS (Design Framework):** Demonstrates how QAIS primitives and modules serve as structural anchors for secure, distributed intelligent systems.
- **CRQC–LLM (Stress-Test Lens):** Highlights how fragile coherence or entanglement could be adversarially stressed, showing why resilience must be engineered into the substrate itself.

By positioning QAIS as the disciplinary scaffold for the future of AI, this spotlight prepares the transition to Section 3.9.12, QAIS as the Logic of Next-Generation AI.

Whereas Section 3.9.11 emphasizes the substrate-level constants of hybrid intelligence, the next section contrasts them directly with the logical foundations of classical AI, highlighting how QAIS establishes the new "grammar" of intelligent systems.

3.9.12 QAIS as the Logic of Next-Generation AI

In the same way that Boolean logic, probability theory, and linear algebra defined the mathematical foundations of classical AI, QAIS now establish the corresponding substrate for intelligent systems in the quantum era. QAIS is not an add-on to classical formalisms; it redefines the primitives of representation, learning, and computation by grounding them directly in physics. Superposition, entanglement, coherence, and no-cloning emerge as the new operational constants, shaping how information can be encoded, transformed, and secured in hybrid intelligent architectures.

Abstract Formalism to Physics-Embedded Logic

Figure 3.9.10 provides a comparative framing of classical and QAIS foundations. On the classical side, Boolean logic, probability theory, and linear algebra are shown as the abstract formalisms that historically supported symbolic reasoning, uncertainty modeling, and vector-based computation. On the QAIS side, physically embodied primitives, superposition, entanglement, coherence, and no-cloning, anchor Quantum AI in the laws of quantum mechanics, reframing limits as resources and embedding resilience directly into the substrate of intelligent systems.

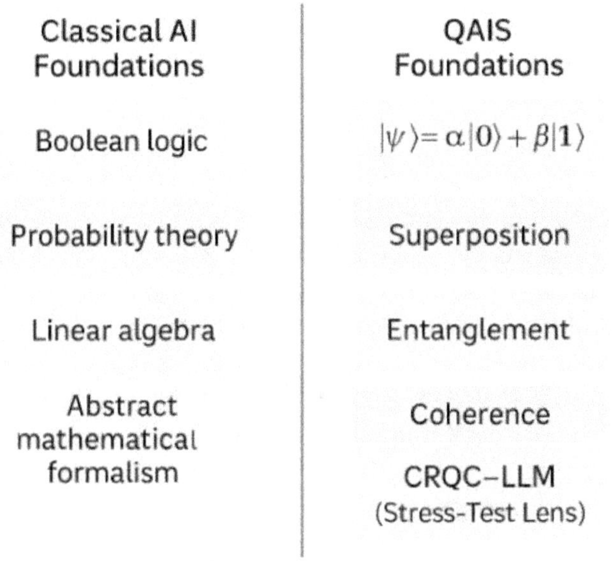

Figure 3.9.10 — QAIS as the Logic of Next-Generation AI.

This side-by-side diagram contrasts classical and QAIS foundations. The left panel shows Boolean logic, probability theory, and linear algebra as the building blocks of classical AI. The right panel highlights QAIS primitives, superposition, entanglement, coherence, and no-cloning, as the physical constants of Quantum AI. Together, the comparison emphasizes how QAIS functions as the new logic of intelligent systems, anchoring design Frameworks such as QALIS and resilience stress-tests such as CRQC–LLM.

Technical Spotlight — QAIS as the Logic of Next-Generation AI

Conceptual Grounding

In classical AI, abstract mathematical formalisms provided the scaffolding for intelligent behavior. QAIS redefines this foundation by embedding logic directly in the physics of quantum systems. Superposition expands representational capacity by enabling states to exist in linear combinations. Entanglement encodes correlations that cannot be described classically, binding distributed elements into coherent wholes. Coherence defines usable lifetimes for computation and inference, setting temporal boundaries for reliable operation. No-cloning and measurement disturbance transform fundamental limits into structural integrity rules, ensuring that information security and verification are embedded at the substrate level.

System Relevance

The implications of QAIS logic are profound: trust and resilience are no longer add-ons but structural properties of intelligent systems. Quantum key distribution enforces confidentiality as a law of physics rather than an assumption of complexity. Entanglement-based correlations may support specified authentication or synchronization protocols, but those functions still require classical communication, timing, control, and verification. Disturbance itself becomes intrusion detection, turning fragility into a defensive signal. Together, these mechanisms establish QAIS as both the computational and security substrate of hybrid AI.

Cross-Disciplinary Orientation

For machine learning specialists, quantum states act as feature maps that expand generalization power beyond classical kernels. Engineers recognize that gate fidelities, coherence windows, and entanglement depth function not only as performance targets but as design constraints. Cybersecurity experts encounter systems where adversarial probing cannot remain hidden, because disturbance becomes an unavoidable alert. Physicists interpret QAIS as the natural maturation of quantum theory into deployable logic for intelligent systems.

Case Study Anchor — QAIS as the Logic of Next-Generation AI Systems

Intent: To close the Chapter 3 case-study sequence by framing QAIS as the governing logic of next-generation AI systems, where ecosystem readiness determines whether quantum capability becomes durable intelligence or systemic risk.

- **QALIS (Design Framework):** Embeds QAIS primitives into operational architectures, demonstrating how foundational physics becomes functional system design.
- **CRQC–LLM (Stress-Test Lens):** Highlights how fragile coherence or entanglement can be adversarially probed, underscoring the need for resilience at the foundational layer.

Section 3.9 Summary — QAIS and Quantum AI

Section 3.9 established Quantum AI as a modular hybrid architecture rather than a simple coupling of quantum processors and classical AI. Quantum states provide physically structured representations; controlled transformations, entanglement, and interference propagate computational relationships; measurement produces classical evidence; and classical optimization, learning, inference, and control coordinate iterative system behavior.

Within this architecture, quantum primitives are organized into reusable computational structures, including variational circuits, quantum kernels, hybrid quantum–classical algorithms, and related learning and inference modules. These structures connect quantum-native representation and transformation with classical processing while remaining bounded by coherence, calibration, verification, measurement, and interface constraints.

QALIS evaluates how these primitives and modules can be assembled into stable learning, reasoning, and decision capabilities. CRQC-LLM examines the complementary risks arising from fragile state preparation, coherence loss, insecure interfaces, unreliable measurement evidence, calibration weakness, and uncontrolled propagation across hybrid workflows. Together, these lenses establish that Quantum AI capability depends on coordinated architecture, verifiable execution, and disciplined management of physical and operational constraints.

With this architectural foundation established, Section 3.10 turns to practical QAIS deployment contexts, where sector requirements, infrastructure conditions, and integration constraints shape scalability, reliability, and resilience.

3.10 Practical Applications — QAIS Deployment

Architectural Framing: *Applied System Realization*

Section Framing: *This section treats application as an execution control surface through which emerging QAIS architectures are evaluated in realistic operational environments. Representation is translated into implementable system configurations, while propagation appears as system behavior under actual workloads, constraints, and performance demands. Stability is maintained through bounded propagation, with environmental noise, deployment conditions, and integration challenges defining practical limits. Applications therefore provide the settings in which QAIS theory, prototypes, and architectural claims can be tested against operational evidence.*

The ecosystem dynamics examined throughout Chapter 3—historical evolution, sectoral adoption, infrastructure coordination, barrier interdependence, and workforce development—are becoming visible in quantum-AI research programs, experimental platforms, testbeds, and proofs of concept. Although complete QAIS architectures remain at an early stage of development, these initiatives already demonstrate how ecosystem readiness affects the feasibility and reliability of quantum-enabled capabilities.

At the national and institutional levels, quantum testbeds and research consortia provide shared environments for benchmarking, interoperability testing, workforce training, and coordinated infrastructure development. These initiatives translate ecosystem coordination into operational settings in which hardware calibration, communication protocols, software interfaces, and distributed workflows can be evaluated under common procedures. They illustrate the constructive dimension of ecosystem maturation: coordinated investment aligned with verification and stabilization discipline.

In the commercial sector, enterprises are exploring quantum-enhanced optimization, secure communication, sensing, and hybrid inference through research programs and limited proofs of concept across logistics, finance, energy, telecommunications, and related domains. These efforts reveal how cost, regulatory obligations, interoperability demands, hardware access, and resilience requirements influence architectural design. Commercial experimentation does more than test technical feasibility; it exposes the practical boundary conditions that future scalable QAIS systems would have to satisfy.

Academic–industry partnerships and distributed training frameworks provide another expression of ecosystem maturity. Cloud-accessible quantum platforms, interdisciplinary curricula, and collaborative research pipelines expand workforce capability while supporting shared benchmarking and verification practices. Workforce preparation therefore functions not as a peripheral activity but as a scaling mechanism that directly influences the quality, reproducibility, and reliability of experimental deployments.

At the same time, the barrier networks identified earlier in the chapter remain active operational constraints. Interoperability gaps, supply-chain dependencies, verification overhead, heterogeneous hardware environments, and uneven institutional readiness influence how quantum-enabled systems are designed, evaluated, and secured. These constraints define realistic operating envelopes and shape long-term architectural choices. From a resilience perspective, ecosystem weaknesses form distributed stress points through which instability may be amplified as systems scale.

Section 3.10 Summary — Practical QAIS Deployment

These practical contexts demonstrate that Chapter 3's ecosystem analysis is not merely conceptual. The ecosystem acts as a sociotechnical substrate that shapes where, how, and under what conditions QAIS capabilities may be developed and evaluated. Coordinated infrastructure, standards, investment, and workforce preparation support stable integration, while misalignment across these areas propagates fragility.

Practical deployment therefore reinforces the chapter's guiding principle: systemization must precede scale. This application-oriented perspective prepares the case study advancement in Section 3.11, where ecosystem maturation is evaluated as a structural commitment governing the future trajectory of QAIS architecture.

3.11 Case Study Advancement — QALIS and CRQC–LLM at the Ecosystem Layer

Architectural Framing: *Validation Across Full Ecosystem*

Section Framing*: This section establishes ecosystem-level validation as a control surface for evaluating QAIS frameworks under full-system conditions. Representation extends across distributed components and interacting subsystems, while propagation emerges through multilayer coordination, dependency, and feedback. Stability is maintained through bounded propagation, with cross-system dependencies and integration constraints determining robustness. The case study therefore evaluates QAIS architecture within complex sociotechnical environments.*

Intent: To advance the QAIS case study from localized realization to ecosystem-scale architectural obligation under distributed scaling.

Chapter 3 advances the dual-case trajectory by extending the QALIS and CRQC–LLM lenses from physical and operational mechanisms to the sectoral, infrastructural, and institutional conditions governing QAIS maturity. Earlier chapters established the physical substrate and operational grammar of quantum systems. This chapter demonstrated how those foundations develop through policy coordination, communication architectures, adoption pathways, infrastructure design, barrier management, workforce formation, and hybrid convergence. The ecosystem itself now becomes the architectural object of evaluation.

Through the constructive lens of QALIS, ecosystem maturation appears as disciplined systemization at scale. Workforce-development pipelines provide the human capability needed for sustained architectural growth. Research and adoption efforts across business, academia, government, and critical infrastructure demonstrate pathways from experimental capabilities toward mission-relevant systems. Standards development, interoperable benchmarks, policy alignment, and infrastructure coordination form stabilizing scaffolds that preserve constraint awareness during distributed expansion.

From this perspective, the historical progression from foundational algorithms to hybrid quantum–AI architectures reflects increasing coordination among capability, verification, infrastructure, and governance. Ecosystem strength does not guarantee successful scaling, but it improves the conditions under which coherent and verifiable development can occur.

Through the resilience lens of CRQC–LLM, the same ecosystem reveals systemic fragility surfaces. Workforce shortages, uneven institutional readiness, and concentrated expertise can restrict long-term stability. Interoperability gaps, heterogeneous platforms, and supply-chain dependencies introduce infrastructural vulnerabilities analogous to constraint propagation at the physical and operational layers.

Governance asymmetries, inconsistent standards, and misaligned policy incentives create additional pathways for institutional drift. Adoption across domains may signal progress, but it also exposes developing systems to operational stress and expanding verification obligations. Ecosystem maturity is therefore not a passive enabler of capability; it is a verification landscape in which distributed misalignment may propagate into system-level instability.

Section 3.11 Summary — Ecosystem-Layer Case Study

The case study advances QAIS from localized technical realization to distributed architectural commitment. Scientific and engineering progress creates requirements for communication infrastructure, institutional coordination, standards, workforce readiness, verification regimes, and supply-chain stability. Ecosystem maturation is therefore structural rather than merely technological.

Under QALIS, this trajectory reflects constructive coordination across distributed layers of the Quantum Stack. Under CRQC–LLM, it reveals how ecosystem weaknesses can magnify adversarial exposure and systemic fragility. Together, these lenses establish that QAIS resilience depends on sociotechnical infrastructure as well as on the physical principles and operational mechanisms developed in earlier chapters.

QAIS must consequently be understood not only as a collection of quantum devices or algorithms but as a coordinated ecosystem whose viability depends on disciplined systemization before scale. This advancement prepares the transition to Chapter 4, where the discussion turns from ecosystem readiness to representational discipline and encoding architecture.

Chapter 3 Conclusion

Ecosystem Closure and Architectural Readiness

Chapter 3 established ecosystem maturation as a precondition for scalable architecture within the Quantum Stack. Historical development, infrastructure, governance, workforce capability, and institutional coordination form interconnected control surfaces through which QAIS readiness must be evaluated. Stability depends on bounded propagation: alignment across these domains can preserve architectural integrity, while misalignment can amplify fragility.

The movement from isolated demonstrations toward coordinated research infrastructures marks an important transition in QAIS development. Testbeds, shared platforms, standards efforts, educational programs, and public–private initiatives increasingly shape how quantum-enabled capabilities are developed, verified, integrated, and sustained. The ecosystem therefore influences whether expansion reinforces disciplined architecture or propagates unresolved constraints.

Through QALIS, coordinated development across policy, infrastructure, standards, and workforce supports constructive pathways toward coherent scaling. Through CRQC–LLM, weaknesses across those same dimensions reveal distributed fragility surfaces capable of transmitting instability across organizational and technical boundaries. Ecosystem maturity should therefore be evaluated not simply by growth, investment, or activity, but by the degree of coordination achieved across coupled constraint networks.

The central conclusion is structural: QAIS resilience depends as much on sociotechnical infrastructure as on the physical and operational principles established in earlier chapters. Scale is neutral; it amplifies the discipline—or the weakness—already embedded within system design, institutional coordination, and information propagation. Systemization therefore remains the precondition for responsible expansion.

With ecosystem maturation defined as bounded propagation across distributed control surfaces, the architectural trajectory now turns inward. Chapter 4 advances to representational discipline, examining how encoding strategies and information architectures operate within the stabilization envelope established here. Ecosystem coordination governs scalability; representational precision governs expressive capacity. The next chapter therefore moves from ecosystem readiness to the encoding structures through which quantum information becomes usable within QAIS.

Quick Self-Check Questions

Evolution, Ecosystem, and Workforce of QAIS

1. What does Layer 3 of the Quantum Stack primarily represent?

A) Quantum gate operations
B) Mathematical formalism
C) Ecosystem realization and scaling
D) Classical preprocessing

2. In Chapter 3, representation is best understood as:

A) Data storage format
B) Control surface across distributed systems
C) Classical encoding method
D) Algorithm output

3. Propagation at the ecosystem-level refers to:

A) Signal amplification
B) Data duplication
C) System behavior across coordinated domains
D) Hardware execution speed

4. Stability in Chapter 3 is defined as:

A) Error-free computation
B) System uptime
C) Bounded propagation under constraints
D) Hardware reliability only

5. Constraint propagation at Layer 3 primarily extends into:

A) Quantum circuits only
B) Classical algorithms
C) Institutional and infrastructural systems
D) Compiler optimization

6. What is the primary role of QALIS in Chapter 3?

A) Detect system failures
B) Enable constructive system scaling
C) Replace classical AI
D) Perform error correction

7. What does CRQC–LLM emphasize?

A) Performance optimization
B) Hardware acceleration
C) Identification of fragility and instability
D) Data compression

8. Ecosystem maturity is primarily determined by:

A) Investment levels
B) Public visibility
C) Synchronization across constraint systems
D) Number of qubits

9. The transition from NISQ to scalable QAIS requires:

A) Faster processors
B) More data
C) System-level coordination
D) Classical optimization

10. Communication and measurement function as:

A) Data storage tools
B) Control primitives for coordination and extraction
C) Error correction methods
D) Optimization routines

11. Entanglement at the ecosystem-level implies:

A) Hardware coupling
B) Distributed relational dependency
C) Data redundancy
D) Faster execution

12. Infrastructure in QAIS acts as:

A) Passive support
B) Control surface for system deployment
C) Data storage layer
D) Security module

13. Workforce architecture contributes to QAIS by:

A) Increasing computation speed
B) Acting as a human control surface
C) Replacing algorithms
D) Reducing data needs

14. Barriers to maturity are best described as:

A) Temporary inefficiencies
B) Constraint surfaces limiting propagation
C) Hardware defects only
D) Market delays

15. Adoption across domains represents:

A) Marketing strategy
B) Diffusion of control surfaces into applications
C) Data migration
D) Hardware replication

16. Benchmarking in QAIS primarily ensures:

A) Faster execution
B) Alignment of system behavior
C) Data accuracy
D) Hardware compatibility

17. Governance in QAIS ecosystems functions as:

A) Regulatory overhead
B) Constraint enforcement mechanism
C) Data management
D) Hardware scheduling

18. Interoperability challenges affect:

A) Hardware cost
B) System behavior across platforms
C) Data storage
D) Algorithm design only

19. Hybrid QAIS–AI systems enable:

A) Replacement of quantum systems
B) Integration of classical and quantum control surfaces
C) Data elimination
D) Hardware independence

20. Measurement disturbance impacts:

A) Data storage
B) Representation integrity
C) Network speed
D) Workforce training

21. Scaling without coordination leads to:

A) Faster systems
B) Reduced cost
C) Amplified instability
D) Improved accuracy

22. The ecosystem acts as a:

A) Data repository
B) Distributed control system
C) Hardware platform
D) Algorithm library

23. Systemization in QAIS refers to:

A) Automation
B) Structured coordination across layers
C) Data processing
D) Hardware scaling

24. The primary function of standards in QAIS is to:

A) Increase speed
B) Ensure compatibility and alignment
C) Reduce cost
D) Improve hardware

25. Chapter 3 ultimately prepares the transition to:

A) Hardware optimization
B) Algorithm design
C) Representational discipline
D) Data analytics

Reader Guidance

For answers to these questions, see Appendix B for expanded reflection.

References

[1] R. P. Feynman, "Simulating physics with computers," *International Journal of Theoretical Physics*, vol. 21, no. 6/7, pp. 467–488, 1982.

[2] D. Deutsch, "Quantum theory, the Church–Turing principle and the universal quantum computer," *Proceedings of the Royal Society of London A*, vol. 400, pp. 97–117, 1985.

[3] P. W. Shor, "Algorithms for quantum computation: Discrete logarithms and factoring," in *Proceedings of the 35th Annual Symposium on Foundations of Computer Science (FOCS)*, pp. 124–134, 1994.

[4] L. K. Grover, "A fast quantum mechanical algorithm for database search," in *Proceedings of the 28th Annual ACM Symposium on the Theory of Computing (STOC)*, pp. 212–219, 1996.

[5] European Commission, Quantum Flagship Strategic Agenda 2024–2030, Brussels, 2024.

[6] J. Preskill, "Quantum Computing in the NISQ era and beyond," *Quantum*, vol. 2, p. 79, 2018.

[7] K. Bharti et al., "Noisy Intermediate-Scale Quantum (NISQ) Algorithms," *Reviews of Modern Physics*, vol. 94, p. 015004, 2022.

[8] A. S. Holevo, *Quantum Coding Theorems*, Russian Academy of Sciences, 1998.

[9] National Institute of Standards and Technology (NIST), *Post-Quantum Cryptography Standardization Process*, Gaithersburg, MD, 2023.

[10] A. Marchisio et al., "Cutting is All You Need: Execution of Large-Scale Quantum Neural Networks on Limited-Qubit Devices," *arXiv:2412.04844*, 2024.

[11] M. Zaman et al., "TunnElQNN: A Hybrid Quantum-Classical Neural Network for Efficient Learning," *arXiv:2505.00933*, 2025.

[12] J. Bowles et al., "Better than Classical? The Subtle Art of Benchmarking Quantum Machine Learning Models," *arXiv:2403.07059*, 2024.

[13] R. Meyer et al., "QUARK: A Modern Framework for Quantum Computing Application Benchmarking," Linux Foundation Quantum Initiative, 2024. [Online]. Available: https://www.linkedin.com/pulse/quark-modern-framework-quantum-computing-application-m1upf

[14] U.S. National Quantum Initiative Act, Washington, DC: U.S. Government, 2018.

[15] EuroQCI (European Quantum Communication Infrastructure), *Strategic Roadmap*, Brussels, 2025.

[16] Quantum Economic Development Consortium (QED-C), *Workforce Development Strategic Report*, 2025. [Online]. Available: https://quantumconsortium.org

[17] National Academies of Sciences, Engineering, and Medicine, *Advancing Workforce Development in Quantum Information Science and Technology*, Washington, DC: National Academies Press, 2023.

[18] M. Cerezo et al., "Variational Quantum Algorithms," *Nature Reviews Physics*, vol. 3, pp. 625–644, 2021.

[19] M. Schuld, A. Bocharov, K. Svore, and N. Wiebe, "Circuit-Centric Quantum Classifiers," *Physical Review A*, vol. 101, no. 3, p. 032308, 2020.

[20] R. Groisman, S. Popescu, and A. Winter, "Quantum, Classical, and Total Amount of Correlations in a Quantum State," *Physical Review A*, vol. 72, p. 032317, 2005.

[21] M. Berta, M. Christandl, R. Colbeck, J. Renes, and R. Renner, "The Uncertainty Principle in the Presence of Quantum Memory," *Nature Physics*, vol. 6, no. 9, pp. 659–662, 2010.

Validation and System Resources

The appendices in this volume extend the core chapters into a structured system of reference, application, and instructional support. Rather than functioning solely as supplemental material, these appendices serve as integrated components of the QAIS framework, supporting foundational study, mathematical formalization, experimental exploration, and systems-oriented learning.

Appendix A — Mathematical Foundations

Provides the mathematical foundations underlying quantum state representation, transformation, and measurement, consolidating key equations, operators, and formal relationships used throughout the volume.

Appendix B — Quick Self-Check Answers

Provides concise answers to the Quick Self-Check exercises appearing throughout the chapters, supporting independent validation of conceptual understanding while maintaining continuity within the main text.

Expanded derivations, guided solutions, and instructional materials are available through the QuSciTech Labs instructional environment for verified users and participating instructional programs.

Appendix C — QAIS Glossary

Serves as a centralized glossary defining the core terminology, operational concepts, and foundational vocabulary used throughout the QAIS framework.

Appendix D — Curriculum and Systems Framework

Introduces the instructional and architectural framework supporting foundational QAIS education, including curriculum mappings, systems-layer relationships, and instructional progression models aligned with the Quantum Stack architecture.

Appendix E — Laboratory Framework

Provides a sequence of hands-on laboratories and guided exercises designed to translate foundational theoretical concepts into observable system behavior using hybrid quantum–classical workflows and applied computational experimentation.

Appendix F — The Quantum Stack Reference Architecture

Provides a foundational reference framework for the Quantum Stack introduced throughout Volume I. The appendix consolidates the architectural relationships among Quantum Physics, Quantum Mechanics, and Quantum Technology while illustrating how these foundational disciplines collectively support the emergence of higher-layer capabilities throughout the QAIS framework.

Within the context of Volume I, Appendix F emphasizes the progression from physical quantum phenomena to mathematical representation and engineered implementation. It examines how quantum behavior is discovered, described, controlled, and measured, while introducing the dependency relationships that connect foundational science to future developments in Quantum Information Science, Quantum Computing, and Quantum AI Systems.

The appendix serves as both a conceptual roadmap and an architectural reference model, helping readers understand how the foundational principles developed throughout Chapters 1–3 contribute to the broader QAIS discipline. It also establishes the systems-oriented perspective that will guide subsequent volumes as the Quantum Stack progresses from Quantum Physics toward information processing, computation, learning, communication, governance, and intelligent system deployment.

Collectively, these appendices provide an integrated resource layer supporting the transition from conceptual understanding to mathematical formalization, operational analysis, experimental validation, architectural comprehension, and foundational QAIS system development. Together they reinforce the systems-oriented progression of the QAIS framework, connecting the physical, mathematical, and technological foundations established in Volume I to the broader Quantum Stack architecture that underpins the complete QAIS ecosystem.

Appendix A — Mathematical Foundations

Single-Qubit State Representation

Source: Equation (1.4.1)

$|\psi\rangle = \alpha|0\rangle + \beta|1\rangle$

Explanation: This equation defines a single-qubit quantum state as a linear superposition of computational basis states, where α and β are complex amplitudes governing probability and phase.

QAIS Interpretation
- Representation: Encodes information into amplitudes within Hilbert space
- Propagation: Evolves under unitary transformations, measurement, and noise
- Stability: Amplitude structure determines sensitivity to perturbations
- System Role: Foundation for quantum representation in QAIS architectures

Normalization Condition

Source: Equation (1.4.2)

$|\alpha|^2 + |\beta|^2 = 1$

Explanation: This condition ensures that the total probability of all measurement outcomes equals one, preserving physical validity of quantum states.

QAIS Interpretation
- Representation: Constrains admissible encoding space
- Propagation: Maintains probabilistic consistency across transformations
- Stability: Preserves valid probability normalization under properly implemented quantum operations
- System Role: Defines valid state configurations in QAIS

Expectation Value (Variational Objective)

Source: Equation (1.4.3)

$E(\theta) = \langle\psi(\theta)|\ H\ |\psi(\theta)\rangle$

Explanation: This defines the expectation value of a Hamiltonian with respect to a parameterized quantum state, serving as an objective function.

QAIS Interpretation
- Representation: Links quantum states to measurable observables
- Propagation: Reflects evolution of system behavior
- Stability: Governs optimization landscape sensitivity
- System Role: Drives variational learning and adaptive optimization

Amplitude Representation (Equivalent Form)

Source: Equation (1.4.4)

$\psi = \alpha|0\rangle + \beta|1\rangle$

Explanation: This expresses a qubit state emphasizing amplitude structure without explicit Dirac notation, reinforcing encoding relationships.

QAIS Interpretation

- Representation: Highlights amplitude-based encoding
- Propagation: Evolves identically under quantum operations
- Stability: Determines robustness to perturbations
- System Role: Reinforces representation-level modeling

Von Neumann Entropy

Source: Equation (1.4.5)

$S(\rho) = -\mathrm{Tr}(\rho \log \rho)$

Explanation: This defines entropy of a quantum state, measuring uncertainty or mixedness.

QAIS Interpretation

- Representation: Quantifies information content
- Propagation: Tracks information loss or gain
- Stability: Indicates susceptibility to noise
- System Role: Provides a diagnostic measure of mixedness and uncertainty relevant to learning stability and interpretability

Single-Qubit State Formalism

Source Equation (2.9.1)

$|\psi\rangle = \alpha|0\rangle + \beta|1\rangle$

Explanation: This equation defines the general form of a single-qubit quantum state, where α and β are complex amplitudes associated with the computational basis states $|0\rangle$ and $|1\rangle$.

QAIS Interpretation

- Representation: Encodes quantum information through amplitude and phase structure
- Propagation: Evolves under unitary transformations and measurement operations
- Stability: Requires normalization to preserve valid probability behavior
- System Role: Provides the foundational state model for QAIS computation, inference, and verification pipelines

Mutual Information (Subsystem Correlation)

Source: Equation (2.9.2)
$I(A:B) = S(A) + S(B) - S(AB)$

Explanation: This defines mutual information between subsystems, measuring shared information and correlation.

QAIS Interpretation
- Representation: Captures interdependence between states
- Propagation: Tracks correlation evolution
- Stability: Reflects robustness of relationships
- System Role: Enables distributed learning and coordination

Bell State (Φ^+)

Source: Equation (2.9.3)

$|\Phi^+\rangle = (|00\rangle + |11\rangle)/\sqrt{2}$

Explanation: This defines a maximally entangled two-qubit state exhibiting non-separable correlations.

QAIS Interpretation
- Representation: Encodes non-factorizable joint states
- Propagation: Enables entanglement-based information flow
- Stability: Sensitive to decoherence
- System Role: Supports correlation-aware computation

Bell State Representation (Alternate Context)

Source: Equation (3.9.1)
$|\psi\rangle = (|00\rangle + |11\rangle) / \sqrt{2}$

Explanation: This expresses a maximally entangled state emphasizing amplitude structure in propagation contexts.

QAIS Interpretation
- Representation: Encodes strong subsystem correlation
- Propagation: Enables coordinated transformations
- Stability: Sensitive to noise and decoherence
- System Role: Supports protocols that explicitly use entanglement-based correlations; the Bell state alone does not establish distributed inference or quantum advantage

Appendix B — Quick Self-Check Answers

Appendix B provides concise answers to the Quick Self-Check questions to support independent study and conceptual validation. Answers are presented in a structured format emphasizing correctness and clarity, while avoiding full derivations within the printed volume.

Expanded solutions, derivations, and instructional materials are available through the QuSciTech Labs platform for verified users and instructors.

B.1 Chapter 1 — Foundations of Quantum AI Systems

Q1 — C: As a physically embodied, coherence-sensitive property of quantum systems
Explanation: Quantum information is inseparable from its physical substrate and governed by superposition, entanglement, coherence, and measurement.

Q2 — B: Superposition, entanglement, coherence, and governed measurement
Explanation: These primitives define the structure, correlation, and operational constraints of QAIS.

Q3 — C: It frames computation, representation, and trust as system-level architectural properties
Explanation: QAIS integrates physics, architecture, and verification into a unified discipline.

Q4 — B: It functions as a runtime budget that bounds usable quantum evolution
Explanation: Coherence determines how long quantum effects remain usable before decoherence.

Q5 — C: They define encoding limits, integrity thresholds, and distinguishability bounds prior to deployment claims
Explanation: These metrics constrain system validity before scaling.

Q6 — C: As a constructive design lens and a stress-test verification lens
Explanation: QALIS builds capability; CRQC–LLM exposes fragility.

Q7 — C: It creates non-decomposable joint states that cannot be factored into independent parts
Explanation: Entanglement encodes inseparable relational structure.

Q8 — B: The measurement interface between quantum evolution and classical interpretation
Explanation: Measurement defines the trust boundary.

Q9 — B: Pure single-qubit states represented as points on the unit sphere under normalization and relative phase constraints
Explanation: The Bloch sphere geometrically represents qubit states.

Q10 — B: It governs interference behavior under unitary evolution
Explanation: Relative phase affects evolution, not direct measurement probabilities.

Q11 — C: States that cannot be expressed as tensor products of individual subsystems
Explanation: This defines entanglement mathematically.

Q12 — C: As coherence-dependent representational geometry enabling interference
Explanation: Superposition encodes structured amplitude relationships.

Q13 — B: It enables coordinated evolution across distributed subsystems within a shared relational state space
Explanation: Entanglement acts as infrastructure.

Q14 — B: It bounds distinguishability and therefore error probability
Explanation: Trace distance limits decision confidence.

Q15 — C: Co-engineer capability and resilience under physical constraint
Explanation: Advantage and fragility co-emerge.

Q16 — B: Because it collapses coherent amplitude structure into classical output
Explanation: Measurement is irreversible.

Q17 — C: It constrains executable sequence length before decoherence dominates
Explanation: Coherence limits circuit depth.

Q18 — C: They are inseparable throughout the information lifecycle
Explanation: Quantum information depends on its substrate.

Q19 — B: A layered architectural ascent from physics to intelligent systems
Explanation: The Quantum Stack propagates constraints upward.

Q20 — B: As connective structure enabling coordinated inference pipelines
Explanation: QALIS interprets entanglement constructively.

Q21 — C: As a fragility surface under adversarial or cryptographic stress
Explanation: CRQC–LLM reveals vulnerabilities.

Q22 — C: The transition from quantum phenomena as theory to architectural doctrine
Explanation: Physics becomes system design.

Q23 — C: Across layers 4–6, where infrastructure integration, architectural composition, learning and inference, and hybrid deployment converge.
Explanation: QAIS emerges at higher layers.

Q24 — C: Because disturbance of shared correlation alters observable statistics
Explanation: Enables authentication mechanisms.

Q25 — C: Intelligent quantum systems must be engineered within physically bounded, verification-aware architectures
Explanation: Core takeaway of Chapter 1.

B.2 Chapter 2 — Operations and Scientific Framework of QAIS

Q1 — C: A physical and context-dependent property of quantum systems
Explanation: Information is embodied in amplitudes, phase, and coherence.

Q2 — B: Superposition, entanglement, coherence, unitary evolution, and measurement
Explanation: These define operational behavior in QAIS.

Q3 — C: It redefines computation, representation, and trust as system-level properties
Explanation: QAIS is not just faster computing.

Q4 — B: They reinforce one another, demanding coordinated solutions
Explanation: Constraints are interdependent.

Q5 — B: Linear algebra, quantum mechanics, quantum information theory, quantum programming tools, systems-level thinking
Explanation: QAIS requires interdisciplinary mastery.

Q6 — C: Integrating trust into the substrate via no-cloning, entanglement, and measurement disturbance
Explanation: Trust is physically embedded.

Q7 — B: QALIS demonstrates design-aligned behavior; CRQC–LLM stress-tests resilience
Explanation: Complementary analytical roles.

Q8 — B: Interleaving classical preprocessing and optimization with quantum-native operations
Explanation: Hybrid architecture is essential.

Q9 — B: Correlations through physically embodied amplitudes, entanglement, and phase
Explanation: Representation is physical, not symbolic.

Q10 — B: By establishing qubits, gates, and circuits as the operational substrate of QAIS
Explanation: Chapter 1 enables Chapter 2.

Q11 — D: Deutsch's universal model
Explanation: Established programmable quantum computation.

Q12 — B: They demonstrated quantum advantage as an operational resource
Explanation: Proof of advantage.

Q13 — C: Boundary condition for extracting classical information
Explanation: Measurement defines output.

Q14 — B: It enforces limits on duplication and supports verification integrity
Explanation: No-cloning ensures security.

Q15 — C: Establishing non-separable correlations across qubits
Explanation: Core role of entanglement.

Q16 — C: Measurement outcomes depend on operational context
Explanation: Contextuality.

Q17 — B: Geometric evolution of amplitudes and phases
Explanation: Computation is physical evolution.

Q18 — C: Preserve coherence without copying states
Explanation: Goal of QEC.

Q19 — B: Fidelity
Explanation: Measures similarity.

Q20 — C: Distinguishability between quantum states
Explanation: Trace distance.

Q21 — B: Quantifies uncertainty and mixed-state structure
Explanation: Entropy.

Q22 — C: Integrating primitives into a unified lifecycle
Explanation: Operational flow model.

Q23 — B: It preserves phase relationships required for interference
Explanation: Role of coherence.

Q24 — C: Transfer state information via entanglement and classical communication
Explanation: Teleportation.

Q25 — D: Capability and fragility arise from the same physical primitives
Explanation: Core insight.

Q26 — C: Feynman's quantum simulation proposal
Explanation: First recognition of quantum simulation need.

B.3 Chapter 3 — Evolution, Ecosystem, and Workforce of QAIS

Q1 — C: Ecosystem realization and scaling
Explanation: Layer 3 represents the transition from isolated systems to coordinated, scalable ecosystems across infrastructure and institutions.

Q2 — B: Control surface across distributed systems
Explanation: Representation governs how information propagates and is controlled across distributed environments.

Q3 — C: System behavior across coordinated domains
Explanation: Propagation reflects how system dynamics emerge across organizations and platforms.

Q4 — C: Bounded propagation under constraints
Explanation: Stability requires limiting how signals and deviations propagate across systems.

Q5 — C: Institutional and infrastructural systems
Explanation: Constraints extend beyond physics into governance and workforce structures.

Q6 — B: Enable constructive system scaling
Explanation: QALIS ensures alignment for scalable system growth.

Q7 — C: Identification of fragility and instability
Explanation: CRQC–LLM exposes constraint violations and failure modes.

Q8 — C: Synchronization across constraint systems
Explanation: Ecosystem maturity depends on coordinated alignment across domains.

Q9 — C: System-level coordination
Explanation: Scaling requires coordinated infrastructure, governance, and workforce alignment.

Q10 — B: Control primitives for coordination and extraction
Explanation: Communication and measurement regulate system behavior.

Q11 — B: Distributed relational dependency
Explanation: Entanglement implies non-local coordination dependencies.

Q12 — B: Control surface for system deployment
Explanation: Infrastructure enables representation and propagation.

Q13 — B: Acting as a human control surface
Explanation: Workforce capabilities shape system design and execution.

Q14 — B: Constraint surfaces limiting propagation
Explanation: Barriers define limits to scalability.

Q15 — B: Diffusion of control surfaces into applications
Explanation: Adoption spreads QAIS capabilities into real-world systems.

Q16 — B: Alignment of system behavior
Explanation: Benchmarking ensures consistency across deployments.

Q17 — B: Constraint enforcement mechanism
Explanation: Governance bounds system propagation.

Q18 — B: System behavior across platforms
Explanation: Interoperability affects coordinated execution.

Q19 — B: Integration of classical and quantum control surfaces
Explanation: Hybrid systems unify computational domains.

Q20 — B: Representation integrity
Explanation: Measurement disturbance affects system state validity.

Q21 — C: Amplified instability
Explanation: Uncoordinated scaling increases fragility.

Q22 — B: Distributed control system
Explanation: Ecosystems regulate system-wide behavior.

Q23 — B: Structured coordination across layers
Explanation: Systemization aligns components for coherence.

Q24 — B: Ensure compatibility and alignment
Explanation: Standards maintain system integrity.

Q25 — C: Representational discipline (Chapter 4)
Explanation: Chapter 3 prepares the transition to encoding and representation.

Appendix C — Foundational Glossary of Quantum Terms

This appendix provides a foundational glossary of key terms used throughout Volume I — Foundations of Quantum AI Systems. The glossary is scoped specifically to Chapters 1–3 and focuses on the physical, mathematical, operational, and infrastructural principles underlying Quantum AI Systems (QAIS).

C.1 PHYSICAL QUANTUM FOUNDATIONS

Bell State

A maximally entangled two-qubit state exhibiting nonclassical correlations between the two subsystems.

Coherence

The preservation of phase relationships within a quantum state during evolution.

Decoherence

The degradation of quantum coherence caused by interaction with external environments, resulting in loss of quantum information fidelity.

Entanglement

A non-separable relationship between quantum subsystems in which the state of one subsystem cannot be described independently of another.

Hilbert Space

The mathematical vector space in which quantum states are represented and evolved.

Interference

The constructive or destructive combination of quantum amplitudes during state evolution.

Measurement (Quantum)

A physical process that produces classical outcomes from a quantum state, described generally by a POVM and, when the postmeasurement state is relevant, a quantum instrument. Projective measurement is a special case.

No-Cloning Theorem

A fundamental principle stating that arbitrary unknown quantum states cannot be copied exactly.

Probability Amplitude

A complex-valued coefficient associated with a quantum basis state that determines measurement probabilities and phase behavior.

Qubit

The fundamental unit of quantum information capable of superposition and entanglement.

Relative Phase

The phase relationship between components of a quantum superposition state.

Superposition

The representation of a quantum state as a linear combination of basis states.

Tensor Product

The mathematical operation used to combine multiple quantum systems into a composite state space.

C.2 QUANTUM OPERATIONS AND DYNAMICS

Circuit Depth

The number of sequential gate layers within a quantum circuit.

Coherence Budget

The operational time or computational complexity available before decoherence significantly degrades system reliability.

Composite Hilbert Space

The combined state space formed by multiple interacting quantum subsystems.

Computational Basis

The standard orthonormal basis states used to represent qubits, typically $|0\rangle$ and $|1\rangle$.

Contextuality

A property of quantum systems in which measurement outcomes depend on the experimental configuration.

Gate Fidelity

A measure of how accurately a physical quantum gate approximates its intended ideal transformation.

Measurement Backaction

The disturbance introduced to a quantum system as a consequence of measurement.

Quantum Circuit

An ordered sequence of quantum gates and measurements acting on qubits.

Quantum Gate

A unitary transformation applied to one or more qubits.

Reversible Evolution

Quantum evolution governed by unitary transformations that preserve information.

State Vector

The mathematical representation of a quantum state within Hilbert space.

Unitary Evolution

The deterministic evolution of a closed quantum system governed by unitary operators.

Unitary Operator

A linear operator that preserves normalization and reversibility during quantum evolution.

C.3 INFORMATION METRICS AND MEASUREMENT

Born Rule

The probabilistic rule stating that measurement probabilities are determined by squared probability amplitudes.

Coherence Time

The duration over which a quantum system maintains usable coherence.

Density Matrix

A matrix representation describing pure or mixed quantum states.

Distinguishability

The degree to which two quantum states can be differentiated through measurement.

Entropy (Quantum)

A measure of uncertainty or informational content within a quantum system.

Fidelity

A measure of similarity between quantum states.

Mixed State

A statistical ensemble of quantum states represented by a density matrix.

Normalization

The requirement that total measurement probability equals one.

Observable

A measurable physical quantity represented mathematically by an operator.

Pure State

A fully specified quantum state represented by a single state vector.

Trace Distance

A metric quantifying the distinguishability between quantum states.

Von Neumann Entropy

A measure of quantum informational uncertainty defined using density operators.

C.4 FOUNDATIONAL QAIS CONCEPTS

Noise-Aware Architecture

A system design approach that explicitly incorporates decoherence, gate error, and operational instability into computational modeling and execution planning.

Quantum AI Systems (QAIS)

A systems-level framework integrating quantum information science, computation, artificial intelligence, and operational system design.

Quantum Information Science (QIS)

The interdisciplinary field studying quantum computation, communication, measurement, and information processing.

Quantum Stack

A layered architectural framework organizing quantum systems from physical hardware through computation and higher-level system integration.

Stability Constraint

A physical or operational limitation governing whether quantum system evolution remains reliable under noise, scaling, and interaction effects.

System Infrastructure

The hardware, software, control, and execution environment supporting quantum computational operation.

Verification

The process of evaluating whether a quantum system behaves according to specified operational or computational expectations.

Appendix D — QAIS Curriculum Design Framework

D.1 PURPOSE AND SCOPE

This appendix provides a high-level curriculum and institutional design framework supporting the broader *QAIS* series. The framework establishes how the concepts developed throughout the QAIS volumes align with academic program development, workforce preparation, interdisciplinary instruction, and emerging standards in quantum information science and intelligent systems engineering.

The purpose of this appendix is not to prescribe specific course implementations, but rather to provide institutions, instructors, and independent learners with a structured reference model for organizing QAIS-related educational pathways and instructional progression.

Because the QAIS series is organized across multiple volumes, this appendix is included in each volume as a shared curricular reference supporting both standalone and integrated instructional use.

D.2 QAIS AS AN INTERDISCIPLINARY ACADEMIC FRAMEWORK

Quantum AI Systems (QAIS) represent the convergence of:

- Quantum Information Science
- Artificial Intelligence
- Systems Engineering
- Computational Modeling
- Information Theory

As an interdisciplinary framework, QAIS emphasizes:

- Integration of physical and computational models
- Hybrid quantum–classical architectures
- Representation, learning, communication, and control
- Verification, evaluation, and operational stability
- Systems-level architectural reasoning

This positioning supports the development of QAIS as a structured field of study spanning scientific, computational, and engineering disciplines.

D.3 CURRICULUM DESIGN MODEL

The QAIS curriculum framework is organized as a layered progression aligned with increasing system complexity, abstraction, and operational integration.

D.3.1 Foundational Layer

Introduces the mathematical, physical, and computational foundations underlying QAIS systems, including:

- Linear algebra and probability
- Quantum mechanics fundamentals
- Quantum information science
- Classical computation and algorithms
- Foundations of artificial intelligence

This layer aligns primarily with:

- Volume I — Foundations of Quantum AI Systems

D.3.2 Representation and Computational Layer

Focuses on information representation, reasoning, and computational transformation within quantum-enabled systems, including:

- Quantum encoding and representation
- Computational inference models
- Quantum reasoning architectures
- Hybrid representational systems
- Information propagation mechanisms

This layer aligns primarily with:

Volume II — Quantum Representation and Reasoning

D.3.3 Learning and Adaptive Systems Layer

Addresses adaptive computation, optimization, and trainable intelligent architectures, including:

- Quantum machine learning
- Variational optimization
- Hybrid quantum–classical learning systems
- Adaptive feedback and control
- Stability-preserving computational architectures

This layer aligns primarily with:

- Volume III — Quantum Learning and System Architectures

D.3.4 Communication and Deployment Layer

Examines large-scale system integration, communication infrastructures, governance considerations, and operational deployment environments, including:

- Quantum communication systems
- Distributed architectures
- Sensing and system integration
- Verification and resilience
- Governance and deployment frameworks

This layer aligns primarily with:

- Volume IV — Quantum Information and Communication Systems
- Volume V — Quantum Governance, Resilience, and Deployment

D.4 ALIGNMENT WITH EDUCATIONAL PROGRAM LEVELS

The QAIS curriculum framework supports multiple academic and instructional levels, including:

- Undergraduate instruction emphasizing foundational principles and introductory systems concepts
- Graduate instruction emphasizing architectural design, integration, and applied methodologies
- Doctoral and research-level study emphasizing innovation, theoretical advancement, and systems-scale investigation

This layered structure enables institutions to integrate QAIS concepts into existing programs or develop specialized interdisciplinary pathways.

D.5 WORKFORCE AND PROFESSIONAL COMPETENCIES

QAIS education supports the development of interdisciplinary competencies including:

- Quantum systems literacy
- Computational and algorithmic reasoning
- Hybrid system integration
- Information representation and evaluation
- Adaptive-system analysis
- Systems-level architectural thinking
- Verification and operational assessment

These competencies align with emerging workforce needs across academia, industry, national laboratories, and advanced technology sectors.

D.6 INSTITUTIONAL IMPLEMENTATION MODELS

Institutions may implement QAIS-related curricula through multiple pathways, including:

- Integration into existing computer science, engineering, physics, or data science programs
- Development of interdisciplinary degree concentrations or certificates
- Establishment of dedicated research initiatives or instructional laboratories
- Collaboration with industry, government, and national research organizations

The framework is designed to support both incremental adoption and comprehensive program development.

D.7 STANDARDS AND ACCREDITATION ALIGNMENT

The QAIS curriculum framework supports alignment with emerging educational and technical standards, including:

- IEEE technical standards
- NIST-aligned emerging technology frameworks
- Outcome-based accreditation models
- Systems-engineering instructional practices

Key alignment objectives include:

- Clearly defined learning outcomes
- Measurable competency development
- Integration of theory and application
- Systems-oriented reasoning and evaluation
- Interdisciplinary technical fluency

D.8 LONG-TERM DEVELOPMENT OF QAIS EDUCATION

Quantum AI Systems are expected to evolve into a foundational domain supporting next-generation computational and intelligent infrastructures.

The QAIS curriculum framework supports:

- Development of interdisciplinary educational ecosystems
- Expansion of quantum-enabled workforce capabilities
- Research innovation and technical specialization
- Integration of emerging quantum technologies into operational environments

As the field evolves, QAIS education will continue adapting to advances in computation, information science, intelligent systems, and quantum technologies.

D.9 SUMMARY

This appendix establishes a flexible curriculum and institutional framework supporting the broader QAIS series. The framework:

- Organizes QAIS as a structured interdisciplinary field
- Aligns educational progression with systems architectures
- Supports instructional and workforce development
- Encourages integration across scientific and engineering domains
- Provides a scalable foundation for future QAIS educational programs

Together, these principles support the transition of QAIS from an emerging theoretical framework into a mature instructional and operational discipline.

Appendix E — Quantum AI Labs

Laboratory Access and Code Redemption

Companion laboratories for the *QAIS* series are available through **QuSciTech Labs**:

https://quscitech.com/labs.php

The laboratory environment is organized into progressive instructional tracks aligned with the objectives of *Volume I — Foundations of Quantum AI Systems*:

- **E.1 Beginner Laboratories** — Foundational quantum phenomena and state behavior
- **E.2 Intermediate Laboratories** — Control, optimization, and system evaluation
- **E.3 Advanced Laboratories** — Integrated QAIS architectures and hybrid-system analysis

Together, these laboratories provide a progression from single-qubit dynamics to system-level QAIS evaluation, illustrating how representation, propagation, optimization, and stability emerge within quantum-enabled intelligent systems.

Public users may access the E.1 Beginner Labs through QuSciTech Labs, GitHub, and Google Colab. The E.2 and E.3 laboratory tracks are available to verified users and extend the material presented in the text through guided implementation exercises and hybrid quantum–classical workflows.

Purchasers of *QAIS: Architectures for Quantum AI Systems* may request access to restricted laboratory environments using the access phrase:

QAIS-BOOK-BUYER-2026

Access requests:

https://quscitech.com/labs/request-code.php

Verified users receive individual, non-transferable access credentials. Instructor and institutional access options, including expanded laboratory and assessment resources, are available through approved academic channels:

https://quscitech.com/instructors/access/

Reproducibility and Execution Environment

Laboratories are supported by companion repositories containing executable code, datasets, visualizations, and instructional resources.

Archival DOI

https://doi.org/10.5281/zenodo.18955801

Additional Resources

https://quscitech.com

Minimum Software Requirements

- Python 3.10+
- Qiskit 1.0+
- NumPy
- Matplotlib
- SciPy
- scikit-learn

Optional Frameworks

- PennyLane

All laboratories are compatible with Google Colab and may be executed without local installation.

Implementation and Reproducibility Notes

All coded laboratories include complete executable implementations designed to demonstrate correct methodology, reproducible execution, and expected QAIS system behavior. Emphasis is placed on:

- Reproducible experimentation
- Interpretable system behavior
- Hybrid quantum–classical workflows
- Architecture-aware analysis

Results are reproducible under simulation using the specified software environment. Variations may occur on physical quantum hardware due to noise, decoherence, calibration differences, queue conditions, and other operational constraints. These effects form part of the practical systems-analysis considerations examined throughout the QAIS framework.

Visual Attribution Statement

Unless otherwise noted, all figures, diagrams, illustrations, architectural models, and visualizations in this volume were prepared specifically for QAIS: Architectures for Quantum AI Systems, Volume I — Foundations of Quantum AI Systems.

Visual materials are intended to support the conceptual, mathematical, operational, and architectural presentation of QAIS throughout the text.

In select cases, computational and generative visualization tools may be used to assist with figure refinement and presentation; however, all conceptual design, technical interpretation, and scientific content remain under the direct supervision and responsibility of the author.

Hands-On Quantum AI Labs

E.1 Beginner Quantum Laboratories

The E.1 series establishes the physical and informational foundations of QAIS through direct observation of quantum behavior.

E.1.1 Superposition — Probability Distribution

Source: Chapter 1

Objective: Observe probabilistic behavior of a qubit under superposition.

Methodology: Prepare a single qubit in $|0\rangle$, apply a Hadamard gate, and perform repeated measurements across increasing shot counts.

Expected Results: Figure E.1.1a shows convergence of $P(|1\rangle)$ toward 0.5 as the number of shots increases. Figure E.1.1b shows a balanced histogram with equal counts for $|0\rangle$ and $|1\rangle$.

Caption: Figure E.1.1a–b. Probability convergence and measurement histogram for a single-qubit superposition state.

QAIS Interpretation: Superposition encodes probabilistic representation intrinsically, forming the basis for quantum information processing.

E.1.2 Bell State — Correlated Outcomes

Source: Chapter 1

Objective: Observe entanglement through correlated measurement outcomes.

Methodology: Prepare a Bell state using Hadamard and CNOT gates.

Expected Results: Figure E.1.2a shows diagonal dominance ($|00\rangle$, $|11\rangle$). Figure E.1.2b shows divergence between marginal and joint probabilities.

Caption: Figure E.1.2a–b. Entanglement correlation heatmap and marginal/joint distribution comparison.

QAIS Interpretation: Entanglement produces non-separable relational structure.

E.1.3 Quantum Teleportation — State Transfer

Source: Chapter 1

Objective: Demonstrate quantum state transfer using entanglement and classical communication.

Methodology: Prepare an arbitrary single-qubit state and teleport the state using a Bell-pair resource and conditional correction operations.

Expected Results: The reconstructed output state matches the original input state with high probability.

QAIS Interpretation: Teleportation demonstrates that quantum information transfer depends upon entanglement structure and coordinated measurement operations rather than direct physical transport of the underlying quantum state.

E.1.4 Depth and Noise Sensitivity — Probability Decay

Source: Chapter 2 — Operations and Scientific Framework of QAIS

Objective: Evaluate the impact of circuit depth under noise.

Methodology: Execute circuits of increasing depth under both noiseless and depolarizing noise conditions.

Expected Results: Figure E.1.4a shows stable success probability under noiseless execution. Figure E.1.4b shows decay in success probability under noise as depth increases.

Caption: Figure E.1.4a–b. Comparison of noiseless and noisy circuit depth scaling.

QAIS Interpretation: Circuit depth directly consumes coherence, reducing reliability under realistic conditions.

E.2 Intermediate Quantum Laboratories

The E.2 laboratories extend foundational quantum operations through state evolution analysis, entanglement behavior, circuit composition, and experimental measurement interpretation.

E.2.1 Bloch Trajectories Under Composite Gates

Source: Chapter 2

Objective: Analyze state evolution under gate composition.

Methodology: Apply multiple composite gate sequences to a single qubit and visualize state evolution on the Bloch sphere.

Expected Results: Distinct gate orderings produce unique Bloch-sphere trajectories, illustrating how gate composition governs quantum-state geometry and evolution.

Caption: Figure E.2.1a–b. Distinct Bloch trajectories under different gate sequences.

E.2.2 CHSH Correlation Sweep

Source: Chapter 1

Objective: Demonstrate Bell inequality violation.

Methodology: Execute CHSH correlation measurements across varying detector angle configurations using entangled Bell states.

Expected Results: Correlation sweeps exceed the classical Bell inequality bound of 2 and approach the quantum limit of approximately 2.828 under optimal measurement settings.

Caption: Figure E.2.2a–e. CHSH correlation exceeding classical bound.

Appendix F — The Quantum Stack

Appendix Overview

Appendix F provides a foundational reference framework for the Quantum Stack introduced in Chapter 1 and developed throughout Volume I of the *QAIS* series. The appendix consolidates the architectural progression connecting Quantum Physics, Quantum Mechanics, and Quantum Technology while providing a forward-looking view of the higher layers that will be explored in subsequent volumes.

The purpose of this appendix is not to re-derive the mathematical formalisms presented throughout the volume, nor to duplicate detailed discussions already developed within the main chapters. Instead, it serves as a systems-oriented reference that places the foundational concepts of Volume I within the broader architectural context of the Quantum Stack.

Throughout Chapters 1–3, readers examined the physical principles governing quantum phenomena, the mathematical frameworks used to describe quantum behavior, and the technological systems developed to control and exploit quantum effects. While each of these topics may be studied independently, their true significance emerges when viewed as interconnected layers within a larger architectural progression.

The Quantum Stack provides a structured framework for understanding how fundamental physical phenomena evolve into increasingly sophisticated forms of capability. At the foundational level, the stack describes how quantum behavior arises within nature, how that behavior is represented mathematically, and how it becomes accessible through engineering and technological implementation. In later volumes, these foundations support the emergence of information-processing systems, programmable computation, and intelligent quantum architectures.

This appendix therefore serves as both a summary of the foundational layers developed in Volume I and a roadmap for understanding how those layers contribute to the broader QAIS discipline.

Accordingly, this appendix should be read as:

- a conceptual hierarchy describing the progression from physical phenomena to increasingly sophisticated quantum capabilities,
- an architectural framework illustrating relationships among foundational quantum disciplines,
- a reference model connecting Quantum Physics, Quantum Mechanics, and Quantum Technology,
- a systems perspective on capability inheritance and constraint propagation,
- and a roadmap for understanding how future layers of the Quantum Stack emerge from the foundations established in this volume.

Readers are encouraged to use this appendix as a reference while reviewing the material presented throughout Volume I and as a preparatory guide for the information-centric disciplines introduced in subsequent volumes of the QAIS series.

F.1 The Quantum Stack as an Architectural Framework

The foundational structure of QAIS is most effectively understood through the layered framework of the Quantum Stack, a conceptual and architectural model that traces the disciplined progression from physical law to intelligent system deployment. The framework provides a structured explanation of how quantum phenomena become transformed into operational capabilities through successive layers of abstraction, engineering, and coordination.

The Quantum Stack should not be interpreted as a simple taxonomy of technologies or a catalog of independent disciplines. Instead, it represents an ordered dependency hierarchy in which each layer contributes specialized functionality while remaining constrained by the layers beneath it. Every architectural decision made at higher levels must ultimately respect limitations originating within lower-level physical and operational realities.

Within this hierarchy:

- each layer inherits constraints from the layer below,
- each layer abstracts complexity to support higher-level functionality,
- each layer introduces new operational capabilities,
- and each layer serves as an enabling foundation for subsequent architectural development.

Physics defines the fundamental possibility space within which all quantum behavior occurs. Quantum mechanics formalizes that behavior into predictive models and mathematical structures. Engineering transforms theoretical possibility into stable and controllable systems. Information science organizes engineered behavior into meaningful representations and protocols. Computation activates these protocols through programmable execution. Architecture coordinates all layers into coherent systems capable of achieving operational objectives.

Viewed through this perspective, intelligence does not emerge independently of the stack. Rather, intelligent behavior represents the highest level of organized capability supported by the cumulative interaction of all preceding layers.

Traditional quantum computing models often terminate at the level of programmable execution, focusing primarily on algorithms, circuits, and computational outcomes. While such frameworks remain valuable, they do not fully capture the architectural requirements of intelligent quantum systems operating within realistic environments.

The QAIS Quantum Stack extends beyond conventional computational boundaries into domains that include:

- representational architectures,
- learning systems,
- hybrid quantum–classical orchestration,
- verification and validation workflows,
- resilience engineering,
- adaptive control frameworks,
- security-aware operational models,
- and intelligent decision-support systems.

This broader perspective recognizes that future quantum platforms will function not as isolated computational devices but as components within larger ecosystems of interconnected technologies, organizations, and decision processes.

The stack therefore describes how structured physical interference evolves into programmable computation and subsequently into organized adaptive behavior operating under physical, operational, economic, and adversarial constraints. It provides a disciplined framework for understanding how intelligence emerges from coordinated interactions among multiple layers of scientific and engineering abstraction.

F.1.1 Conceptual Hierarchy of the Quantum Stack

The conceptual Quantum Stack consists of seven architectural layers:

- Quantum Physics
- Quantum Mechanics
- Quantum Technology
- Quantum Information Science
- Quantum Computing
- Hybrid Quantum–Classical Computing & AI
- QAIS

These layers collectively represent an ascending hierarchy of abstraction, capability, and organizational complexity. Each successive layer introduces new mechanisms for representing, controlling, and exploiting quantum behavior while remaining fundamentally conditioned by lower-layer realities.

Quantum Physics establishes the foundational possibility space governing the behavior of matter, energy, fields, and interactions at microscopic scales. It defines the physical conditions under which coherence, interference, uncertainty, and probabilistic behavior become possible.

Quantum Mechanics formalizes these phenomena through mathematical structures describing state evolution, observables, measurement processes, and state-space geometry. This layer provides the theoretical language necessary for predicting and analyzing quantum behavior.

Quantum Technology transforms theoretical models into engineered systems capable of practical operation. Through advances in fabrication, control electronics, calibration procedures, cryogenic infrastructure, photonic systems, and error management techniques, this layer stabilizes fragile quantum dynamics sufficiently for operational use.

Quantum Information Science organizes engineered quantum behavior into structured representations supporting information processing. At this level, quantum states become informational resources capable of supporting:

- computation,
- communication,
- sensing,
- cryptographic operations,
- and distributed information exchange.

Quantum Computing operationalizes these informational structures through programmable transformations, executable circuits, runtime environments, and algorithmic workflows. This layer converts informational capability into computational utility.

Hybrid Quantum–Classical Computing & AI integrates programmable quantum computation with classical processing, data systems, optimization, simulation, machine learning, inference, and orchestration. This layer provides the computational integration bridge through which specialized quantum capabilities participate in larger hybrid workflows. Quantum processors contribute capabilities suited to selected computational tasks, while classical processors, AI models, data services, and orchestration mechanisms coordinate preparation, optimization, interpretation, feedback, and control. The layer therefore converts quantum computational utility into coordinated hybrid computational capability suitable for integration within intelligent-system architectures.

At the highest level, Quantum AI Systems integrate all preceding layers into coordinated architectures supporting:

- learning,
- inference,
- optimization,
- orchestration,
- adaptation,
- resilience,
- decision support, and

- intelligent autonomous behavior

Rather than viewing intelligence as a standalone capability, QAIS treats intelligent behavior as an emergent property arising from the coordinated interaction of all lower architectural layers. As systems ascend the stack, they gain increasing levels of abstraction, integration, flexibility, and autonomy while simultaneously inheriting deeper dependencies on their physical, informational, computational, and operational foundations.

Figure F.1 presents the canonical Quantum Stack framework introduced in Chapter 1 and subsequently refined throughout the QAIS series. The figure serves as a visual synthesis of the architectural progression connecting foundational quantum science to advanced intelligent system deployment.

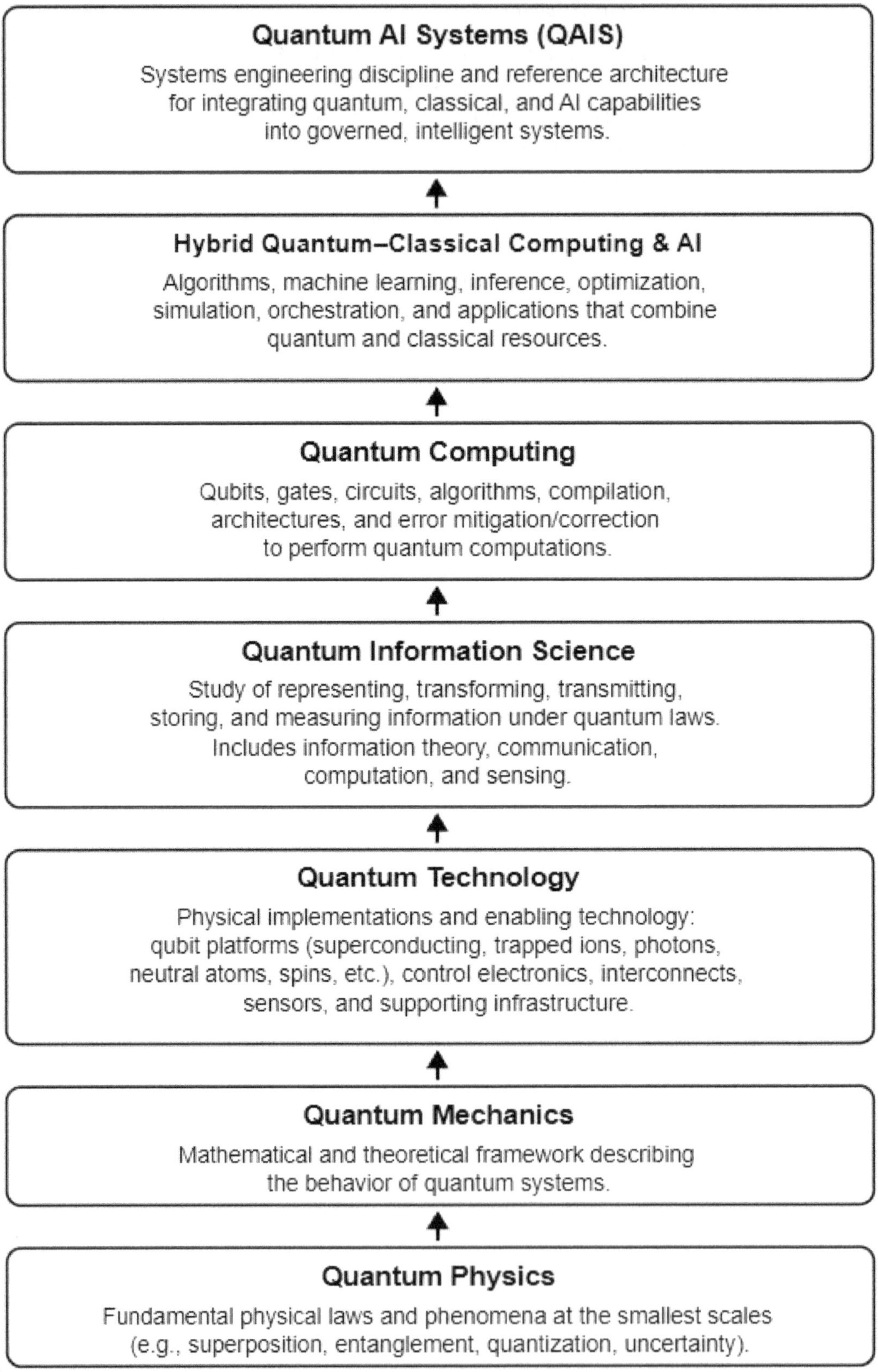
Quantum AI Systems (QAIS)
Systems engineering discipline and reference architecture for integrating quantum, classical, and AI capabilities into governed, intelligent systems.
Hybrid Quantum–Classical Computing & AI
Algorithms, machine learning, inference, optimization, simulation, orchestration, and applications that combine quantum and classical resources.
Quantum Computing
Qubits, gates, circuits, algorithms, compilation, architectures, and error mitigation/correction to perform quantum computations.
Quantum Information Science
Study of representing, transforming, transmitting, storing, and measuring information under quantum laws. Includes information theory, communication, computation, and sensing.
Quantum Technology
Physical implementations and enabling technology: qubit platforms (superconducting, trapped ions, photons, neutral atoms, spins, etc.), control electronics, interconnects, sensors, and supporting infrastructure.
Quantum Mechanics
Mathematical and theoretical framework describing the behavior of quantum systems.
Quantum Physics
Fundamental physical laws and phenomena at the smallest scales (e.g., superposition, entanglement, quantization, uncertainty).

Figure F.1 — The Quantum Stack: From Physics to QAIS.

The seven-layer Quantum Stack illustrates the architectural ascent from quantum physics and quantum mechanics through quantum technology, quantum information science, quantum computing, hybrid quantum–classical computing and AI, and ultimately Quantum AI Systems (QAIS). The framework demonstrates how fundamental quantum phenomena become physically realizable resources, structured information processes, programmable computational capabilities, hybrid quantum–classical and AI workflows, and ultimately governed intelligent architectures supporting learning, inference, optimization, reasoning, resilience, and deployment.

F.1.2 Structural Logic of the Stack

The Quantum Stack should be interpreted as a structured architectural progression rather than a simple list of scientific topics or technological achievements. Although each layer may be associated with a particular discipline, the stack is fundamentally a framework for understanding how increasingly sophisticated capabilities emerge from deeper Quantum Physics.

In Volume I, the first three layers of the stack—Quantum Physics, Quantum Mechanics, and Quantum Technology—form the foundational portion of this progression. Together, they explain how fundamental physical phenomena become transformed into engineered systems capable of controlled operation.

The layers represent ordered transitions in:

- understanding,
- abstraction,
- mathematical description,
- experimental control,
- and technological realization.

Each layer builds upon the one below it while introducing new methods for describing, predicting, and manipulating quantum behavior.

The progression is cumulative rather than substitutive. Higher layers do not replace lower layers. Instead, they depend upon them and extend their capabilities. Quantum Technology does not replace Quantum Mechanics, and Quantum Mechanics does not replace Quantum Physics. Rather, each layer provides additional structure that allows the capabilities of the preceding layer to be understood and utilized more effectively.

This relationship is important because quantum systems cannot be fully understood from any single perspective.

For example:

- Physics identifies the existence of quantum phenomena.
- Quantum Mechanics provides the mathematical language needed to describe those phenomena.
- Quantum Technology transforms those descriptions into controllable physical systems.
- Each stage therefore answers a different question.

Quantum Physics asks: *"What behaviors occur in nature?"*

Quantum Mechanics asks: *"How can those behaviors be described and predicted?"*

Quantum Technology asks: *"How can those behaviors be engineered and controlled?"*

Viewed from this perspective, the stack encodes a disciplined transformation:

- from observation of natural phenomena,
- to mathematical formalization,
- to experimental verification,
- to engineering implementation,
- and ultimately toward future information-processing systems.

The first layer establishes the physical possibility space. The second layer creates the theoretical framework needed to understand that possibility space. The third layer converts theoretical understanding into practical capability.

This progression mirrors the historical development of quantum science itself. Early experimental observations revealed behaviors that could not be explained by classical physics. The development of quantum mechanics provided a mathematical framework capable of explaining those observations. Advances in engineering later enabled researchers to manipulate and control quantum systems with increasing precision.

The stack therefore reflects both a conceptual hierarchy and a historical progression in scientific understanding.

Another important feature of the stack is that capability and constraint emerge together. Every new capability gained at a higher layer introduces additional challenges that must be addressed.

Examples include:

- Superposition increases representational possibilities but requires preservation of coherence.
- Interference enables useful quantum effects but requires precise control of phase relationships.
- Quantum state manipulation creates opportunities for computation and sensing but introduces sensitivity to environmental disturbances.
- Greater experimental control enables more sophisticated applications while increasing engineering complexity.

These examples illustrate a recurring principle throughout quantum science: enhanced capability is accompanied by increased responsibility for control and stability.

As systems move upward through the stack, they become more powerful but also more dependent upon successful management of lower-layer constraints.

A quantum device, for example, cannot outperform the physical limitations imposed by the quantum states from which it is constructed. Likewise, engineering solutions cannot eliminate the fundamental probabilistic behavior described by quantum mechanics. Instead, successful technologies must be designed to operate within those constraints.

For this reason, the Quantum Stack functions not only as a capability model but also as a dependency model. It helps readers understand how developments at one level influence every level above it.

Within the context of Volume I, the stack provides a unifying framework for understanding the relationship between:

- physical quantum phenomena,
- mathematical quantum theory,
- and engineered quantum systems.

These foundational layers establish the scientific and technological base upon which later developments in quantum information science, quantum computing, and intelligent quantum systems will eventually be constructed. Understanding the structural logic of the stack at this foundational level is therefore essential for understanding the broader architectural progression developed throughout the remainder of the Quantum AI Systems series.

F.2 From Quantum Physics to Engineered Quantum Systems

The first three layers of the Quantum Stack describe a progression from natural quantum behavior to practical technological implementation. While these layers may initially appear as separate domains of study, they are more accurately understood as interconnected stages within a continuous scientific and engineering process.

Quantum Physics establishes the existence of phenomena that differ fundamentally from classical expectations. Quantum Mechanics provides the mathematical framework needed to describe and predict those phenomena. Quantum Technology transforms this theoretical understanding into physical systems capable of controlled operation.

Together, these layers explain how abstract scientific principles become usable technological capabilities.

This progression represents one of the defining characteristics of modern science and engineering. Scientific discovery alone is insufficient to create practical systems. Likewise, engineering cannot proceed effectively without an accurate understanding of the underlying physical principles. Meaningful technological advancement emerges only when scientific understanding and engineering implementation develop together.

Within the Quantum Stack, this relationship forms the bridge between theory and practice.

The transition from Quantum Physics to Quantum Mechanics represents the movement from observation to explanation. Early experimental discoveries revealed behaviors that challenged classical assumptions about matter, energy, and measurement. Phenomena such as discrete energy levels, wave-particle duality, and probabilistic measurement outcomes required entirely new mathematical descriptions. Quantum Mechanics emerged as the framework capable of providing those descriptions.

The transition from Quantum Mechanics to Quantum Technology represents a second transformation. Once quantum behavior could be modeled mathematically, researchers began exploring methods for controlling and exploiting those behaviors in physical systems. This effort required advances in experimental techniques, materials science, precision instrumentation, electronics, and environmental control.

As a result, quantum phenomena that were once observed only in laboratory experiments gradually became the basis for practical technologies.

Examples include:

- atomic clocks utilizing controlled quantum transitions,
- semiconductor devices exploiting quantum behavior within solid-state materials,
- laser systems based upon stimulated emission,
- superconducting circuits supporting coherent quantum states,
- trapped-ion platforms capable of precise quantum control,
- photonic systems that manipulate individual quantum particles of light,
- and quantum sensing technologies capable of detecting extremely small environmental variations.

These technologies illustrate an important principle of the Quantum Stack: practical capability emerges only when theoretical understanding is matched by engineering control.

Control is particularly important because quantum systems are inherently sensitive to their environments. Quantum states can be disrupted by thermal fluctuations, electromagnetic interference, material imperfections, and measurement interactions. Consequently, engineering quantum systems often involves protecting delicate quantum behavior from external disturbances while simultaneously maintaining sufficient access for useful operation.

This challenge creates a continual balance between isolation and control.

A quantum system must be isolated enough to preserve coherent behavior, yet accessible enough to permit manipulation, observation, and measurement.

Maintaining this balance remains one of the central engineering challenges throughout the development of quantum technologies.

Another important aspect of this progression is the gradual increase in abstraction that occurs as systems advance through the stack. At the level of Quantum Physics, attention focuses on fundamental phenomena. At the level of Quantum Mechanics, emphasis shifts toward mathematical models and predictive frameworks. At the level of Quantum Technology, these models become embedded within physical devices and operational systems.

As abstraction increases, the complexity visible to users often decreases.

For example, a scientist studying quantum state evolution may focus on wavefunctions, probability amplitudes, and measurement operators. An engineer designing a quantum device may instead focus on fabrication methods, calibration procedures, and environmental stability. An end user of a quantum-enabled technology may never directly encounter the underlying mathematics at all.

Each layer therefore hides portions of lower-level complexity while preserving essential functionality.

This process of abstraction is a defining characteristic of engineering systems. Complex underlying mechanisms become encapsulated within structures that enable higher-level operation. The same principle that allows modern users to operate computers without understanding semiconductor physics also enables future users to employ quantum technologies without mastering the full mathematical formalism of quantum mechanics.

The Quantum Stack provides a framework for understanding these relationships.

Rather than viewing Physics, Mechanics, and Technology as isolated disciplines, the stack reveals them as interconnected stages in the transformation of knowledge into capability. Physical phenomena establish what is possible. Mathematical theory explains why it is possible. Engineering determines how it can be implemented.

Together, these foundational layers create the scientific and technological infrastructure upon which later developments in quantum information science, quantum computing, and advanced intelligent systems will ultimately be constructed.

F.3 Quantum Physics Layer

The Quantum Physics Layer forms the foundational level of the Quantum Stack and serves as the starting point for all subsequent developments in quantum science, engineering, and technology. It is within this layer that the fundamental behaviors of matter and energy are observed, investigated, and characterized. Every higher layer of the stack ultimately derives its capabilities from the physical principles established here.

Unlike higher layers that focus on mathematical modeling, system design, or technological implementation, Quantum Physics is concerned primarily with understanding how nature behaves at microscopic scales. It seeks to identify the physical laws governing particles, fields, interactions, and energy exchange under conditions where classical descriptions become insufficient.

Historically, the emergence of quantum physics represented one of the most significant transformations in scientific thought. Experimental observations such as blackbody radiation, atomic emission spectra, the photoelectric effect, and electron diffraction revealed behaviors that could not be adequately explained using classical physics. These discoveries established the need for an entirely new understanding of physical reality.

F.3.1 Fundamental Quantum Phenomena

Several foundational phenomena emerge within the Quantum Physics Layer:

- quantization,
- wave-particle duality,
- superposition,
- interference,
- uncertainty,
- entanglement,
- and probabilistic measurement outcomes.

These phenomena collectively define the physical possibility space from which all subsequent quantum capabilities emerge.

F.3.2 Quantization and Discrete Physical Behavior

One of the earliest discoveries of quantum physics was that certain physical quantities occur in discrete increments rather than continuous values.

Energy exchange, atomic transitions, and many microscopic interactions exhibit quantized behavior. This discovery fundamentally altered scientific understanding and established one of the defining characteristics of quantum systems.

F.3.3 Wave–Particle Duality

Quantum entities exhibit both particle-like and wave-like characteristics.

Photons, electrons, and other quantum particles may behave as localized objects under some conditions while simultaneously exhibiting interference and diffraction behavior typically associated with waves.

This duality remains one of the most distinctive features of quantum physics.

F.3.4 Superposition and Interference

Quantum systems may exist in combinations of multiple possible states prior to measurement.

This property, known as superposition, enables probability amplitudes to evolve simultaneously. Interference emerges when these amplitudes combine constructively or destructively, producing behaviors that differ fundamentally from classical statistical systems.

F.3.5 Entanglement as a Physical Phenomenon

Entanglement describes a relationship in which multiple quantum systems become correlated beyond what is possible through classical descriptions. These correlations persist regardless of spatial separation and represent one of the most powerful and intriguing resources within quantum science.

F.3.6 The Role of Quantum Physics Within the Stack

Within the broader Quantum Stack:

- Quantum Physics establishes what is physically possible.
- Quantum Mechanics explains how those possibilities behave.
- Quantum Technology determines how those behaviors can be controlled.

Consequently, Quantum Physics serves as the foundational source of capability, constraint, and possibility throughout the entire stack.

F.4 Quantum Mechanics Layer

The Quantum Mechanics Layer occupies the second foundational position within the Quantum Stack and serves as the formal bridge between observed quantum phenomena and engineered quantum systems. Whereas the Quantum Physics Layer focuses on the discovery and characterization of quantum behavior, the Quantum Mechanics Layer provides the mathematical language required to describe, predict, and analyze that behavior.

Historically, the development of quantum mechanics represented a profound shift in scientific reasoning. Experimental observations revealed behaviors that could not be adequately explained using classical models of matter and energy. Phenomena such as quantized energy levels, wave-particle duality, and probabilistic measurement outcomes required an entirely new framework capable of describing microscopic systems.

F.4.1 Quantum mechanics emerged as that framework.

Within the Quantum Stack, quantum mechanics performs a unique role. It transforms observation into prediction and converts physical intuition into mathematical structure. Without quantum mechanics, the phenomena identified within quantum physics would remain isolated experimental curiosities. Through mathematical formalism, however, these phenomena become predictable, reproducible, and ultimately exploitable for technological purposes.

Consequently, the Quantum Mechanics Layer can be viewed as the theoretical engine of the Quantum Stack. It supplies the conceptual tools that make higher-layer engineering and technology possible.

F.5 Quantum Technology Layer

The Quantum Technology Layer occupies the third foundational position within the Quantum Stack and represents the transition from theoretical understanding to practical implementation. Whereas Quantum Physics establishes the underlying phenomena and Quantum Mechanics provides the mathematical framework for describing those phenomena, Quantum Technology focuses on engineering systems capable of controlling, manipulating, and measuring quantum behavior.

This layer transforms scientific knowledge into operational capability.

Through advances in materials science, precision instrumentation, control electronics, cryogenic engineering, photonics, and fabrication technologies, researchers have developed increasingly sophisticated platforms capable of harnessing quantum phenomena for practical applications.

Quantum Technology therefore represents the point at which quantum science becomes engineering.

F.5.1 Engineering Quantum Behavior

Engineering quantum systems requires precise control over fragile physical processes.

Unlike many classical systems, quantum states are highly sensitive to environmental disturbances. Consequently, engineering solutions must carefully balance accessibility, stability, and operational control.

F.5.2 State Preparation and Control Systems

Before useful quantum operations can occur, quantum states must be prepared in known configurations.

State preparation systems establish initial conditions for experimentation, sensing, communication, and future computational applications.

Control systems then manipulate these states through carefully designed interactions.

F.5.3 Representative Quantum Technologies

Examples of contemporary quantum technologies include:

- superconducting circuits,
- trapped-ion systems,
- photonic architectures,
- neutral-atom platforms,
- atomic clocks,
- quantum sensors,
- and spin-based devices.

Each technology provides unique advantages while introducing distinct engineering challenges.

F.5.4 Measurement Technologies

Quantum measurement requires specialized hardware capable of detecting extremely small signals while minimizing environmental interference.

Examples include:

- photodetectors,
- microwave readout systems,
- ion-state detection platforms,
- superconducting measurement circuits,
- and precision timing systems.

These technologies provide the observational capabilities necessary for practical quantum operation.

F.5.5 Stability, Noise, and Environmental Management

One of the primary responsibilities of the Quantum Technology Layer is maintaining operational stability.

Important engineering considerations include:

- coherence preservation,
- thermal management,
- electromagnetic shielding,
- calibration control,
- vibration isolation,
- and noise reduction.

Without these protections, useful quantum behavior rapidly degrades.

F.5.6 The Role of Quantum Technology Within the Stack

- Quantum Technology serves as the implementation layer of the foundational stack.
- Quantum Physics establishes possibility.
- Quantum Mechanics establishes predictability.

• Quantum Technology establishes controllability.

Together, these three foundational layers transform quantum phenomena from scientific observations into practical technological resources capable of supporting future developments in Quantum Information Science, Quantum Computing, and Quantum AI Systems.

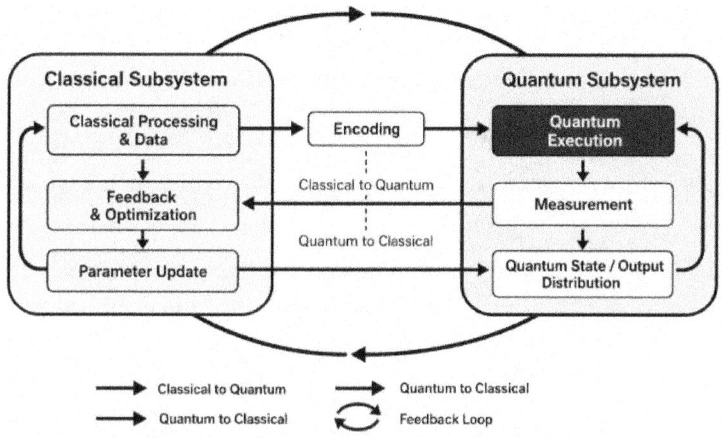

Figure F.2 — Hybrid Quantum-Classical Interaction Model for Foundational Quantum Information Processing.

Foundational interaction between classical computational resources and quantum execution systems. Classical subsystems perform data preparation, parameter specification, and result interpretation, while quantum subsystems execute state preparation, quantum evolution, and measurement. The bidirectional exchange between the two subsystems illustrates the fundamental operational pattern that underlies many contemporary quantum computing workflows and serves as a precursor to more advanced QAIS architectures developed in later volumes.

F.6 Interlayer Dependency Relationships

The Quantum Stack is often presented as a sequence of layers progressing from physical phenomena toward increasingly sophisticated technological capabilities. While this representation is useful for visualization, it can create the mistaken impression that each layer operates independently. In reality, the layers of the Quantum Stack form a tightly coupled dependency hierarchy in which capabilities emerge through the coordinated interaction of multiple levels of abstraction.

Understanding these dependencies is essential because higher-layer achievements cannot be fully understood without reference to the foundational layers that support them. Every technological capability, experimental platform, and future information-processing system ultimately depends upon principles established at lower levels of the stack.

The concept of interlayer dependency therefore serves as one of the central architectural themes of the Quantum Stack.

Rather than viewing the stack as a collection of separate disciplines, readers should view it as a system of inherited capability and inherited constraint. Each layer receives both opportunities and limitations from the layers beneath it. Consequently, advances at higher levels remain fundamentally connected to developments occurring at lower levels.

F.6.1 Vertical Dependency Structure

The Quantum Stack exhibits a vertical dependency structure in which each layer builds upon the resources established by preceding layers.

At the foundational level, Quantum Physics identifies the behaviors that nature permits. These behaviors include superposition, interference, quantization, uncertainty, and entanglement. Without these physical phenomena, the remaining layers of the stack would have no basis for existence.

Quantum Mechanics builds directly upon this foundation by providing mathematical descriptions capable of representing and predicting quantum behavior. Theoretical models, state representations, and measurement frameworks depend entirely upon the physical phenomena they seek to describe.

Quantum Technology then relies upon both physics and mechanics. Engineering solutions require an understanding of physical constraints as well as mathematical models capable of guiding system design.

This relationship can be summarized as follows:

- Physics establishes possibility.
- Mechanics establishes predictability.
- Technology establishes controllability.

The removal of any one layer would compromise the functionality of those above it.

For example, technological innovation cannot proceed without mathematical understanding. Likewise, mathematical models cannot exist without underlying physical phenomena. The stack therefore functions as a chain of dependency rather than a collection of isolated domains.

F.6.2 Capability Propagation

One of the most important consequences of interlayer dependency is the concept of capability propagation.

Capabilities discovered at lower layers often reappear at higher layers in increasingly sophisticated forms.

Consider the phenomenon of superposition.

- At the Quantum Physics Layer, superposition represents a fundamental physical behavior describing the existence of multiple potential outcomes within a quantum system.
- At the Quantum Mechanics Layer, superposition becomes a mathematical construct represented through state vectors and probability amplitudes.
- At the Quantum Technology Layer, superposition becomes an operational resource that engineers attempt to preserve and manipulate within physical devices.

Although the interpretation changes from layer to layer, the underlying capability remains the same.

A similar pattern appears with:

- interference,
- coherence,
- entanglement,
- quantum state evolution,
- and measurement.

These capabilities originate at the physical level but propagate upward through the stack, acquiring new forms of representation and application as abstraction increases.

The concept of capability propagation helps explain why foundational quantum phenomena remain relevant even when discussing highly engineered systems.

Every advanced capability ultimately traces its origins to lower-layer physical behavior.

F.6.3 Constraint Inheritance

Capabilities are not the only properties that propagate through the stack.

Constraints propagate as well.

In many cases, limitations established at lower layers become increasingly important as systems grow more sophisticated.

Examples include:

- decoherence,
- environmental noise,
- thermal instability,
- measurement uncertainty,
- fabrication imperfections,
- and control precision requirements.

These limitations originate at the physical level but continue to influence every higher layer.

For example, a technological system may incorporate advanced control mechanisms and sophisticated calibration procedures. However, those mechanisms cannot eliminate the fundamental sensitivity of quantum states to environmental interactions.

Similarly, mathematical models may provide accurate descriptions of quantum behavior, but they cannot remove the probabilistic nature of measurement outcomes.

This principle demonstrates an important architectural reality:

Higher layers may manage lower-layer constraints, but they cannot eliminate them.

Consequently, successful quantum technologies are designed not to ignore constraints but to operate effectively within them.

F.6.4 Dependency Versus Abstraction

As systems move upward through the stack, abstraction increases.

Abstraction simplifies interaction by hiding lower-level complexity from users and designers.

For example, an engineer working with a quantum sensor may not need to explicitly calculate every wavefunction associated with the device. Similarly, an end user of a quantum-enabled technology may never encounter the underlying mathematics governing system operation.

This concealment of complexity creates an important distinction between abstraction and dependency. A higher layer may abstract away lower-level details while remaining completely dependent upon them. Modern computer systems provide a useful analogy.

Most users operate software applications without understanding semiconductor physics. Nevertheless, the software remains entirely dependent upon the hardware and physical principles supporting computation. The same relationship exists within the Quantum Stack.

Technological systems may abstract away quantum mechanical complexity, but they remain fundamentally dependent upon both quantum mechanics and quantum physics. Understanding this distinction helps prevent the misconception that higher levels somehow replace lower levels. They do not. They simply provide additional layers of organization and usability.

F.6.5 Architectural Consequences

The interlayer dependency structure of the Quantum Stack has several important architectural consequences.

First, system performance can often be traced downward through the stack to identify foundational causes of success or failure. Second, advances at lower layers frequently create opportunities for innovation at higher layers. Third, limitations encountered at higher layers are often rooted in unresolved challenges originating below.

These observations highlight the importance of maintaining a systems perspective when studying quantum technologies.

Engineers, scientists, and system architects must recognize that no layer exists in isolation. Effective design requires an understanding of how capabilities, constraints, and dependencies interact throughout the stack.

Within the context of Volume I, this perspective reinforces the importance of mastering the foundational layers before progressing toward more advanced topics. Quantum Information Science, Quantum Computing, and Quantum AI Systems all depend upon the relationships established here.

Consequently, interlayer dependency should be viewed as a defining characteristic of the Quantum Stack itself. The stack is not merely a sequence of topics arranged for convenience. It is a structured architectural framework describing how physical reality, mathematical understanding, and technological implementation cooperate to create increasingly sophisticated quantum capabilities.

Summary

The dependency relationships explored in this section establish the foundation for understanding how information emerges from physical systems. Once quantum phenomena can be represented mathematically and controlled technologically, they become candidates for information storage, transmission, transformation, and interpretation. This transition from physical behavior to informational significance forms the subject of the next section and provides the conceptual bridge toward the information-centric disciplines that will be developed in subsequent volumes.

F.7 Information Emergence Within the Quantum Stack

One of the most important insights provided by the Quantum Stack is that information does not emerge independently of physical systems. Before information can be stored, transmitted, processed, or interpreted, it must first exist within a physical substrate capable of representing distinguishable states. For this reason, the study of information ultimately begins with physics.

Within Volume I, the Quantum Stack has been examined through three foundational layers:

- Quantum Physics,
- Quantum Mechanics,
- and Quantum Technology.

Collectively, these layers establish the conditions under which information can eventually become a meaningful operational resource. Although the formal study of Quantum Information Science is reserved for later volumes, the foundations of information emerge naturally from the concepts already introduced.

Understanding this progression is important because future quantum information systems, quantum computers, and intelligent quantum architectures all depend upon the ability to represent and manipulate information through physical quantum states.

The emergence of information within the Quantum Stack therefore represents the transition from understanding quantum behavior to utilizing that behavior for purposeful representation and communication.

F.7.1 Physical States as Information Resources

At the most fundamental level, information requires distinguishable physical conditions.

A classical switch, for example, can occupy an ON state or an OFF state. These distinct physical configurations permit information to be represented as binary values.

Quantum systems also possess physical states that may serve as information resources. However, quantum states introduce capabilities that extend beyond traditional binary representations.

Examples include:

- superposition,
- interference,
- phase relationships,
- polarization states,
- spin states,
- and energy-level transitions.

Each of these physical properties can be used to represent information under appropriate conditions. Importantly, the information itself is not separate from the physical system. Rather, information exists because the physical system can occupy distinguishable states that may be observed, manipulated, and interpreted.

This principle is sometimes summarized by the statement: Information is physical.

Within the Quantum Stack, this means that every future information-processing capability originates from physical phenomena established within the foundational layers.

F.7.2 From Physical Observation to Mathematical Representation

While physical states provide the substrate for information, information cannot be effectively utilized without a means of description.

This requirement introduces the role of Quantum Mechanics.

Quantum mechanics provides mathematical representations capable of describing quantum states and their evolution. Concepts such as state vectors, wavefunctions, probability amplitudes, and operators allow researchers to model quantum systems in ways that support prediction and analysis.

From an informational perspective, these mathematical descriptions perform an important function. They transform physical behavior into representable structure.

A physical quantum state may exist within a laboratory device, but its informational significance emerges only when that state can be described, compared, and interpreted.

Consequently, the Quantum Mechanics Layer serves as an intermediary between physical reality and informational meaning.

Without mathematical representation, physical states remain difficult to organize into useful information-processing frameworks.

F.7.3 Measurement and Information Extraction

The emergence of information depends not only upon representation but also upon extraction. Information that cannot be observed or communicated has limited practical value.

Measurement therefore occupies a central role within the Quantum Stack. Through measurement, quantum systems produce observable outcomes that can be interpreted by classical observers and computational systems.

Measurement converts:

- quantum possibilities,

into

- classical observations.

This conversion process creates one of the most important boundaries in quantum science.

Prior to measurement, information exists within quantum state descriptions characterized by probability amplitudes and evolving relationships among possible outcomes.

Following measurement, information becomes available as experimentally observable data.

The ability to extract information from quantum systems establishes the operational foundation upon which future communication, sensing, and computational technologies depend.

Consequently, measurement can be viewed as one of the earliest information-generating processes within the Quantum Stack.

F.7.4 Information Preservation and Physical Stability

Information is useful only when it can be preserved long enough to support meaningful operation.

This requirement introduces the importance of stability within quantum systems.

Quantum states are sensitive to environmental interactions. External disturbances may alter state evolution, reduce coherence, or disrupt the relationships necessary for preserving information.

Examples include:

- thermal fluctuations,
- electromagnetic interference,
- material defects,
- environmental noise,
- and unintended measurement interactions.

These influences demonstrate that information preservation is fundamentally linked to physical control.

The challenge of maintaining stable quantum states therefore becomes one of the first major engineering problems encountered within the Quantum Stack.

Quantum Technology addresses this challenge through:

- environmental isolation,
- precision control,
- calibration,
- error mitigation,
- and hardware design.

These efforts enable quantum states to remain sufficiently stable for future informational applications.

In this sense, Quantum Technology serves as the first layer in which information preservation becomes an explicit engineering objective.

F.7.5 Correlation, Relationship, and Meaning

Information is often associated with individual states, but many of the most important quantum phenomena arise from relationships among states.

- Interference patterns emerge from relationships among probability amplitudes.
- Entanglement produces correlations among multiple systems.
- Measurement outcomes frequently depend upon interactions occurring throughout an entire system rather than within isolated components.

These observations suggest that information within quantum systems is not always localized.

Instead, meaningful information may reside within relationships, correlations, and patterns extending across multiple physical components.

This idea represents an important departure from many classical perspectives and foreshadows concepts that will later become central to Quantum Information Science.

At this stage, it is sufficient to recognize that information may emerge not only from individual states but also from the relationships connecting them.

F.7.6 Preparing for Quantum Information Science

The first three layers of the Quantum Stack establish the conditions necessary for information to emerge as a meaningful resource.

- Quantum Physics provides the phenomena.
- Quantum Mechanics provides the representation.
- Quantum Technology provides the control.

Together, these layers create the foundation upon which future information-processing disciplines can be constructed.

The transition to Quantum Information Science occurs when quantum states cease to be viewed solely as physical phenomena and begin to be treated as informational resources capable of supporting communication, computation, sensing, and structured information processing.

This transition represents one of the most significant developments within the broader Quantum Stack because it transforms quantum behavior from a subject of study into a resource for purposeful operation.

Summary

The emergence of information marks a critical turning point in the architectural progression of the Quantum Stack. Once physical states can be represented mathematically, controlled technologically, preserved operationally, and interpreted meaningfully, they become candidates for organized information processing. The next section examines how constraints originating within the foundational layers continue to influence every subsequent capability built upon them, establishing the principle of constraint propagation throughout the Quantum Stack.

F.8 Constraint Propagation Across the Stack

One of the most important architectural principles within the Quantum Stack is that constraints propagate upward through layers in much the same way that capabilities do. While much attention is often given to the opportunities created by quantum phenomena, an equally important consideration involves the limitations inherited from lower levels of the stack.

Every layer of the Quantum Stack receives not only enabling resources from the layers beneath it but also the restrictions, uncertainties, and operational boundaries associated with those resources. As a result, higher-level systems must continually operate within constraints established at lower levels.

*This principle is referred to as **constraint propagation**.*

Constraint propagation explains why advanced technological systems remain fundamentally connected to the physical realities from which they emerge. No matter how sophisticated a higher-layer architecture becomes, it cannot completely escape the limitations imposed by the layers supporting it.

Within Volume I, understanding constraint propagation is particularly important because it reveals the close relationship between:

- physical reality,
- mathematical representation,
- and engineering implementation.

The principle also provides a foundation for understanding why future information-processing systems, quantum computers, and intelligent quantum architectures remain dependent upon lower-layer performance and stability.

F.8.1 Constraints Within Quantum Physics

The origin of constraint propagation begins within the Quantum Physics Layer.

Quantum phenomena provide remarkable capabilities, but these capabilities are accompanied by inherent limitations.

Examples include:

- probabilistic measurement outcomes,
- uncertainty relationships,
- environmental sensitivity,
- finite coherence duration,
- state disturbance during observation,
- and interaction-dependent behavior.

These characteristics are not engineering deficiencies. They are fundamental properties of nature itself.

For example, measurement uncertainty cannot be eliminated through improved technology because it originates from the physical principles governing quantum systems. Likewise, the probabilistic nature of quantum measurement reflects an intrinsic feature of quantum behavior rather than an imperfection in experimental design.

Consequently, all higher layers of the Quantum Stack inherit these realities.

- Technology may manage them.
- Mathematics may describe them.
- Engineering may mitigate their practical effects.

However, none of these approaches can remove the underlying physical constraints themselves.

F.8.2 Constraints Within Quantum Mechanics

The Quantum Mechanics Layer introduces a second category of constraints associated with representation and prediction.

Although quantum mechanics provides a powerful mathematical framework for describing quantum behavior, that framework imposes its own limitations.

Examples include:

- probabilistic outcome prediction,
- measurement dependence,
- state-space complexity,
- computational scaling challenges,
- and representational abstraction.

As quantum systems increase in size, the complexity of their mathematical description grows rapidly. This growth can make analysis and simulation increasingly difficult, even when the governing equations remain well understood.

In this sense, mathematical understanding does not automatically translate into computational simplicity. A system may be theoretically describable while remaining extraordinarily challenging to model in practice. This distinction becomes increasingly important as systems progress toward technological implementation.

F.8.3 Constraints Within Quantum Technology

Quantum Technology introduces engineering constraints that arise during the transition from theory to implementation. At this layer, abstract concepts must operate within physical devices.

Examples include:

- fabrication tolerances,
- control precision,
- calibration requirements,
- thermal management,
- hardware reliability,
- measurement accuracy,
- and environmental isolation.

These constraints reflect the practical realities of constructing systems capable of manipulating delicate quantum states.

For example, maintaining coherence requires extensive environmental control. Achieving precise state preparation often requires sophisticated instrumentation and calibration procedures. Reliable measurement may depend upon highly specialized detection systems.

As a result, engineering challenges become inseparable from system performance.

Even when physical principles are well understood and mathematical models are accurate, technological implementation remains constrained by the practical realities of hardware design and operation.

F.8.4 Constraint Amplification Across Layers

An important characteristic of the Quantum Stack is that lower-layer constraints may become amplified as systems increase in complexity.

A small physical limitation at one layer can produce significant operational consequences at higher layers.

For example:

- minor calibration errors may influence measurement reliability,
- measurement inaccuracies may affect information extraction,
- information degradation may influence computational outcomes,
- and computational errors may affect application-level performance.

This amplification process illustrates why seemingly minor physical effects often receive significant attention within quantum engineering.

The importance of a constraint is not determined solely by its immediate impact. Its significance also depends upon how that constraint propagates through the stack.

Consequently, successful system design requires consideration of both local effects and long-range architectural consequences.

F.8.5 Managing Constraints Through Engineering

Although constraints cannot be eliminated entirely, they can often be managed effectively.

A central objective of engineering is therefore not the removal of constraints but the development of strategies that allow useful operation despite their presence.

Examples include:

- environmental shielding,
- calibration protocols,
- precision control systems,
- redundancy mechanisms,
- error-mitigation techniques,
- and measurement optimization procedures.

Each of these approaches seeks to reduce the practical impact of lower-layer limitations.

Importantly, these strategies do not alter the underlying physical laws.

Instead, they improve the ability of systems to operate within those laws.

This distinction reinforces one of the central themes of the Quantum Stack:

Engineering succeeds not by overcoming physics, but by working effectively within physical reality.

F.8.6 Constraints as Architectural Design Drivers

Constraints are often viewed negatively because they restrict system behavior.

From an architectural perspective, however, constraints can be valuable design drivers.

Constraints encourage:

- disciplined engineering,
- efficient resource allocation,
- realistic performance expectations,
- robust system design,
- and improved operational reliability.

Many important technological innovations emerge specifically because engineers seek ways to manage difficult constraints.

Consequently, constraints should not be viewed solely as obstacles. They are also catalysts for creativity and innovation.

Throughout the history of science and engineering, limitations have frequently inspired new approaches that ultimately expand technological capability.

The same pattern continues throughout the development of quantum technologies.

F.8.7 Architectural Implications for Future Layers

The principle of constraint propagation becomes increasingly important as the Quantum Stack expands into future layers.

- When quantum states become informational resources, physical constraints influence information quality.
- When information becomes computation, physical constraints influence computational performance.
- When computation supports intelligent behavior, physical constraints continue to influence learning, adaptation, and decision-making.

Thus, the influence of foundational layers never disappears.

Instead, it remains embedded within every subsequent capability built upon them.

Understanding this relationship helps explain why the study of foundational quantum science remains relevant even when discussing advanced technological systems.

No higher layer can be fully understood without reference to the constraints inherited from below.

Summary

The propagation of constraints reveals that quantum systems must be analyzed as interconnected architectures rather than isolated components. Capabilities emerge through layered cooperation, while limitations propagate through the same dependency pathways. Understanding these relationships prepares the reader for examining one of the most important interfaces within the Quantum Stack: measurement. Measurement not only extracts information from quantum systems but also establishes a boundary across which physical behavior, mathematical description, and technological operation converge.

F.9 Measurement as a Cross-Layer Boundary

Among all concepts introduced within the foundational layers of the Quantum Stack, measurement occupies a uniquely important position. Few processes influence as many layers simultaneously, and few concepts serve as such a critical bridge between physical phenomena, mathematical representation, and technological implementation.

Measurement is not merely an experimental procedure. Within the Quantum Stack, it functions as a cross-layer boundary through which quantum behavior becomes observable, interpretable, and operationally useful.

- At the level of Quantum Physics, measurement reveals the outcomes of quantum processes.
- At the level of Quantum Mechanics, measurement is represented mathematically through probability distributions and measurement operators.
- At the level of Quantum Technology, measurement becomes an engineering activity requiring specialized hardware, calibration procedures, and signal-processing systems.

Consequently, measurement serves as one of the most important connecting mechanisms within the entire Quantum Stack.

Understanding measurement therefore requires more than a physical perspective or a mathematical perspective alone. It requires an appreciation of how multiple layers cooperate to transform quantum behavior into usable information.

F.9.1 Measurement Within Quantum Physics

Within the Quantum Physics Layer, measurement provides the primary means through which quantum phenomena become observable.

Many quantum properties cannot be directly perceived through ordinary human experience. Instead, their existence must be inferred through carefully designed experiments and observations.

Examples include:

- energy-level transitions,
- interference patterns,
- spin orientations,
- photon polarization states,
- and quantum correlations.

Measurement provides the mechanism through which these phenomena become accessible to scientific investigation.

Importantly, measurement in quantum systems differs from measurement in many classical systems.

In classical physics, observation is often treated as a passive process that reveals properties already possessed by an object.

- Quantum systems exhibit more complex behavior.
- The act of measurement may influence the system being observed and can affect the subsequent evolution of that system.
- As a result, measurement becomes an active participant in quantum behavior rather than a purely external observation process.

This characteristic represents one of the defining features of quantum science and distinguishes quantum measurement from many familiar forms of classical observation.

F.9.2 Measurement Within Quantum Mechanics

While Quantum Physics establishes the existence of measurable phenomena, Quantum Mechanics provides the mathematical framework used to describe measurement outcomes.

Within this framework, quantum systems are represented through mathematical states that evolve according to well-defined rules.

Prior to measurement, these states may exist in superpositions representing multiple possible outcomes.

Measurement introduces a transition from potential outcomes to observed results.

Quantum mechanics describes this process through concepts such as:

- observables,
- operators,
- probability amplitudes,
- expectation values,
- and measurement probabilities.

These mathematical tools allow researchers to predict the likelihood of various outcomes even when individual measurements remain probabilistic.

One of the most important lessons provided by quantum mechanics is that prediction and certainty are not identical concepts.

The mathematical framework often predicts distributions of outcomes rather than single deterministic results.

Consequently, measurement becomes the mechanism through which probability is translated into observation.

This relationship between prediction and observation forms one of the central themes of the Quantum Mechanics Layer.

F.9.3 Measurement Within Quantum Technology

The transition from theory to implementation introduces a new perspective on measurement.

Within the Quantum Technology Layer, measurement becomes an engineering problem.

Physical systems must be designed to:

- detect quantum states,
- capture measurement signals,
- process observations,
- reduce noise,
- and generate reliable outputs.

Accomplishing these objectives often requires sophisticated instrumentation.

Examples include:

- photon detectors,
- superconducting readout circuits,
- ion-state detection systems,
- microwave control electronics,
- and precision timing mechanisms.

These technologies transform theoretical measurement concepts into operational capabilities.

The engineering challenge is particularly significant because quantum signals are often extremely delicate.

Small disturbances can influence measurement accuracy and reduce confidence in observed outcomes.

Consequently, successful measurement systems require careful attention to:

- calibration,
- environmental isolation,
- signal integrity,
- detector efficiency,
- and operational stability.

Measurement therefore becomes one of the most demanding aspects of practical quantum system design.

F.9.4 Measurement and Information Generation

One of the most important functions of measurement is the generation of information.

Prior to measurement, quantum systems contain physical states and mathematical representations describing possible outcomes.

Measurement transforms those possibilities into observable results.

This process creates information that can be:

- recorded,
- interpreted,
- communicated,
- analyzed,
- and utilized.

From an architectural perspective, measurement serves as the primary mechanism through which quantum systems contribute useful information to broader technological environments.

Without measurement, quantum behavior would remain inaccessible to users, applications, and decision-making systems.

Measurement therefore acts as an information-generation boundary.

It is the point at which physical behavior becomes operational knowledge.

This concept provides an important bridge toward the information-centric disciplines that will be explored in later volumes.

F.9.5 Measurement as a Boundary Between Quantum and Classical Domains

Measurement also defines one of the most important interfaces within modern quantum systems.

The outcome of measurement is typically expressed in classical form.

Consequently, measurement functions as a boundary between:

- quantum state evolution,

and

- classical information processing.

Before measurement, system behavior is described through quantum states and probability amplitudes.

After measurement, results become available for interpretation through conventional computational and analytical methods.

This boundary appears repeatedly throughout modern quantum technologies.

Whether the system involves sensing, communication, or computation, useful outcomes generally emerge only after measurement converts quantum behavior into accessible information.

For this reason, measurement occupies a central role in virtually every practical quantum architecture.

F.9.6 Measurement and Experimental Verification

Measurement performs another critical function within the Quantum Stack: verification.

Scientific theories must be tested against observation.

Engineering systems must be validated against performance requirements.

Technological implementations must be evaluated against expected behavior.

Measurement enables all three activities.

Through measurement, researchers can determine whether:

- theoretical predictions are accurate,
- experimental systems behave as intended,
- engineering objectives have been achieved,
- and technological performance satisfies operational requirements.

Without reliable measurement, neither science nor engineering can progress effectively.

Consequently, measurement serves not only as an information-generation process but also as a verification mechanism supporting continual refinement of knowledge and technology.

F.9.7 Architectural Significance of Measurement

From the perspective of the Quantum Stack, measurement occupies a unique architectural position.

It participates simultaneously in:

- physical observation,
- mathematical interpretation,
- technological implementation,
- information generation,
- and system verification.

Few concepts span as many layers or contribute to as many operational functions.

Measurement therefore serves as one of the most important cross-layer processes within the foundational portion of the Quantum Stack.

Its significance extends far beyond the act of observation itself.

- Measurement connects theory to experiment.
- Measurement connects physical systems to information.
- Measurement connects possibility to evidence.

Most importantly, measurement provides the mechanism through which quantum behavior becomes useful within practical technological environments.

Summary

The first three layers of the Quantum Stack reveal a recurring pattern. Physical phenomena establish what is possible. Mathematical frameworks explain how those phenomena behave. Technology enables those behaviors to be controlled, observed, and measured. Through these interactions, the foundations emerge for treating quantum states not merely as physical entities but as resources capable of supporting organized information processing. The next section examines how these foundational layers collectively prepare the transition toward Quantum Information Science, the subject that becomes the primary focus of Volume II.

F.10 From Physics to Information: Preparing for Volume II

The foundational layers of the Quantum Stack examined throughout Volume I establish far more than a collection of scientific principles and technological capabilities. Together, they create the conditions necessary for the emergence of information as a structured operational resource. While Quantum Physics, Quantum Mechanics, and Quantum Technology are often studied as distinct disciplines, their combined significance becomes most apparent when viewed through the broader architectural perspective of the Quantum Stack.

- The first layer establishes the physical phenomena from which quantum behavior emerges.

- The second layer provides the mathematical framework required to represent and predict that behavior.
- The third layer transforms theoretical understanding into engineered capability.

Collectively, these developments prepare the transition from the study of quantum systems to the study of quantum information.

This transition represents one of the most important conceptual developments within the broader Quantum Stack because it changes the primary question being asked.

Throughout much of Volume I, the central concern has been:

"How do quantum systems behave?"

As the stack progresses toward Quantum Information Science, the question gradually evolves into:

"How can quantum systems represent, preserve, transform, and communicate information?"

This change in perspective marks the architectural boundary between the foundational sciences and the information-centric disciplines that follow.

F.10.1 Information as a Physical Resource

One of the most important insights emerging from modern science is that information is not independent of physical reality.

Information must always exist within some physical medium.

- A written page stores information through ink patterns.
- A computer stores information through electronic states.
- A communication network transfers information through physical signals.

Similarly, quantum information must be represented through physical quantum systems.

This observation reinforces a recurring theme throughout the Quantum Stack:

- Information is inseparable from the physical systems that represent it.
- Quantum states therefore possess a dual significance.
- They are physical entities governed by the laws of quantum mechanics.
- They are also potential carriers of information.

The ability of quantum states to support superposition, interference, and correlation suggests that they may represent information in ways unavailable to purely classical systems.

Recognizing this dual role establishes the conceptual bridge between foundational quantum science and Quantum Information Science.

F.10.2 The Emergence of Information-Carrying States

Not every physical state automatically functions as an information resource.

For a state to support information processing, it must possess characteristics that allow it to be:

- distinguished,
- represented,
- manipulated,
- preserved,
- and interpreted.

The foundational layers introduced throughout Volume I provide the mechanisms necessary for these capabilities.

Quantum Physics establishes the existence of distinguishable states.

Quantum Mechanics provides mathematical representations of those states.

Quantum Technology creates methods for preparing, controlling, and measuring them.

Together, these developments transform physical states into information-carrying resources.

Examples include:

- spin states,
- polarization states,
- energy-level configurations,
- superconducting current states,
- trapped-ion internal states,
- and photonic quantum states.

Each of these may eventually serve as a foundation for information representation and processing.

The significance of these examples lies not in their technological implementation alone but in their ability to bridge physical reality and informational meaning.

F.10.3 Preservation of Information

The usefulness of information depends heavily upon its preservation.

Information that cannot be maintained long enough for interpretation or utilization possesses limited practical value.

This principle applies equally to classical and quantum systems.

Within quantum environments, preservation presents unique challenges because quantum states are often highly sensitive to environmental interactions.

Information preservation therefore depends upon many of the engineering concepts introduced throughout Volume I, including:

- coherence preservation,
- environmental isolation,
- control precision,
- calibration stability,

- noise reduction,
- and measurement integrity.

These requirements demonstrate that information processing cannot be separated from the physical realities of system operation.

Before information can be computed, communicated, or analyzed, it must first survive.

Consequently, information preservation becomes one of the foundational objectives linking Quantum Technology to Quantum Information Science.

F.10.4 Transformation of Information

Information becomes useful when it can be transformed in purposeful ways.

In classical systems, transformations may involve logical operations, arithmetic computation, communication protocols, or signal processing.

Quantum systems introduce additional possibilities because quantum states evolve according to uniquely quantum principles.

Examples include:

- superposition evolution,
- interference-based transformations,
- correlated state interactions,
- and measurement-driven state transitions.

Although Volume I does not yet explore the formal information-processing frameworks associated with these behaviors, the underlying mechanisms have already been introduced.

The foundational layers therefore establish the physical and mathematical basis for future information transformations.

This observation highlights an important architectural principle:

The study of information processing begins long before information science formally appears in the stack.

Its roots are embedded within the physical and mathematical foundations established at lower layers.

F.10.5 Communication and Information Exchange

Information acquires additional significance when it can be exchanged among systems.

Communication requires:

- representation,
- transmission,
- reception,
- interpretation,

- and verification.

Each of these activities depends upon capabilities originating within the foundational layers.

- Quantum phenomena provide the states.
- Quantum mechanics provides the descriptions.
- Quantum technology provides the control infrastructure necessary for preparation, transmission, and measurement.

Consequently, the foundations of communication are already visible within the first three layers of the Quantum Stack.

The formal study of quantum communication will emerge in later volumes, but its conceptual roots are firmly established within the material presented thus far.

F.10.6 Information as the Next Architectural Layer

The progression from Physics to Mechanics to Technology naturally leads toward a new architectural perspective.

Instead of focusing primarily on:

- physical behavior,
- mathematical representation,
- or engineering implementation,

attention shifts toward:

- information representation,
- information transformation,
- information preservation,
- information exchange,
- and information interpretation.

This shift marks the emergence of Quantum Information Science as the next major layer of the Quantum Stack.

Importantly, Quantum Information Science does not replace the foundational layers.

Rather, it builds upon them.

Every information-processing capability remains dependent upon the physical, mathematical, and technological infrastructure established below.

The stack therefore continues its pattern of cumulative development.

New capabilities emerge while foundational dependencies remain intact.

F.10.7 Architectural Significance of the Transition

The transition from foundational science to information science represents one of the most consequential developments within the entire Quantum Stack.

Once quantum states become recognized as information resources, entirely new categories of capability become possible.

Future layers of the stack will explore:

- information encoding,
- quantum communication,
- computational frameworks,
- algorithmic structures,
- sensing architectures,
- and eventually intelligent systems.

Each of these developments traces its origins to the foundations established within Volume I.

Consequently, the significance of Physics, Mechanics, and Technology extends far beyond their immediate subject matter.

They collectively provide the infrastructure required for every subsequent layer of the Quantum Stack.

F.11 Foundational Architectural Lessons of the Quantum Stack

The Quantum Stack is more than a conceptual hierarchy describing the progression from quantum phenomena to future information-processing systems. It also serves as a framework for understanding how scientific knowledge evolves into practical capability through successive layers of abstraction, formalization, and implementation.

Throughout Volume I, three foundational layers have been examined in detail:

- Quantum Physics,
- Quantum Mechanics,
- and Quantum Technology.

Although these layers address different questions and employ different methodologies, together they reveal several important architectural lessons that extend beyond the immediate subject matter of quantum science.

These lessons provide insight into how complex systems emerge from simpler foundations and why higher-level capabilities remain dependent upon the layers that support them.

F.11.1 Capability Emerges Through Layered Development

One of the most important lessons of the Quantum Stack is that sophisticated capabilities rarely emerge from a single discovery or technological breakthrough.

Instead, capability develops progressively through multiple layers of understanding and implementation.

- Quantum Physics establishes the phenomena.
- Quantum Mechanics explains the phenomena.

- Quantum Technology enables controlled utilization of the phenomena.
- Each layer contributes functionality that would be insufficient if operating alone.
- Physics without mathematical description limits predictive understanding.
- Mathematical theory without engineering implementation limits practical application.
- Technology without physical understanding limits innovation and reliability.

Meaningful capability therefore emerges only through the interaction of all three layers.

This principle extends beyond quantum science and appears throughout engineering, computing, and systems architecture.

F.11.2 Abstraction Enables Complexity

As systems become more sophisticated, abstraction becomes increasingly important.

Abstraction allows higher layers to utilize lower-layer capabilities without requiring continual engagement with every underlying detail.

For example, a researcher operating a quantum sensing platform may not need to derive the complete mathematical formalism governing every quantum interaction occurring within the device. Likewise, a future user of quantum communication technologies may not require expertise in hardware fabrication techniques.

This ability to hide complexity while preserving functionality is one of the defining characteristics of layered architectures.

The Quantum Stack demonstrates how abstraction enables increasingly advanced forms of capability while maintaining connections to foundational principles.

F.11.3 Constraints Never Disappear

A recurring theme throughout Volume I has been the relationship between capability and constraint.

Every quantum capability emerges alongside limitations that must be understood and managed.

Examples include:

- coherence limitations,
- environmental sensitivity,
- probabilistic measurement outcomes,
- noise,
- calibration requirements,
- and engineering complexity.

Although higher layers introduce new forms of organization and control, they do not eliminate these foundational constraints.

Instead, constraints propagate upward through the stack and continue to influence system behavior.

Understanding this relationship encourages realistic expectations regarding future quantum technologies and highlights the importance of systems-level thinking.

F.11.4 Measurement Connects Theory and Practice

Few concepts illustrate the unity of the Quantum Stack more effectively than measurement.

- Measurement begins as a physical interaction.
- It is described mathematically through the formalism of quantum mechanics.
- It is implemented technologically through specialized hardware and instrumentation.
- It produces information that may subsequently support communication, computation, and decision-making.

Consequently, measurement serves as one of the most important examples of cross-layer interaction.

Its presence throughout the foundational layers demonstrates how scientific concepts often span multiple levels of abstraction simultaneously.

F.11.5 Engineering as the Bridge Between Knowledge and Capability

The progression from Physics to Mechanics to Technology illustrates another important lesson.

Scientific understanding alone does not create operational systems.

Knowledge must be translated into engineered capability.

This translation requires:

- experimentation,
- design,
- fabrication,
- calibration,
- verification,
- and continual refinement.

Engineering therefore functions as the bridge between theoretical possibility and practical utility.

Many of the quantum technologies explored today exist not because the underlying science is new, but because advances in engineering have made it possible to control quantum behavior with increasing precision.

F.11.6 Preparing for the Information-Centric Layers

The foundational layers examined in Volume I establish the prerequisites for the next stage of the Quantum Stack.

- Physical phenomena provide the resources.
- Mathematical frameworks provide the descriptions.
- Technological systems provide the control.

Together, these elements create the conditions necessary for treating quantum states as information resources.

This transition represents the architectural threshold separating foundational quantum science from the information-centric disciplines that follow.

As the stack progresses into Quantum Information Science, attention increasingly shifts toward representation, communication, transformation, and preservation of information.

Yet these future developments remain inseparable from the foundations established here.

F.11.7 Architectural Summary

The Quantum Stack demonstrates that complex systems emerge through the accumulation of layered capability rather than through isolated scientific or technological advances. Each layer contributes specialized functionality while simultaneously inheriting constraints, assumptions, and dependencies from the layers beneath it. As a result, the stack should be viewed not as a collection of independent disciplines, but as a coordinated architectural progression through which increasingly sophisticated forms of capability become possible.

At the foundation of this progression lies Quantum Physics. This layer establishes the fundamental possibility space governing quantum behavior. Phenomena such as superposition, interference, entanglement, uncertainty, and measurement define the physical realities within which all higher-layer capabilities must operate. Without these foundational principles, the remainder of the stack would have no basis upon which to develop.

Quantum Mechanics provides the mathematical and conceptual framework required to interpret these phenomena. Through state vectors, operators, probability amplitudes, Hilbert spaces, and measurement theory, quantum mechanics transforms observation into predictive understanding. It enables researchers and engineers to describe quantum behavior systematically, analyze system evolution, and establish the theoretical foundations necessary for practical implementation.

Quantum Technology then converts theoretical understanding into operational capability. Through advances in fabrication, control systems, calibration techniques, sensing infrastructure, and experimental engineering, technology transforms fragile quantum phenomena into controllable and measurable systems. This layer provides the practical mechanisms through which quantum behavior becomes accessible for scientific investigation, information processing, and future computational applications.

The relationship among these layers can be summarized as follows:

- Physics establishes possibility.
- Mechanics establishes understanding.
- Technology establishes controllability.

Together, these layers create the foundational architecture upon which all subsequent developments within the QAIS framework depend.

An important lesson emerging from this progression is that capability and dependency increase simultaneously. As systems ascend the stack, they gain additional functionality, abstraction, and operational sophistication. At the same time, however, they remain dependent upon lower-layer realities. Advanced architectures cannot escape the constraints imposed by physical law, mathematical structure, or technological implementation. Instead, they inherit and manage those constraints while introducing new capabilities.

This dependency structure serves as one of the defining characteristics of the Quantum Stack. Every higher-layer capability ultimately rests upon lower-layer foundations. Information processing depends upon engineered quantum systems. Computation depends upon information representation. Learning depends upon computation. Communication depends upon learning and information exchange. Governance and deployment depend upon all preceding layers. Understanding this progression helps establish a systems-oriented perspective that remains essential throughout the QAIS series.

The architectural significance of Volume I therefore extends far beyond its immediate subject matter. Although the volume focuses primarily on physics, mechanics, and technology, these topics collectively establish the conditions necessary for every subsequent layer of the stack. They provide the physical resources, theoretical understanding, and operational control mechanisms required for the emergence of information science, quantum computing, learning architectures, communication systems, governance frameworks, and intelligent system deployment.

For this reason, Volume I should be viewed not simply as an introduction to quantum concepts, but as the foundational architectural layer of the entire QAIS framework. Understanding the relationships established here is essential because every future layer of the stack ultimately traces its origins to the principles, constraints, and capabilities introduced within these foundational disciplines.

F.12 Common Misconceptions About the Quantum Stack

As readers encounter the Quantum Stack for the first time, it is common to interpret the framework through assumptions inherited from classical computing, traditional engineering, or popular descriptions of quantum technologies. While these perspectives may provide useful intuition, they can also introduce misconceptions that obscure the architectural purpose of the stack.

This section addresses several common misunderstandings and clarifies the intended interpretation of the Quantum Stack as presented throughout Volume I.

F.12.1 Misconception: The Quantum Stack Is a Technology Catalog

One common misunderstanding is to view the Quantum Stack as merely a list of technologies arranged in hierarchical order.

While technologies certainly appear within the stack, the framework is not intended to function as a product inventory or classification system.

Instead, the stack represents a dependency hierarchy describing how capabilities emerge through successive layers of scientific understanding and engineering implementation.

The purpose of the stack is not simply to identify what technologies exist, but to explain how those technologies become possible.

For example, a quantum sensor is not merely a device occupying the technology layer. Its existence depends upon physical phenomena identified within Quantum Physics and mathematical principles established within Quantum Mechanics.

The stack therefore emphasizes relationships rather than inventories.

F.12.2 Misconception: Higher Layers Replace Lower Layers

Another frequent misunderstanding is the belief that higher layers make lower layers obsolete.

The opposite is true.

Higher layers depend continuously upon the layers beneath them.

Quantum Technology does not replace Quantum Mechanics.

Quantum Mechanics does not replace Quantum Physics.

Instead, each layer adds new forms of abstraction while preserving dependence upon foundational principles.

This relationship is similar to modern computing systems. Application software does not eliminate operating systems, and operating systems do not eliminate hardware. Each layer remains dependent upon those supporting it.

The Quantum Stack follows the same architectural principle.

F.12.3 Misconception: Quantum Systems Escape Physical Constraints

Popular discussions occasionally create the impression that quantum technologies possess unlimited capability.

Although quantum phenomena enable powerful new forms of behavior, they remain governed by physical law.

Quantum systems must contend with:

- decoherence,
- environmental noise,
- control limitations,
- measurement uncertainty,

- and engineering constraints.

These realities do not diminish the significance of quantum technologies. Rather, they define the conditions under which those technologies must operate.

Successful engineering requires understanding constraints as well as capabilities.

F.12.4 Misconception: Quantum Mechanics Is Only Mathematics

Because Quantum Mechanics relies heavily upon mathematical formalism, some readers view it as separate from physical reality.

In practice, the mathematical structures of quantum mechanics exist precisely because they describe experimentally observed behavior.

- The mathematics and the physics are therefore inseparable.
- Physical observations motivate the theory.
- The theory predicts experimental outcomes.
- Experiments verify the theory.
- Together, these elements form a continuous cycle of scientific understanding.

F.12.5 Misconception: Quantum Technology Begins With Quantum Computers

Public attention frequently focuses on quantum computing platforms, leading some readers to assume that quantum technology begins with quantum computers.

Historically, this is not the case.

Many important quantum technologies existed long before modern quantum computing initiatives emerged.

Examples include:

- lasers,
- atomic clocks,
- magnetic resonance systems,
- semiconductor devices,
- and advanced sensing technologies.

These systems demonstrate that quantum technology encompasses a far broader domain than computation alone.

Understanding this broader perspective helps place future developments within their proper historical and architectural context.

F.12.6 Misconception: Information Appears Only in Later Layers

Another misconception is that information becomes relevant only when Quantum Information Science is introduced.

- In reality, information-related concepts begin emerging much earlier.
- Measurement generates information.

- State representation supports information description.
- Control systems enable information preservation.

Consequently, the foundations of information processing are already visible within the first three layers of the stack.

Quantum Information Science does not create information from nothing; it formalizes and extends capabilities whose roots are already present within the foundational layers.

F.12.7 Lessons Learned

The Quantum Stack is best understood as:

- an architectural framework,
- a dependency hierarchy,
- a capability model,
- a constraint model,
- and a roadmap for future development.

While these descriptions are individually useful, the true value of the Quantum Stack emerges when they are considered collectively. The framework does more than organize topics into convenient categories. It reveals the relationships that connect physical phenomena, mathematical description, engineered implementation, information processing, computation, and intelligent system behavior into a coherent architectural progression.

One of the most important lessons established throughout Volume I is that higher-level capabilities do not emerge independently of lower-level foundations. Every capability observed within a quantum system ultimately depends upon physical principles governing matter, energy, probability, and measurement. As systems become increasingly sophisticated, these foundational constraints do not disappear. Instead, they continue to influence behavior throughout every subsequent layer of the stack.

This observation reinforces a central principle of systems engineering:

Complexity does not eliminate dependency; it often increases dependency.

The Quantum Stack provides a structured mechanism for visualizing and understanding these dependencies. Each layer introduces new forms of capability while simultaneously inheriting limitations from the layers beneath it. Consequently, successful system design requires understanding not only what a layer can accomplish, but also what conditions are required to sustain that capability.

A second lesson concerns the relationship between abstraction and understanding. As systems progress upward through the stack, they become increasingly abstract. Physical phenomena become mathematical models. Mathematical models become engineered systems. Engineered systems become informational resources. Informational resources become computational processes. Computational processes ultimately support intelligent behavior.

Although abstraction simplifies complexity, it can also obscure foundational dependencies. The Quantum Stack serves as a reminder that every abstraction rests upon lower-layer realities.

Intelligent systems may appear far removed from quantum physics, yet their operation ultimately depends upon physical principles established at the foundational layers of the architecture.

A third lesson involves the role of information within the broader QAIS framework. Throughout Volume I, readers encounter physical states, measurements, probabilities, and engineered technologies. While these concepts initially appear distinct, they collectively establish the conditions necessary for information to emerge as a meaningful resource. This transition from physical behavior to informational capability represents one of the most significant architectural developments within the entire Quantum Stack and serves as the foundation for the volumes that follow.

The framework also highlights the importance of interdisciplinary thinking. Quantum systems cannot be understood solely through physics, mathematics, engineering, computer science, or artificial intelligence. Each discipline contributes a different perspective on the same architectural reality. The Quantum Stack provides a common language through which these perspectives can be integrated into a unified systems-oriented understanding.

Another important lesson concerns capability and constraint. Popular discussions often emphasize the potential advantages of quantum technologies while overlooking the operational realities that accompany them. The Quantum Stack encourages a more balanced perspective. Every new capability introduces corresponding challenges involving control, measurement, scalability, reliability, verification, and resource management. Understanding both opportunity and limitation is therefore essential for realistic system evaluation.

Finally, Volume I demonstrates that quantum technologies should not be viewed as isolated computational tools. They are components within a much larger architectural ecosystem extending from physical law to intelligent system deployment. Appreciating this broader perspective allows readers to understand how seemingly independent topics contribute to a unified developmental progression.

F.13 Additional Resources for Quantum AI Systems

As QAIS continue to mature as an interdisciplinary field, practitioners increasingly require access to authoritative references, research communities, software platforms, and emerging standards initiatives. While the preceding chapters established the foundational principles of quantum information, computation, and intelligent system behavior, continued learning requires engagement with the broader ecosystem supporting quantum technologies. The resources presented in this section provide readers with pathways for further study, experimentation, professional development, and participation in the evolving QAIS community.

F.13.1 Recommended Reading for Continued Study

The study of QAIS spans multiple disciplines, including quantum information science, artificial intelligence, systems engineering, mathematics, cybersecurity, and computational theory. Readers seeking deeper understanding may benefit from consulting foundational texts and advanced references that expand upon the concepts introduced throughout this volume.

Quantum Information Science Foundations

Several works have become foundational references within the broader quantum computing community. These texts provide rigorous treatments of quantum states, quantum gates, information theory, entanglement, measurement, and quantum algorithms.

Recommended topics for continued study include:

- Quantum computation theory
- Quantum information theory
- Quantum communication systems
- Quantum cryptography
- Quantum error correction
- Quantum complexity theory

Together, these subjects provide the mathematical and Quantum Physics upon which QAIS architectures are ultimately constructed.

Artificial Intelligence and Machine Learning

Because QAIS integrates quantum technologies with intelligent computational systems, readers may also benefit from strengthening their understanding of artificial intelligence, machine learning, probabilistic reasoning, optimization, and adaptive control systems.

Particularly relevant areas include:

- Machine learning architectures
- Neural networks
- Probabilistic inference
- Knowledge representation
- Reinforcement learning
- Optimization theory
- Explainable artificial intelligence

These topics establish the conceptual basis for the adaptive behaviors explored in later QAIS volumes.

Quantum Machine Learning

Quantum machine learning remains an active area of research. Readers interested in emerging approaches should monitor developments related to:

- Variational quantum algorithms
- Quantum-enhanced optimization
- Quantum kernel methods
- Hybrid quantum–classical learning
- Quantum neural networks
- Quantum feature representations
- Quantum data encoding strategies

Although many of these techniques remain experimental, they provide insight into future directions for QAIS development.

F.13.2 Research Institutions and Scientific Communities

The advancement of QAIS depends upon collaboration across universities, national laboratories, government agencies, and industrial research organizations. Participation in these communities enables practitioners to remain informed regarding new discoveries, technological capabilities, and emerging applications.

Academic Research Centers

Leading universities continue to contribute substantially to quantum information science, quantum engineering, and quantum-enabled artificial intelligence. These institutions often publish open research papers, technical reports, educational materials, and conference proceedings that provide valuable learning opportunities.

Areas commonly supported by academic research centers include:

- Quantum hardware development
- Quantum algorithms
- Quantum communication
- Quantum sensing
- Quantum machine learning
- Hybrid quantum–classical systems

National Laboratories and Government Programs

Government-sponsored research initiatives continue to play a central role in advancing quantum technologies. National laboratories frequently support large-scale efforts involving:

- Quantum networking
- Post-quantum cryptography
- Quantum sensing
- Quantum benchmarking
- Verification methodologies
- Quantum workforce development

These programs often produce publicly accessible reports that help shape future directions within the field.

Professional Communities

Professional organizations provide opportunities for education, collaboration, certification, standards participation, and technical exchange. Participation in conferences, workshops, and technical working groups can significantly accelerate professional development within QAIS-related disciplines.

F.13.3 Open-Source Quantum Computing Platforms

Practical experimentation remains one of the most effective ways to develop intuition regarding quantum information processing. Open-source software platforms provide accessible environments for constructing circuits, simulating quantum behavior, and exploring algorithmic concepts.

Quantum Development Frameworks

Several widely adopted frameworks support quantum programming, simulation, and experimentation.

Platform	Primary Focus	Typical Use Cases
Qiskit	Quantum Computing	Circuit Design and Simulation
Cirq	NISQ Systems	Quantum Algorithms
PennyLane	Quantum Machine Learning	Hybrid Learning Systems
ProjectQ	Education and Simulation	Instructional Development
Qibo	Quantum Simulation	Research and Experimentation

These platforms continue to evolve and serve as valuable environments for developing practical quantum computing skills.

Cloud-Based Quantum Services

Many organizations now provide cloud access to quantum processors and simulation environments. These services allow users to experiment with quantum circuits without requiring direct access to specialized hardware.

Typical capabilities include:

- Circuit execution
- Quantum simulation
- Hybrid workflow development
- Algorithm benchmarking
- Educational experimentation

Cloud-accessible environments have significantly lowered barriers to entry for students and researchers entering the field.

Educational Quantum Tools

Numerous browser-based and instructional platforms support introductory learning through visual circuit construction, simulation, and interactive experimentation. Such tools provide useful supplements to the conceptual material presented throughout Volume I.

F.13.4 Standards, Governance, and Emerging Guidance

As quantum technologies mature, increasing attention is being directed toward standardization, interoperability, benchmarking, verification, security, and governance. These efforts are particularly relevant to QAIS because future deployments will require coordination across diverse computational, informational, and operational environments.

Standards Organizations

Several organizations contribute to the development of standards and best practices relevant to quantum technologies.

Examples include:

- Institute of Electrical and Electronics Engineers (IEEE)
- International Organization for Standardization (ISO)
- International Electrotechnical Commission (IEC)
- National Institute of Standards and Technology (NIST)

These organizations support activities related to measurement, benchmarking, interoperability, security, and systems engineering.

Emerging Quantum Standards Areas

Current standards efforts increasingly focus on:

- Quantum benchmarking
- Quantum networking
- Quantum communication protocols
- Post-quantum cryptography
- Hybrid quantum–classical interoperability
- Verification methodologies
- Trust and assurance frameworks

Many of these initiatives remain under active development but are expected to become increasingly important as operational quantum systems mature.

Implications for QAIS

Future QAIS architectures will likely require standardized approaches for:

- System verification
- Performance evaluation
- Interoperability assessment
- Security validation
- Governance compliance
- Operational resilience measurement

Consequently, standards development should be viewed not as a peripheral activity but as a foundational component of long-term QAIS deployment and sustainability.

F.13.5 Concluding Perspective

The resources presented throughout this section extend the educational objectives of Volume I beyond the boundaries of the printed text. Through engagement with foundational literature, research communities, open-source platforms, and emerging standards initiatives, readers can continue developing the knowledge and practical experience required to participate in the evolving field of QAIS.

More importantly, these resources reinforce a central theme established throughout Appendix F: the development of QAIS is best understood through a systems-oriented perspective rather than through isolated technologies or individual algorithms.

The Quantum Stack provides the architectural framework that connects physical quantum behavior, information representation, adaptive intelligence, communication, governance, and deployment into a coherent progression of capabilities.

The lessons established throughout this appendix therefore extend beyond the specific content of Volume I. They provide an interpretive framework for understanding all subsequent volumes in the QAIS series. As readers progress into representation, reasoning, learning, communication, governance, and deployment architectures, the Quantum Stack will continue to serve as the reference model linking increasingly sophisticated system behaviors to their foundational origins in quantum information science.

Recognizing these relationships establishes the systems perspective that becomes increasingly important throughout the remainder of the series. More importantly, it provides the conceptual foundation required for understanding how physical quantum phenomena ultimately support the emergence of scalable, adaptive, resilient, and intelligent quantum-enabled systems.

Bibliography

Volume 1

References are organized on a per-chapter basis and retain their original numbering within each chapter.

In-text citations (e.g., [1], [2]) correspond to the reference list at the end of the respective chapter.

The consolidated reference section reproduces all chapter references in a single location for convenience. Identical numbering across chapters reflects independent chapter-level citation sequences.

Chapter 1 — Foundations of Quantum AI Systems

[1] R. P. Feynman, "Simulating physics with computers," *Int. J. Theor. Phys.*, vol. 21, no. 6/7, pp. 467–488, 1982.

[2] D. Deutsch, "Quantum theory, the Church–Turing principle and the universal quantum computer," *Proc. R. Soc. A*, vol. 400, no. 1818, pp. 97–117, 1985.

[3] P. W. Shor, "Algorithms for quantum computation: Discrete logarithms and factoring," in *Proc. 35th Annu. Symp. Found. Comput. Sci. (FOCS)*, 1994, pp. 124–134.

[4] L. K. Grover, "A fast quantum mechanical algorithm for database search," in *Proc. 28th Annu. ACM Symp. Theory Comput. (STOC)*, 1996, pp. 212–219.

[5] M. A. Nielsen and I. L. Chuang, *Quantum Computation and Quantum Information*. Cambridge, U.K.: Cambridge Univ. Press, 2000.

[6] J. Preskill, "Quantum computing in the NISQ era and beyond," *Quantum*, vol. 2, p. 79, 2018.

[7] K. Bharti *et al.*, "Noisy intermediate-scale quantum algorithms," *Rev. Mod. Phys.*, vol. 94, no. 1, p. 015004, 2022.

[8] A. S. Holevo, "The capacity of the quantum channel with general signal states," *IEEE Trans. Inf. Theory*, vol. 44, no. 1, pp. 269–273, Jan. 1998.

[9] F. Arute *et al.*, "Quantum supremacy using a programmable superconducting processor," *Nature*, vol. 574, pp. 505–510, 2019.

[10] National Quantum Initiative Act, Pub. L. No. 115–368, 132 Stat. 5092, Dec. 21, 2018.

[11] European Commission, "Quantum Flagship: Pioneering European Quantum Technologies," 2018. [Online]. Available: https://qt.eu

[12] Ministry of Science and Technology of China, "China National Laboratory for Quantum Information Science," 2020.

[13] C. H. Bennett and G. Brassard, "Quantum cryptography: Public key distribution and coin tossing," in *Proc. IEEE Int. Conf. Computers, Systems and Signal Processing*, 1984.

Chapter 2 — Operations and Scientific Framework of QAIS

[1] R. P. Feynman, "Simulating physics with computers," Int. J. Theor. Phys., vol. 21, no. 6/7, pp. 467–488, 1982.

[2] D. Deutsch, "Quantum theory, the Church–Turing principle and the universal quantum computer," Proc. R. Soc. A, vol. 400, no. 1818, pp. 97–117, 1985.

[3] P. W. Shor, "Algorithms for quantum computation," in Proc. FOCS, 1994.

[4] L. K. Grover, "A fast quantum algorithm for database search," in Proc. STOC, 1996.

[5] M. A. Nielsen and I. L. Chuang, Quantum Computation and Quantum Information. Cambridge Univ. Press, 2000.

[6] A. S. Holevo, "Coding theorems for quantum channels," Russ. Math. Surv., 1998.

[7] J. Preskill, "Quantum computing in the NISQ era and beyond," Quantum, 2018.

[8] K. Bharti et al., "Noisy intermediate-scale quantum algorithms," Rev. Mod. Phys., 2022.

[9] National Quantum Initiative Act, U.S. Congress, 2018.

[10] European Commission, "Quantum Flagship," 2018.

[11] Ministry of Science and Technology of China, 2020.

[12] NIST, "Post-Quantum Cryptography Standardization," 2022.

[13] European Commission, "EuroQCI," 2025.

[14] R. Meyer et al., "QUARK benchmarking framework," Linux Foundation, 2024

Chapter 3 — Evolution, Ecosystem, and Workforce of QAIS

[1] R. P. Feynman, "Simulating physics with computers," *International Journal of Theoretical Physics*, vol. 21, no. 6/7, pp. 467–488, 1982.

[2] D. Deutsch, "Quantum theory, the Church–Turing principle and the universal quantum computer," *Proceedings of the Royal Society of London A*, vol. 400, pp. 97–117, 1985.

[3] P. W. Shor, "Algorithms for quantum computation: Discrete logarithms and factoring," in *Proceedings of the 35th Annual Symposium on Foundations of Computer Science (FOCS)*, pp. 124–134, 1994.

[4] L. K. Grover, "A fast quantum mechanical algorithm for database search," in *Proceedings of the 28th Annual ACM Symposium on the Theory of Computing (STOC)*, pp. 212–219, 1996.

[5] European Commission, Quantum Flagship Strategic Agenda 2024–2030, Brussels, 2024.

[6] J. Preskill, "Quantum Computing in the NISQ era and beyond," *Quantum*, vol. 2, p. 79, 2018.

[7] K. Bharti et al., "Noisy Intermediate-Scale Quantum (NISQ) Algorithms," *Reviews of Modern Physics*, vol. 94, p. 015004, 2022.

[8] A. S. Holevo, *Quantum Coding Theorems*, Russian Academy of Sciences, 1998.

[9] National Institute of Standards and Technology (NIST), *Post-Quantum Cryptography Standardization Process*, Gaithersburg, MD, 2023.

[10] A. Marchisio et al., "Cutting is All You Need: Execution of Large-Scale Quantum Neural Networks on Limited-Qubit Devices," *arXiv:2412.04844*, 2024.

[11] M. Zaman et al., "TunnElQNN: A Hybrid Quantum-Classical Neural Network for Efficient Learning," *arXiv:2505.00933*, 2025.

[12] J. Bowles et al., "Better than Classical? The Subtle Art of Benchmarking Quantum Machine Learning Models," *arXiv:2403.07059*, 2024.

[13] R. Meyer et al., "QUARK: A Modern Framework for Quantum Computing Application Benchmarking," Linux Foundation Quantum Initiative, 2024. [Online]. Available: https://www.linkedin.com/pulse/quark-modern-framework-quantum-computing-application-m1upf

[14] U.S. National Quantum Initiative Act, Washington, DC: U.S. Government, 2018.

[15] EuroQCI (European Quantum Communication Infrastructure), *Strategic Roadmap*, Brussels, 2025.

[16] Quantum Economic Development Consortium (QED-C), *Workforce Development Strategic Report*, 2025. [Online]. Available: https://quantumconsortium.org

[17] National Academies of Sciences, Engineering, and Medicine, *Advancing Workforce Development in Quantum Information Science and Technology*, Washington, DC: National Academies Press, 2023.

[18] M. Cerezo et al., "Variational Quantum Algorithms," *Nature Reviews Physics*, vol. 3, pp. 625–644, 2021.

[19] M. Schuld, A. Bocharov, K. Svore, and N. Wiebe, "Circuit-Centric Quantum Classifiers," *Physical Review A*, vol. 101, no. 3, p. 032308, 2020.

[20] R. Groisman, S. Popescu, and A. Winter, "Quantum, Classical, and Total Amount of Correlations in a Quantum State," *Physical Review A*, vol. 72, p. 032317, 2005.

[21] M. Berta, M. Christandl, R. Colbeck, J. Renes, and R. Renner, "The Uncertainty Principle in the Presence of Quantum Memory," *Nature Physics*, vol. 6, no. 9, pp. 659–662, 2010.

Index

Volume I — Foundations of Quantum AI Systems

Q

R

S

W

About QuSciTech

QuSciTech develops research, instructional, and publication systems focused on quantum information science, artificial intelligence architectures, and advanced computational learning environments. The Quantum AI Systems (QAIS) series forms part of the organization's long-term scholarly and educational research initiative.

Contact and Research Resources

QuSciTech Labs
https://quscitech.com

QuSciTech Labs Repository
https://github.com/jopaneur/quscitech-labs

Permanent Scholarly Record (DOI)
https://doi.org/10.5281/zenodo.18955800

www.ingramcontent.com/pod-product-compliance
Lightning Source LLC
LaVergne TN
LVHW060046210926
848481LV00013B/3966
9798994836903